Acta Universitatis Upsaliensis
Symposia Universitatis Upsaliensis
Annum Quingentesimum Celebrantis: 2

Acta Universitatis Upsaliensis

Symposia Universitatis Upsaliensis
Annum Quingentesimum Celebrantis: 2

The Quaternary History of the North Sea

Uppsala 1979

Distributors: Almqvist & Wiksell International, Stockholm, Sweden and
Societas Upsaliensis pro Geologia Quaternaria, Uppsala, Sweden

Scientific editors:
E. Oele, Rijks Geologische Dienst, Haarlem, the Netherlands
R.T.E. Schüttenhelm, Rijks Geologische Dienst, Haarlem, the Netherlands
A.J. Wiggers, Freie Universiteit, Amsterdam, the Netherlands

Technical editor:
L.-K. Königsson, University of Uppsala, Sweden

The map enclosure was prepared and printed by the Rijks Geologische Dienst, Haarlem, the Netherlands

Typesetting by: Textgruppen i Uppsala AB
Printed by: Borgströms tryckeri, Motala 1979, Sweden

Contents

Chapter I
  Introduction to the volume. By E. Oele, R.T.E. Schüttenhelm & A.J. Wiggers 1

Chapter II
  Tectonic history of the North Sea area 5
  Tectonics of the North Sea. By P.A. Ziegler & C.J. Louwerens 7
  A new isopachyte map of the Quaternary of the North Sea. By V.N.D. Caston 23

Chapter III
  The North Sea in Early and Middle Pleistocene times 29
  Early and Middle Pleistocene coastlines in the southern North Sea basin.
    By W.H. Zagwijn 31

Chapter IV
  The development of ancient shorelines since the Eemian 43
  Late and Post-Weichselian shore level changes in South Norway. By U. Hafsten ... 45
  Late Pleistocene and Holocene shorelines on the Swedish West Coast. By K.G. Eriksson 61
  Late Pleistocene and Holocene shorelines in Western Denmark. By H. Krog 75
  The Quaternary geological development of the German part of the North Sea.
    By K.-E. Behre, B. Menke & H. Streif 85
  Depositional History and coastal development in the Netherlands and the adjacent
    North Sea since the Eemian. By S. Jelgersma, E. Oele & A.J. Wiggers 115
  The Belgian coastal plain during the Quaternary. By R. Paepe & C. Baeteman 143
  Quaternary coastlines in northern France. By J. Sommé 147
  The western (United Kingdom) shore of the North Sea in Late Pleistocene and
    Holocene times. By W.G. Jardine 159
  Late Quaternary sedimentation in the North Sea. By J.H.F. Jansen,
    T.C.E. van Weering & D. Eisma 175

Chapter V
  Summary of present knowledge on the Late Quaternary History of the North Sea 189
  Development of the North Sea after the Saalian glaciation. By E. Oele &
    R.T.E. Schüttenhelm 191
  Sea-floor morphology and recent sediment movement in the North Sea.
    By D. Eisma, J.H.F. Jansen & T.C.E. van Weering 217
  Sea-level changes in the North Sea basin. By S. Jelgersma 233

Chapter I

Introduction to the volume

During the 1970 meeting of the subcommission on shorelines of northwestern Europe, part of the commission on shorelines of the International Union for Quaternary Research (INQUA), it was decided to compile the present information on the subject in order to stimulate further discussions.

It was planned to publish separate volumes on the Quaternary History of the Baltic, on the North Sea and on the Irish Sea, and these were to be presented at the 10th INQUA congress in 1977. In that year the Uppsala University celebrated its 500th anniversary, and kindly offered to print and publish some of these studies. On the occasion of its anniversary, the Uppsala University organized a symposium on the Quaternary development of the mentioned seas. The first volume, on the Irish Sea, was officially presented during the symposium.

The compilation of known information coincided with a rapidly increasing number of data on the European marginal seas as a result of marine exploration for oil and gas. Information on the Quaternary proved valuable in connection with locating platforms and pipelines. Quaternary geological studies, which serve a practical purpose for obtaining sand and gravel, dredging channels, and constructing artificial islands, also provide a better understanding of sea level movements. This is not merely an academic matter, since millions of people are living in coastal areas where a slight rise in sea level would mean disaster. The sea level movements, resulting in transgressions and regressions, are distinctly related to glaciations, and these will be included in the discussions. The central theme, however, is shoreline development.

This volume comprises a series of separate papers, which contain either direct Quaternary geological data or essential geological background information. On numerous subjects there is no consensus of opinion among scientists in northwestern Europe. The sometimes widely conflicting views are reflected in the contributions to this North Sea volume. The first part of the second chapter, on Tectonics of the North Sea, reviews the pre-Quaternary geological framework, which serves as introduction to the second part of the chapter, in which an isopach map of the Quaternary is presented; this map shows that the maximum thickness exceeds 1000 metres. The third chapter deals with the Early to Middle Pleistocene history of the North Sea basin, with special emphasis on the shorelines in the southern North Sea.

In the fourth chapter, the later Quaternary coastal development is discussed in eight contributions by representatives of the various countries bordering the North Sea. The authors are the countries' representatives in the subcommission or they have been invited by the representatives. In some contributions an attempt has been made to integrate the results of marine geological work with data on onshore geology. The articles deal with the countries along the North Sea in a clock-wise order, and the oldest deposits and phenomena are dealt with first. In the last paper of this chapter, the relevant history of the North Sea area proper is reviewed. The information in the various contributions of chapter IV serves as an introduction to the summarizing papers of chapter V, in which an attempt has been made to develop an overall picture of the geological history of the area since the Saalian. In two papers special attention is paid to sea-floor morphology and sediment movements and to sea level changes.

In view of the importance of sea levels, national reference planes are mentioned. Jardine (1976) recommended the use of one reference level, the Normaal Amsterdam Peil (N.A.P.), which is the geodetic reference plane for Europe. He compiled a table (see Fig. I–1) of the individual national reference planes for the countries bordering the North Sea and the Irish Sea.

In the articles, the local stratigraphic terminology is mostly used next to the continental European stratigraphic terminology. As an exception, some authors prefer the term Late Devensian (Late Weichselian) in the British sense to the continental nomenclature. Local unexplained terminology is avoided as much as possible. Lithostratigraphic names are largely absent. Various pollen zonations are mentioned. It has been attempted to correlate most of these with the Dutch zonation (Zagwijn, 1961, 1975), which thus serves more or less as reference.

The writing generally follows the British rules and the AGI dictionary of geology. For abbreviations the international rules have been applied. It has not been attempted to give a complete review of literature on the geological history of the North Sea. In recent years some

Area	National reference plane (N.R.P.)	Present height of reference level in relation to local mean sea level	Height of reference level in relation to Normaal Amsterdams Peil (N.A.P.)
Norway (south)	Normal Null	Zero (± 0.04 m)	? 0.012 m –N.A.P.
Sweden (west coast)	Zero of precision levelling system	?	? 0.007 m –N.A.P.
Denmark	D.N.N.	Zero	0.124 m –N.A.P.
Federal Republik of Germany	Normalnullpunkt (N.N.)	"Nearly identical with MSL of the North Sea"	? 0.016 m +N.A.P.
The Netherlands	Normal Amsterdams Peil (N.A.P.)	Nearly identical with MSL of the North Sea	Zero
Belgium	Provisional datum (O.P.)	2.0.12 m –MSL at Ostend	2.323 m –N.A.P.
	Recent datum	Zero	0.311 m –N.A.P.
France	Nivellement général de la France (N.G.F.)	Zero	? 0.128 m –N.A.P.
U.K. mainland	Ordnance Datum Newlyn (O.D.)	0.086 m below MSL	0.203 m –N.A.P.

Fig. I–1. National reference planes of the North Sea in relation to local mean sea level and Normaal Amsterdams Peil (N.A.P.). After Jardine (1976).

authors presented a summary of previous work on the Quaternary geology of the North Sea (e.g. Maisey 1972). Preference was given to a listing of literature for each separate paper.

In view of the systematic, country-wise treatment of the North Sea region, subject and place indexes have been omitted.

It is hardly possible to thank all those who contributed to the compilation of this volume. We wish to express our appreciations, however, to the University of Uppsala for printing and publishing the book, to the Director and the former Director of the Geological Survey of the Netherlands for their generous permission to use personnel an to allow for printing the colour chart and to Dr. L. Boomgaart and Dr. J.W.C. Doppert for correcting many parts of the manuscript.

References

Jardine, W.G., 1976: Some problems in plotting the mean surface level of the North Sea and the Irish Sea during the last 15,000 years. *Geol. Fören. Stockh. Förh. 98*, 78–82.

Maisey, G.H., 1972: Summary of previous works on Quaternary geology of the North Sea. NTNF's Kontinentalsokkelkontor Oslo, 1–60.

Zagwijn, W.H., 1961: Vegetation, climate and radiocarbon datings in the Late Pleistocene of the Netherlands. Part I: Eemian and Early Weichselian. *Meded. Geol. Sticht., Nieuwe Ser. 14*, 15–45.

Zagwijn, W.H., 1975: Indeling van het Kwartair op grond van veranderingen in vegetatie en klimaat. In W.H. Zagwijn & C.J. van Staalduinen (editors), Toelichting bij geologische overzichtskaarten van Nederland, 109–114, Rijks Geologische Dienst, Haarlem.

Haarlem in April 1978

Erno Oele   *Ruud T.E. Schüttenhelm*   *Albert J. Wiggers*

Chapter II

Tectonic history of the North Sea area

Introduction

As far as crustal movements are concerned, Quaternary geologists were in the past mainly interested in the isostatic uplift of areas previously depressed by overburden of inland ice. Sedimentary basins such as the North Sea basin were assumed to be gradually subsiding for many millions of years. Fortunately, this assumption was not in contradiction with the concepts and models of Quaternary scholars.

However, the intensive and continuous search for hydrocarbons in the North Sea by a multitude of oil companies, initiated by substantial discoveries in the coastal areas, has increased our knowledge of the geology of the North Sea considerably.

It has appeared to be a region with very complex structural and depositional patterns, containing appreciable reserves of oil and gas which are rather unequally distributed over the North Sea.

Naturally the interest of oil geologists in this area is more directed towards the older deposits which may have source rock or reservoir potential.

However, an advantage of the oil boom is the considerable increase of the knowledge on the Pleistocene (and Holocene) in some parts of the North Sea area by a number of foundation borings and subbottom profiler surveys. See Zagwijn (this volume) and some contributions in chapter IV.

Another result of increased interest in the geology of the North Sea basin has been the recognition, first by exploration geologists, that activity along pre-existing tectonic lines has continued for very long periods, even up to the present time.

Recently, it has become clear to some Quaternary geologists that a study of certain problems in their fields of interest could only be satisfactorily solved if the influence of the above-mentioned continuous crustal movements at locally varying rates is taken into consideration. The problems include the understanding of sediment distribution patterns, the establishment and comparison of sea level curves, and the correlation of local transgressions.

The present contribution by P. A. Ziegler and C. J. Louwerens summarizes the structural and sedimentary history of the present North Sea area. In addition, it tries to explain the observed structural phenomena in the North Sea and the North Atlantic by means of the modern concepts on rifting and sea-floor spreading.

The contribution by V.N.D. Caston presents a Quaternary isopach map of the North Sea that stresses the significance and continuing activity of older structural phenomena.

II-*a*

Tectonics of the North Sea

PETER A. ZIEGLER & CAREL J. LOUWERENS

Ziegler, P. A. & Louwerens, C. J., 1979: Tectonics of the North Sea. In E. Oele, R. T. E. Schüttenhelm & A. J. Wiggers (editors), The Quaternary History of the North Sea, 7–22. *Acta Univ. Ups. Symp. Univ. Ups. Annum Quingentesimum Celebrantis: 2*, Uppsala. ISBN 91-554-0495-2.

The North Sea is underlain by a complex sedimentary basin that reaches thicknesses of up to 9 km. The floor of this basin is formed by Precambrian and Caledonian metamorphics and intrusives. In Devonian and Carboniferous times the southern North Sea formed part of the Variscan foredeep. During the Permian large parts of the North Sea were occupied by two east-west striking intracratonic basins.
The Mesozoic and Cainozoic evolution of the North Sea area was intricately linked to the development of the North Atlantic and the Norwegian-Greenland sea. An essentially N-S trending active rift system dominated the Mesozoic devleopment of the North Sea. With the onset of sea-floor spreading in the Norwegian-Greenland sea during the Early Tertiary the North Sea rift became inactive. The Cainozoic North Sea basin is thought to reflect the subsidence pattern of an abortive rift.

Dr. P. A. Ziegler and Drs. C. J. Louwerens, Shell Internationale Petroleum Mij B. V., Carel van Bylandtlaan 30, The Hague, Netherlands.

Introduction

The North Sea occupies large parts of the intracratonic NW-European Basin which extends from eastern Europe to the Atlantic shelf of the Shetland Islands and Western Norway. This basin is flanked to the NE by the Precambrian Fennoscandian shield, which extends eastward

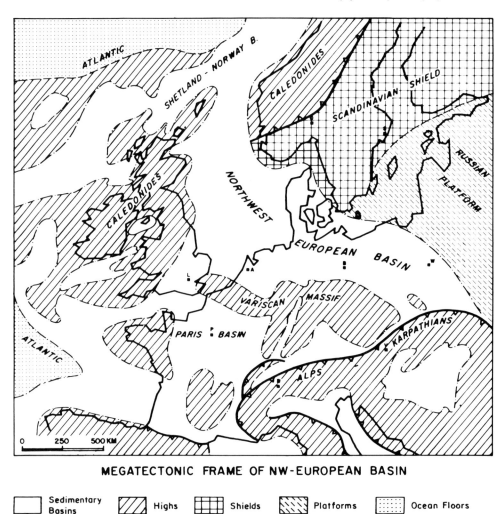

Fig. II-1. Megatectonic frame of the NW-European basin.

under the stable Russian platform, to the south by the Variscan massifs and to the west by the Caledonides of Scotland (Fig. II-1).

The sedimentary sequence underlaying the North Sea reaches maximum thicknesses of up to 9 km with sediments ranging in age from Paleozoic to Quaternary. The North Sea basin came into existence as a structural entity during the Early Tertiary. Cainozoic series reach a maximum thickness of 3500 m; Quaternary sediments alone may attain a thickness of up to 1000 m (Caston this volume). The geometry of the Tertiary and Quaternary North Sea basin is strongly influenced by tectonic patterns established during the Late Paleozoic and the Mesozoic.

During its geologic evolution the North Sea area was occupied by a number of genetically

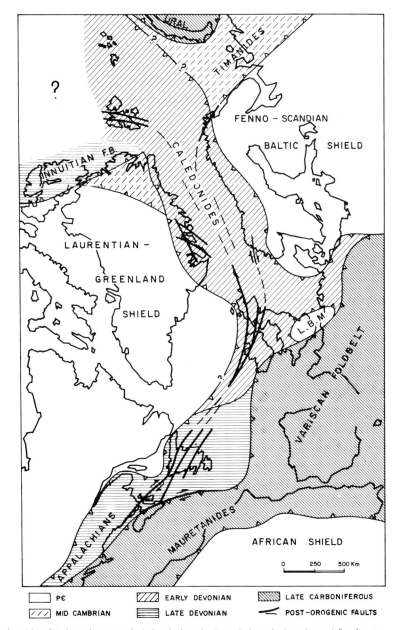

Fig. II-2. Spacial relationship of Paleozoic orogenic belts during the Late Paleozoic (continental fit after Le Pichon 1977). L.B.M. = London — Braband Massif. Legend: regions largely consolidated during:

different sedimentary basins each of which developed in response to a different megatectonic setting (P. A. Ziegler 1975, 1977, W. H. Ziegler 1975). Thus the history of the North Sea can be subdivided into several more or less distinct tectonic as well as despositional cycles. During the Mesozoic and Cainozoic the geologic evolution of the North Sea area was intricately linked with the development of the North Atlantic Ocean and the Norwegian-Greenland Sea.

The development of the North Sea basin

Pre-Permian tectonic framework

The North Sea is underlain by a continental crust consisting of Precambrian and Caledonian basement elements. The front of the Variscan fold belt skirts the North Sea to the south.

The Caledonian orogeny of Late Silurian to Early Devonian age resulted in a fusion of the NW-European — Fennoscandian plate with the Greenland — North American plate along the suture of the Arctic — North Atlantic Caledonides (Dewey 1974).

Following the Caledonian diastrophism the Arctic — North Atlantic Caledonian fold belt was subjected to post-orogenic uplifting resulting in its collapse. However, in the Appalachian-Mauretanides-Variscan geosynclinal system the Caledonian deformation phase was followed during the Devonian and Carboniferous by a number of orogenic events that led to the progressive consolidation of the Appalachians, the Mauretanides and the Variscan fold belt. By Late Carboniferous times the North American — European craton and the African plate were welded together along this fold belt (Fig. II—2). In the Arctic the Innuitian fold belt remained active till the Early Carboniferous.

Viewed against this megatectonic framework the North Sea area occupies a keystone position at the junction between the Fennoscandin-Baltic Precambrian shield, the Caledonides and the Variscides, whereby the Precambrian London-Braband Massif (Fig. II-2) played the role of an intramontane stable platform.

During the Devonian and Carboniferous the southern North Sea formed part of the Variscan foredeep in which a thick wedge of in part marine clastics and minor carbonates were deposited. At the same time the post-orogenic uplift of the Arctic — North Atlantic Caledonides resulted in the deposition of thick continental series in such tensional intramontane basins as the Old Red Orcadian basin of Scotland and the East Greenland Devonian-Carboniferous basin. Rapid subsidence of these fault-bounded basins was locally accompanied by folding; this is thought to be related to transcurrent movements along a complex fault system sub-paralleling the axis of the North Atlantic Caledonides (Harland 1969, Rast & Grant 1973). However, transcurrent movements appear to have abated during the Early Carboniferous and to have given way, at least in the area of the future Norwegian-Greenland sea, to purely extensional movements (Haller 1970). Thus this fault system may be considered as the precursor of the Late Paleozoic and Mesozoic North Atlantic — Arctic rift system.

Permian Variscan foreland-collapse

The post-orogenic uplift of the Variscides during the Permian was paralleled by the subsidence of two large intracratonic basins in the European foreland. These basins displayed an essentially eastwest strike and occupied large parts of the North Sea area. They were separated by the Mid North Sea — Ringkøbing — Fyn trend of highs (Fig. II—3).

After an initial volcanic phase thick Rotliegend clastics were deposited in these two basins. In the central parts of the southern basin these clastics grade laterally into halites deposited in a sabkha. During the Late Permian progressive rifting movements in the area of the future Norwegian-Greenland sea led to the establishment of a presumably narrow sea way between the Arctic Permian basins (Sverdrup, Spitzbergen and Pechora) and those of NW Europe (Halstead 1975); in the latter the thick Zechstein evaporites were deposited. For comparison it is interesting to note that the southern Zechstein basin which encroached in its eastern parts onto the Variscan fold belt had similar dimensions as the Black Sea, which in turn can

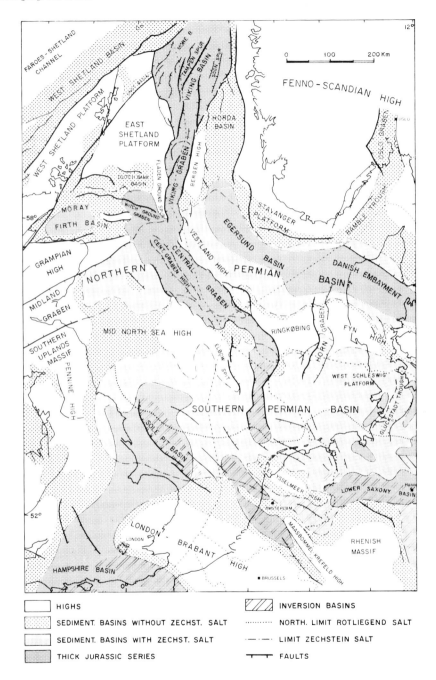

Fig. II-3. Permian and Meozoic tectonic elements of the North Sea.

be considered as an Alpine post-orogenic foreland collapse basin.

Mesozoic rifting stage

The Mesozoic development of the North Sea area in particular and of northwestern Europe in general was dominated by the subsidence of a complex graben system, main elements of which were the North Sea rift, the Danish-Polish trough, the Western Approaches graben and the Celtic Sea trough (Fig. II–4). This graben system formed a part of the northern North Atlantic-Norwegian-Greenland Sea – Eurasian Arctic rift system. This graben system gradually opened up during the Mesozoic leading ultimately to the Early Tertiary crustal separation between the North American-Greenland and the European plate and the onset of sea-floor spreading (Laughton 1975, Burke 1976, Le Pichon 1977). Major tectonic phases in the North Sea apparently correlate with similar phases in the North Atlantic – Arctic rift zone (P. A. Ziegler 1975, Birkelund 1976).

At the transition from the Permian to the Triassic the marine connection between the NW-European Zechstein basins and the Arctic Permian basins was again closed off. In the North Sea this caused at the onset of the Triassic a return to a continental depositional regime. The Triassic subsidence pattern of the North Sea area followed essentially the Permian basin framework but was gradually modified by the development of a N-S oriented graben system that breached the E-W oriented Mid North Sea – Ringkøbing – Fyn high. Main features of this graben system are the Viking and the Central graben, the Horn graben, the Danish-Polish trough as well as the Horda fault system which parallels the Norwegian coast.

In the realm of the Alpine geosyncline, Triassic extension movement led to widespread marine transgressions, that reached during the Middle Triassic as far north as the southern North Sea. The Upper Triassic was marked by a regional regression. The Triassic of the northern North Sea consists entirely of continental red beds. Thus the classical tripartite Germanic subdivision of the Triassic can only be recognised in the southern North Sea area. Triassic series reach in the Viking graben and the Danish trough thicknesses of up to 3000 m; in the Permian basins diapirism of the Zechstein and Rotliegend salts was generally triggered during the Late Triassic.

The "early Kimmerian" tectonic pulse (Keuper-Rhaetian boundary) affected the entire Arctic-North Atlantic-rift system as well as the Tethyan regions; it was followed by the widespread Liassic marine transgression during which much of NW Europe was inundated by a generally shallow sea. At the same time a link was established between the Tethyan seas and those of the Arctic along the Norwegian-Greenland rift zone.

At the Aalenian-Bajocian boundary a major rifting event (mid-Kimmerian phase) accompanied by the uplifting of a large rift dome and the emplacement of volcanic centers in the central North Sea resulted in profound paleogeographic changes (P. A. Ziegler 1977). In the area of this rift dome Lower Jurassic and older series were subjected to extensive truncation with erosion products being shed both northward into the continuously subsiding Viking graben and southward into the Sole Pit-West Netherlands-Lower Saxony basin. During the late Middle Jurassic and the Late Jurassic this rift dome began gradually to founder. This was accompanied by the rapid differential subsidence of the Viking and the Central graben which reached during the Late Jurassic a deep water stage. At the same time volcanic activities in the central North Sea ceased.

At the transition from the Jurassic to the Cretaceous a further rifting movement (late Kimmerian phase) affected the entire North Sea rift system. The Viking graben subsided rapidly giving rise to a submarine block-faulted topography with a relief of some 2000 m. In the Central graben this tectonic event was however in part obscured by the strong diapirism of the Zechstein salts. Southward in the Dutch part of the North Sea the Central graben lost its identity and was eventually cut off by a set of transform faults that led in a SE direction into the Lower Saxony basin. In the areas flanking to the north the London-Brabant and the Variscan massifs

Fig. II-4). Schematic graben pattern of the Mesozoic North Atlantic — Arctic rift system (continental fit approximately in Late Cretaceous times). A — Aegir ridge, C — Celtic sea trough, F — Faeroes, FC — Flemish Cape, G — Galicia Bank, M — Jan Mayen ridge, NS — North Sea rift, O — Orphan Knoll, P — Polish-Danish trough, R-H — Rockall-Hatton bank, W — Western Approaches graben.

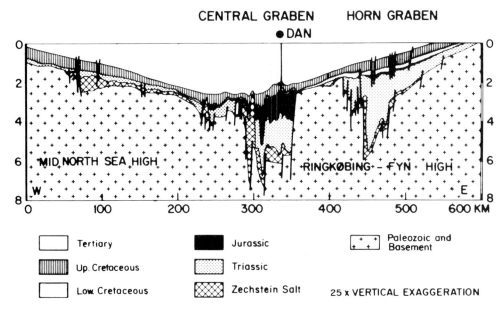

Fig. II-5. Structural cross-section through the central North Sea.

the late Kimmerian phase resulted in the rapid subsidence of so-called Marginal Throughs ("Randtröge", Voight 1962) which all displayed a NW-SE strike (Sole Pit, West Netherlands, Lower Saxony, and Mecklenburg-Brandenburg basin). Subsidence of these basins was probably caused by wrench movements between the Danish-North German block to the North and the Variscan massifs to the South (P. A. Ziegler 1975).

During the Cretaceous a number of albeit minor rifting phases resulted in a repeated temporary uplifting of the flanks of the Viking-Central Graben system thus interrupting the progressive subsidence of the central North Sea rift dome. At the same time the relief of the North Sea rift became gradually infilled by Lower Cretaceous clays and Upper Cretaceous chalks; the latter attain thicknesses of up to 1200 m in the graben itself whereas on the rift flanks only 250 to 300 m were deposited (Hancock & Scholle 1975).

At the end of the Danian a last rifting phase (Laramide pulse) resulted again in a rapid differential subsidence of the North Sea rift and a temporary uplifting of its rift flanks. In contrast, the Dutch parts of the Central graben were inverted, whereby the graben fill was uplifted above the erosional base (Heybroek 1975). Simultaneously the marginal throughs and the Danish-Polish trough were inverted. Inversion of these basins is thought to have taken place in response to compressive and/or transcurrent movements. The contemporaneity of these inversion movements with major orogenic phases in the Alpine geosyncline suggests that the Laramide deformation of the fragmented Alpine foreland was caused by the same forces that led to the progressive consolidation of the Alpine fold belt.

The Cainozoic North Sea basin

During the Tertiary the entire North Sea area was dominated by regional subsidence which apparently is still going on as demonstrated by the great thickness of Quaternary series (Caston 1977). The resultant elongate basin is in cross-

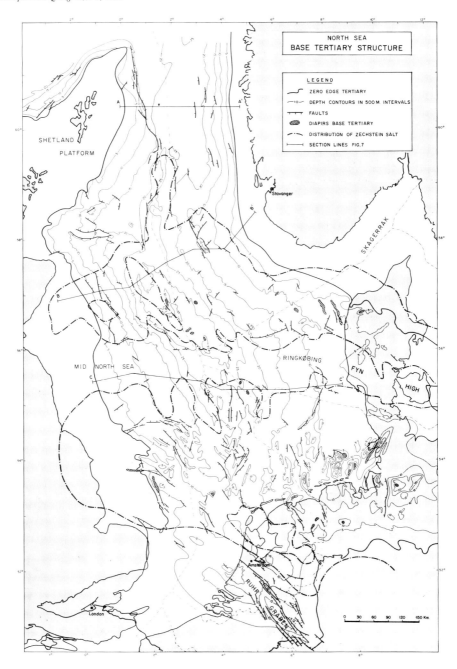

Fig. II-6. Structure map of the base of Tertiary clastics in the North Sea basin.

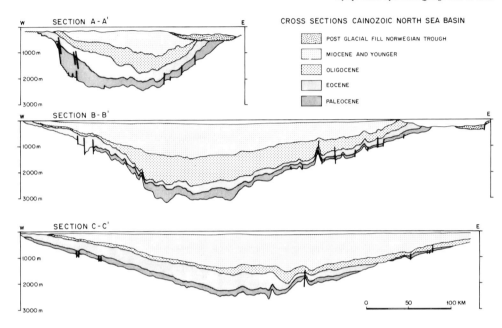

Fig. II-7. Structural corss-sections through the Cainozoic North Sea basin (for location — see Fig. II-6).

section saucer-shaped whereby its axis centres above the Mesozoic rift system (Fig. II-5). This subsidence pattern came into evidence during the Eocene by which time crustal separation between the North American-Greenland and the European plate had been achieved; with this the Norwegian-Greenland rift entered into the sea-floor spreading stage. During the drifting apart of Europe and Greenland-North America extensional stresses had apparently cased to affect the North Sea rift, which by this time had become an inactive yet still subsiding branch of the Arctic-North Atlantic rift system.

The overall structural configuration of the Cainozoic North Sea basin is illustrated on Fig. II-6 by a structure map drawn at the base of the Tertiary clastics. The outline of the Zechstein salt distribution is superimposed on this map which is based on seismic and drilling data.

In areas where Zechstein salts are present the structure of the base Tertiary is considerably complicated by diapirism of the Permian salts. In these areas faulting affecting the base Tertiary was induced by halokinetic movements.

Only diapirs breaking trough the base of the Tertiary are shown on Fig. II—6.

In areas where Zechstein salts are missing, faults affecting the base of the Tertiary were emplaced during the Laramide rifting phase. These faults have generally remained inactive during the subsequent subsidence of the Cainozoic North Sea basin the internal geometry of which is illustrated by the three structural cross-section contained in Fig. II-7. For want of regionally correlative seismic markers a subdivision of the Miocene and younger series is not readily available. In this respect it should be realised that during the recent exploration effort for hydrocarbons only limited attention has been paid to Oligocene and younger series since they proved to be largely non-prospective. Yet from the data available a fair picture can be gained of the Tertiary and Quaternary evolution of the North Sea basin.

In much of the North Sea the Early Tertiary Danian is represented by chalks that are essentially in facies continuity with the Upper Cretaceous sediments. At the end of the Danian the Laramide rifting phase resulted in major paleogeographic changes and a return to a clastic depositional regime that continued throughout the Cainozoic. The base of these clastics gives rise to an excellent seismic marker that can be mapped throughout the North Sea basin. The map given in Fig. II-6 represents the structure of this horizon and excludes therefore the Danian chalks. The Laramide rifting movement along the Shetland-Faeroe trough resulted apparently in the uplifting of the Scottish Highlands and the western Shetland platform. This gave rise to an eastward-directed drainage pattern which during the Paleocene and Eocene resulted in the deposition of an outbuilding clastic foreset wedge on shelf areas east of the Shetland and Orkney Islands. Slope failure of these foreset sequences triggered density currents which supplied sands to the down-faulted rift valleys of the Viking and Central graben (Parker 1975). In the inverted areas of the North Sea a major erosional phase preceeded the deposition of the Tertiary clastics.

Tertiary volcanism. — At the Paleocene-Eocene boundary the Scottish Tertiary volcanicity reached its peak. Tuffaceous material was scattered widely over the North Sea and resulted in the deposition of the Lower Tertiary tuff marker which represents an excellent almost basin-wide seismic marker (Jacqué 1975). During the Eocene the supply of clastics from the Scottish Highlands and the West Shetland platform gradually diminished, probably in response to the subsidence of the Shetland-Faeroe rift system. In the central and southern North Sea Paleocene and Eocene series are mainly represented by marine clays and silts.

Oligocene and younger series. — A regional hiatus marks the base and the top of the Oligocene. These disconformities, which are thought to be caused rather by eustatic sea-level changes than by local tectonic events, are mainly expressed in the marginal areas of the North Sea basin. In the centre of the basin Oligocene series reach thicknesses of up to 1000 m and consist mainly of clays and silts; these massive shales are frequently undercompacted and over-pressured.

Miocene and younger series are made up of shallow marine and paralic sands and clays as well as of Quaternary glacial material. In many parts of the North Sea these series display large scale forset patterns. Although these are readily visible on seismic data they have as yet not been mapped. For this reason no statement can be made on the potential major clastic sources that have contributed to the bulk of the Neogene and Quaternary North Sea basin fill.

Generally, Oligocene and younger time-stratigraphic units expand in thickness from the margins of the North Sea basin towards its centre (Fig. II-7).) Only along its northeastern margin did the basin edge get truncated by the Late Pleistocene uplift of the Fennoscandian shield. The dimensions of this truncation are difficult to assess but appear to be significant. Scouring out of the Skagerrak and the Norwegian trough is attributed to glacial activity (Holtedahl and Bjerkli 1975). The major source of the Late Quaternary sediments in the North Sea is the northwest European river system which gradually infills this basin from the Southeast. Bathymetric maps illustrate that the shorelines of the North Sea are highly susceptible to eustatic sea-level changes.

Geodynamic interpretation

From Fig. II-4 it is evident that the North Sea rift formed during the Mesozoic part of the complex North Atlantic — Arctic megarift system. It is thought that the driving mechanism behind the latter were upwelling mantle convection cells that were at least intermittently active since the Early Mesozoic. However, as long as crustal separation between the North American — Greenland and the European plate was not yet accomplished extensional stresses originating from such a convection system affected a wide area around the eventual ocean spreding centres; this resulted in the development of a complex graben system that followed in part pre-existent lineaments provided by the

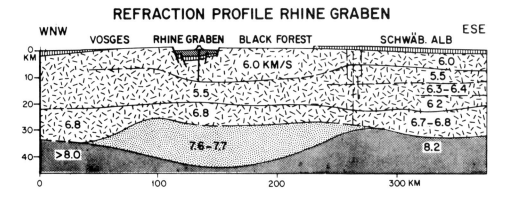

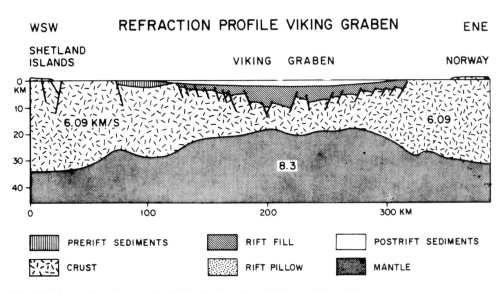

Fig. II-8. Comparison between the Rhine Graben rift and the North Sea rift.

Variscan and Caledonian fold belts. With progressive extension certain grabens became inactive whereas others remained active until crustal separation became effective.

In northwestern Europe a complex graben system started to subside during the Triassic and became accentuated during the intensified Jurassic rifting activities, whereby the North Sea rift and the Polish-Danish through emerged as the dominant fracture zones. Volcanism in the North Sea rift became significant during the Middle Jurassic. At the same time a number of minor Triassic grabens became inactive (i.e. Horn, Glückstadt and Emsland graben). Significantly this volcanic stage of the North Sea rift was accompanied by the development of a major rift dome in the central North Sea.

The volcanic stage of the North Sea rift was, however, relatively shortlived. The Jurassic to Early Tertiary development of the North Sea rift

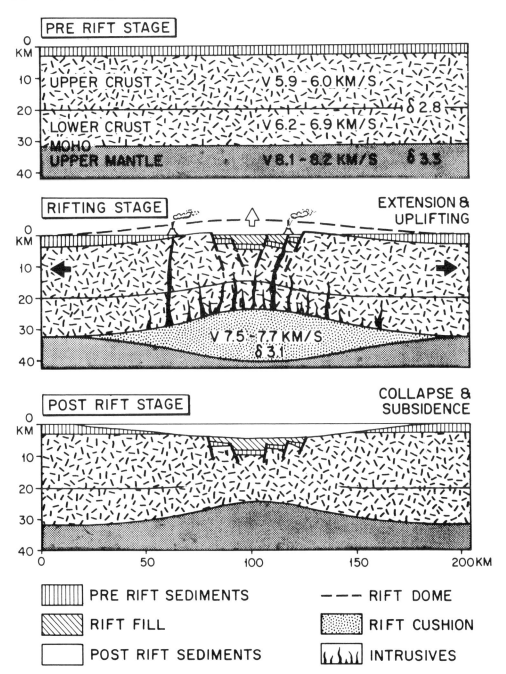

Fig. II-9. Conceptual rift model.

was marked by the rapid differential subsidence of its central graben system and a gradual foundering of the rift dome. By Late Cretaceous times the mid-Jurassic rift dome had essentially disappeared. The Cainozoic isopach map (Fig. II-5), illustrates that the North Sea rift system had ceased to differentially subside during the Tertiary. Instead, a symmetrical, saucer-shaped basin developed the axis of which follows closely the trace of the North Sea rift system (Fig. II-6).

A refraction profile (Fig. II-8) across the northernmost North Sea along a trace from the Shetland Islands to a point on the Norwegian coast located some 120 km north of Bergen shows that the crust thins drastically under the rift zone (Solli 1976). The velocity of the mantle underlaying this thinned crust is in the order of 8.3 km/sec. This refraction profile crosses the Viking graben in an area where it flares out northward and grades into the West Norway Shelf basin. Furthermore in the eastern part of this profile the Tertiary was severely truncated by the Pleistocene Fennoscandian uplift. Despite this and although the southern Viking and the Central graben are generally considerably narrower and show at shallow levels less extension it is still thought that this profile is in essence representative for the crustal configuration of the North Sea rift. However, it should be kept in mind that no refraction profiles are available across the central and southern North Sea graben system.

Fault geometries of the Viking and Central graben indicate that thinning of the crust was accomplished by extension during the rifting stage. Furthermore subcrustal erosion may have played a significant role. It is speculated that development of a rift dome in the central North Sea during the Middle Jurassic was associated with the emplacement of a low density, low velocity rift cushion at the crust/mantle interface. In this respect the model of the currently active Rhine Graben rift is invoked where refraction date permit the mapping of such a rift cushion (Fig. II-8). The width of the Rhine Graben rift dome corresponds closely to the width of the underlaying rift cushion (Ansorge et al. 1970, Müller et al. 1973). The physical properties of such a rift cushion can be explained by melting processes. Magmas intruding the crust from a rift cushion may reach eventually the surface where they display at first the chemical characteristics of a typical alkaline initial rift volcanism (Le Bas 1971). If extension tectonics persist this volcanism changes gradually over to a tholeiitic basalt type. Triple junctions where faulting and thinning of the crust is the most intense are likely places of an early manifestation of a rift volcanism. However, as in the Rhine Graben area volcanic centres are not necessarily restricted only to the actual rift zone (i.e. the Hegau and Urach volcanic centres); yet these lateral centres of volcanicity appear to remain within the areal confines of the rift cushion.

In the North Sea the main Mid Jurassic volcanic centre is located at the triple junction between the Viking, the Central graben and the Moray Firth fault system. Secondary volcanic centres occur in the Egersund basin. The Mid Jurassic volcanics of the North Sea consist essentially of undersaturted porphyritic olivin basalts that display an alkaline affinity (Howitt et al. 1975, Gibb & Kanaris-Sotiriou 1976). The subcrop of Lower Jurassic and older series against the "mid-Kimmerian" unconformity as well as paleogeographical considerations give clear evidence for the existence of a Mid Jurassic rift dome centering over the central North Sea.

It is therefore hypothesized that the North Sea graben system was underlain during its Jurassic rifting stage by a rift cushion related to a rift dome and the associated volcanism. However, refraction data across the northern North Sea indicate that at present the mantle displays a normal velocity under the North Sea rift. Thus it must be assumed that the above postulated rift cushion disappeared during the Late Mesozoic and the Tertiary, presumably by cooling and solidification. This process was apparently paralleled by the gradual subsidence of the bulged-up crust/mantle interface resulting in the development of the broad, saucer-shaped Cainozoic North Sea basin. The width of this basin may approximately correspond to the width of the original rift cushion. However, subsidence of such a basin continues until it has reached isostatic equilibrium. In this respect consideration has to be given to the intrusion of heavy mantle material into the crust during the rifting stage.

Accepting the above hypothesis, the question as to what processes may have led to the emplacement of a rift cushion and its subsequent disappearance remains to be answered. It is believed that emplacement of a rift cushion in the first place may be due to either an increased heat supply from the mantle leading to the partial melting of its upper parts (mantle plume) or the decompression of the upper mantle either due to thinning of the crust in response to tensional stresses or due to buckling of the crust in response to tangential stresses. The regional setting of the North Sea rift permits to rule out the last hypothesis. In the case of such a second-order graben system as the North Sea rift it is most likely that tensional stresses originating from the first-order Arctic — North Atlantic rift triggered the devlopment of a rift cushion. This is supported by the correlation of the major tectonic events affecting both the Arctic — North Atlantic and the North Sea rift system whereby the latter became inactive once crustal separation between Europe and Greenland-America was established and sea-floor spreading started in the Norwegian-Greenland sea.

In this light it is difficult to visualize individual and in part short-lived mantle plumes as the driving mechanism behind the North Sea graben system (Whiteman et al. 1975). Assuming that regional extensional stresses can lead to crustal thinning and can cause the development of a rift cushion it is likely that such a thermally unstable upper-mantle anomaly will regain its requilibrium during periods of tectonic quiescence (Osmaston 1971). Consequently, regional subsidence as observed during the Cainozoic development of the North Sea area is likely to characterise a rift system that has ceased to be under active extension.

The above hypothesis is summarized in the conceptual rift model given in Fig. II-9.

REFERENCES

Ansorge, J., Enter, D., Fuchs, K., Lauer, J. P., Müller, S. T. and Peterschmitt, E., 1970: Structure of the crust and upper mantle in the rift systems around the Rhine graben. In J. H. Illies & S. Müller (editors), Graben problems, *Int. upper mantle proj., sci. rep. 27*, 190–197. Schweizerbartsche Verlagsbuchhandlung, Stuttgart.

Birkelund, T., 1976: Geology of East Greenland: A revies. Offshore North Sea Technology Conference & Exhibition Stavanger 1976, Handout ONS-76-G-W-3.

Burke, K., 1976: Development of grabens associated with initial ruptures of the Atlantic Ocean; *Tectonophys. 36*, 93–112.

Caston, V. N. D., 1977: A new isopachyte map of the Quaternary of the North Sea. Quaternary deposits of the central North Sea 1, *Rep. Inst. Geol. Sci. No. 77/11*, 1–8.

Dewey, J. F., 1974: The Geology of the southern termination of the Caledonides. In A. E. M. Nairn & F. G. Stehli (editors), The Ocean basins and margins, 2, The North Atlantic, 205–231, Plenum press, New York — London.

Gibb, F. G. F. & Kanaris-Sotiriou, R., 1976: Jurassic igneous rocks of the Forties field. *Nature 260*, 23–25.

Haller, J., 1970: Tectonic map of East Greenland (1:500,000), an account of tectonism, plutonism and volcanism in East Greenland *Medd. om Grønland 171*, 286 pp.

Halstead, P. H., 1975: Northern North Sea faulting. Proceedings NPF-Jurassic northern North Sea symposium, Stavanger 1975, JNNS/10, 1–38, Norwegian Petroleum Society, Oslo.

Hancock, J. M. & Scholle, P. A., 1975: Chalk of the North Sea. In A. W. Woodland (editor), Petroleum and the continental shelf of Northwest Europe, 1, Geology, 413–425, Applied Science Publishers, Barking.

Harland, W. B., 1969: Contribution of Spitsbergen to understanding of tectonic evolution of North Atlantic region. In M. Kay (editor), North Atlantic — Geology and Continental Drift, *Am. Assoc. Pet. Geol. Mem. 12*, 817–851.

Heybroek, P., 1975: On the Structure of the Dutch Part of the Central North Sea Graben. In A. W. Woodland (editor), Geology and the continental shelf of North-west Europe, 1, Geology, 339–349, Applied Science Publishers, Barking.

Holtedahl, H. & Bjerkli, K., 1975: Pleistocene and recent sediments of the Norwegian Continental Shelf (62°N-71°N), and the Norwegian Channel area. *Nor. Geol. Unders. 316*, Bull. 29, 241–252.

Howitt, F., Aston, E. R. & Jacqué, M., 1975: The Occurrence of Jurassic Volcanics in the North Sea. In A. W. Woodland (editor), Petroleum and the continental shelf of Northwest Europe, 1, Geology, 379–386, Applied Science Publishers, Barking.

Jacqué, M. & Thouvenin, J., 1975: Lower Tertiary Tuffs and Volcanic Activity in the North Sea. In A. W. Woodland (editor), Petroleum and the Continental shelf of North-west Europe, 1, Geology, 455–465. Applied Science Publishers, Barking.

Laughton, A. S., 1975: Tectonic evolution of the north-eastern Atlantic Ocean, a review. *Nor. Geol. Unders. 316*, Bull. 29, 169–194.

Le Bas, M. J., 1971: Per-alkaline volcanism, crustal swelling and rifting. *Nature 230*, 85–87.

Le Pichon, X., Sibuet J.-C., Francheteau, J., 1977: The fit of the continents around the North Atlantic Ocean. *Tectonophys 38*, 169–209.

Müller, S., Peterschmitt, E., Fuchs, K., Emter, D. & Ansorge, J., 1973: Crustal structure of the Rheingraben area. *Tectonophys, 20,* 381–391.

Osmaston, M. F., 1971: Genesis of ocean ridge median valleys and continental rift vallies. *Tectonophys. 11*, 387–405.

Parker, J. R., 1975: Lower Tertiary sand development in the central North Sea. In A. W. Woodland (editor), Petroleum and the Continental shelf of North-west Europe, 1, Geology, 447–452, Applied Science Publishers, Barking.

Rast, N. & Grant, R., 1973: Transatlantic correlation of the Variscan-Appalachian Orogeny. *Am. J. of Sci. 272*, 542–579.

Solli, M., 1976: En Seismisk skorpeundersøkelse Norges-Shetland. Thesis University of Bergen.

Voigt, E., 1962: Über Randtröge vor Schollenränder und ihre Bedeutung im Gebiet der Mitteleuropäischen Senke und angrenzender Gebiete. *Z. dtsch. Geol. Ges. 114*, 378–418.

Whiteman, A., Naylor, D., Pegrum, R. M. & Rees, C., 1975: North Sea troughs and plate Tectonics. *Tectonophys. 26*, 39–54.

Ziegler, P. A., 1975: Geologic evolution of North Sea and its Tectonic framework. *Bull. Am. Assoc. Pet. Geol. 59,* 7, 1073–1097.

Ziegler, P. A., 1977: Geology and hydrocarbon provinces of the North Sea. *Geo-Journal 1(1),* 7–32.

Ziegler, W. H., 1975: Outline of the geological history of the North Sea. In A. W. Woodland (editor), Petroleum and the Continental Shelf of North-west Europe, 1, Geology, 165–190, Applied Science Publishers, Barking.

A new isopachyte map of the Quaternary of the North Sea

VIVIAN N.D. CASTON

Caston, V. N. D., 1979: A new isopachyte map of the Quaternary of the North Sea. In E. Oele, R. T. E. Schüttenhelm & A. J. Wiggers (editors), The Quaternary History of the North Sea, 23–28. *Acta Univ. Ups. Symp. Univ. Ups. Annum Quingentesimum Celebrantis: 2*, Uppsala. ISBN 91-554-0495-2.

A Quaternary isopach map has been constructed from 188 North Sea wells, supplemented by published work. The most conspicuous feature of the map is the considerable thickness of Quaternary present in a trough trending north-northwest down the centre of the southern North Sea. This trough contains at least two closed basins in which maximum Quaternary thickness exceeds 1,000 m.
   A relatively close correlation exists between the distribution of Quaternary sediments and the Mesozoic and Tertiary structural features, suggesting that Quaternary sedimentation in the North Sea has been influenced by these older structural patterns. Maximum average Quaternary sedimentation rates are 0.3 to 0.5 mm/year, which are up to ten times as high as the comparable figures for the Tertiary.
   This contribution is an extended abstract from: Caston, V. N. D., 1977: A new isopachyte map of the Quaternary deposits of the central North Sea 1, *Rep. Inst. Geol. Sci.* 77/11, 1–8. Crown copyrights reserved.

Dr. V. N. D. Caston, The British Petroleum Company Ltd., Britannic House, Moor Lane, London EC2Y 9BU.

Introduction

An isopach map of total Quaternary thicknesses in the North Sea has been constructed from information obtained from 188 wells drilled in all five sectors (Fig. II–10), supplement by published work.

Methods

Information concerning the thickness in these wells has been provided by the operators concerned; no re-examination of the basic data has been undertaken. Identification of the base Quaternary is dependent upon a number of factors, including lithologic changes and their associated electric log responses. The principal criterion has however been palaeontological. In the southern North Sea identification of the Pleistocene and, where present, the Pliocene, has commonly been based upon assemblages of benthonic foraminifera and, to a lesser extent, ostracods, using established zonations defined primarily in the Netherlands (e.g. van Voorthuysen et al. 1972). Extrapolation seaward of this palaeontologic zonation has provided satisfactory results; in addition, lithological changes, especially in the south and west where Pleistocene sediments overstep the Tertiary and rest directly upon Mesozoic strata, enable positive determinations to be made.

Errors and limitations

Moving northwards into the deeper Tertiary basin, fine grained sedimentation appears to have continued virtually unchanged across the Pliocene/Pleistocene boundary, providing few obvious lithological markers (Caston 1977). Distinct faunal assemblages are also less easy to identify, possibly because of the earlier influence of cold-water conditions with increasing latitude. As a consequence, many operators tend not to distinguish the Pleistocene from the Pliocene. These problems are aggravated by the fact that, for operational reasons, samples are less frequently collected in those initial stages of drilling a well which commonly includes the Quaternary section. Evidence from the northern part of the North Sea, especially to the north of 58° N, is therefore scanty and insufficient for our purpose; the few isopachs based on well data in this area should therefore be treated with considerable caution.

Incorporated published work

The resultant isopach map Fig. II-11) has been prepared principally from this new well data. It however also incorporates the work of Dingle (1970) off the Yorkshire coast, the Institute of Geological Sciences in the Moray Firth and to the east of the Scottish coast between 56° and 58° N (Holmes 1977), Sellevol and Sundvor (1974) off the west coast of Norway between 60° and 63° N, and Sellevoll and Aalstad (1971) in the Skagerrak. Note has also been taken of the contours on the base of the Pleistocene shown on the Sub-Pleistocene Geology map of the British Isles (Institute of Geological Sciences 1972). These contours are however related to a mean sea-level datum and data control points are not given. Although it is therefore difficult to resolve certain inconsistencies between the IGS map and the present data, Quaternary thicknesses shown in the present work are however comparable with the earlier map. Additional individual data control points for Quaternary thickness have been taken from Howitt (1974), Rasmussen (1974), Fowler (1975) and Pennington (1975).

On land, the data for the Netherlands is taken from Keizer and Letsch (1963), De Jong (1967) and Zagwijn (1974). The latter two papers are expecially important as they show those faults known to be active in the Quaternary.

Results

The most conspicuous feature of the isopach map is the considerable thickness of Quaternary present in a linear trough trending north-northwest for over 750 km down the centre of the North Sea, the southern limit of which has been well documented in the Netherlands. This linear trough corresponds approximately in position to the underlying Central graben, which has been identified as a major structural feature of Mesozoic and Tertiary age (Kent 1975), and contains at least two closed basins in which maximum Quaternary thickness exceeds 1000 m. An area exceeding 600 m in thickness centred at about 53° N, 4° E is displaced westwards

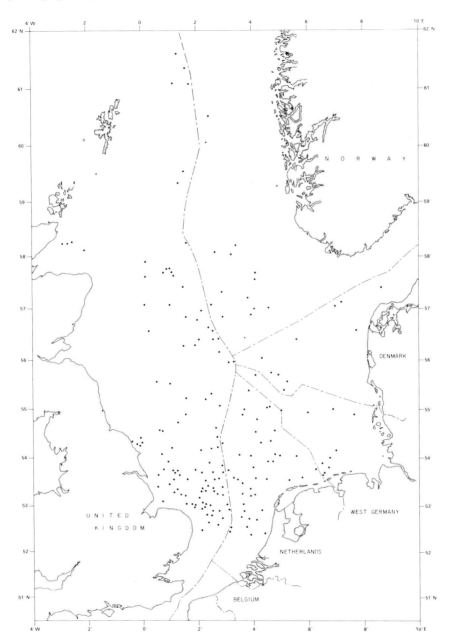

Fig. II-10. Map showing location of wells used for construction of isopach map, together with limits of national sectors.

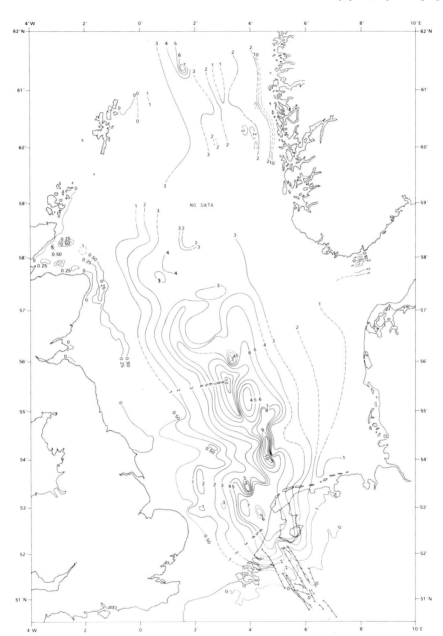

Fig. II-11. Isopach map of total Quaternary thickness in North Sea based primarily upon data from 188 wells. Isopachs at intervals of 100 m, except locally off the UK coast where the 25 m and 50 m contours, in part based upon unpublished data from the Institute of Geologica Sciences, have been added. Sediment thicknesses within the Netherlands and location of faults active in the Quaternary have been taken from published sources given in text.

from the Central graben and appears to be a direct extension of the Quaternary grabens and horsts recognised in the Netherlands. Further north, although there are inadequate data, there appears to be relatively little Quaternary off the east coast of England, and virtually none at all north of about 54°10', which corresponds to the mid-North Sea high.

On little evidence it is suggested that to the East of the Orkneys and Shetlands Quaternary cover is also fairly thin, perhaps less than 100 metres or so out at least as far as 0°, and this corresponds to the East Shetlands platform. There is a small embayment towards the Moray Firth, perhaps associated with the Moray Firth basin. On the eastern side of the North Sea there are little data, but certainly off the Danish and German coasts there appears to be only a fairly gradual westward increase in Quaternary thickness as far as a line that corresponds roughly to the edge of the Central graben. In the central northern North Sea the line of the Viking graben is followed by a poorly-defined linear trough in which the Quaternary is at least 300 m thick, and in one well exceeds 700 m. Moving eastwards, there is a zone in which the Quaternary is locally less than 100 m thick, possibly comparable with the Vestland-Stavanger ridge on Kent's (1975) map, and finally another linear Quaternary deep, which may correspond to the northern part of the Bergen basin.

Conclusions

It is suggested that the relatively close correlation between the distribution of Quaternary sediments and the underlying Mesozic/Tertiary tectonic features suggests that Quaternary sedimentation in the North Sea has been strongly influenced by these older structural patterns. Putting it another way, we have evidence from a study of the Quaternary that the deep-seated structural movements responsible for the North Sea basin development have been active virtually to the present day. Average Quaternary sedimentation rates may have been as high as 0.3 to 0.5 metres per 1000 years, which is up to ten times as high as the comparable rate for the Tertiary, although such averages conceal considerable variations in the deposition rate within the Quaternary.

REFERENCES

Caston, V. N. D., 1977: The Quaternary deposits of the Forties Field, northern North Sea. Quaternary deposits of the central North Sea 2, *Rep. Inst. Geol. Sci. No. 77/11*, 9–22.

De Jong, J. D., 1967: The Quaternary of the Netherlands. In K. Rankama (editor), The Geologic Systems: The Quaternary, 2, 301–426, Interscience, New York.

Dingle, R. V., 1970: Quaternary sediments and erosional features off the north Yorkshire coast, western North Sea. *Mar. Geol. 9*, M17–M22.

Fowler, C., 1975: The geology of the Montrose Field. In A. W. Woodland, (editor), Petroleum and the continental shelf of Norhwest Europe, 1, Geology, 467–476, Applied Science Publishers, Barking.

Holmes, R., 1977: The Quaternary deposits of the central North Sea, 5. The Quaternary geology of the UK sector of the North Sea between 56° and 58° N. *Rep. Inst. Geol. Sci. 77/14*.

Howitt, F., 1974: North Sea oil in a world context. *Nature 249*, 700–703.

Institute of Geological Sciences, 1972: Map of the Sub-Pleistocene geology of the British Isles and the adjacent Continental Shelf. 1st Edition.

Keizer, J. & Letsch, W. J., 1963: Geology of the Tertiary in the Netherlands. *Verh. K. Ned. Geol. Mijnbouwkd. Genoot. Geol. Ser. 21–2*, 147–172.

Kent, P. E., 1975: Review of North Sea Basin development. *J. Geol. Soc. Lond. 131*, 435–468.

Pennington, J. J., 1975: The geology of the Argyll Field. In A. W. Woodland (editor), Petroleum and the continental shelf of Northwest Europe, 1, Geology, 285–291, Applied Science Publishers, Barking.

Rasmussen, L. B., 1974: Some geological results from the first five Danish exploration wells in the North Sea, *Dan. Geol. Unders. III, 42*, 46 pp.

Sellevol, M. A. & Aalstad, I., 1971: Magnetic measurements and seismic profiling in the Skagerrak. *Mar. Geophys. Res. 1*, 284–302.

Sellevol, M. A. & Sundvor, E., 1974: The origin of the Norwegian Channel – a discussion based on seismic measurements. *Can. J. Earth Sci. 11*, 224–231.

Voorthysen, J. H. van, Toering K. & Zagwijn, W. H., 1972: The Plio-Pleistocene boundary in the North Sea Basin. Revision of its position in the Marine Beds. *Geol. Mijnbouw 51*, 627–639.

Zagwijn, W. H., 1974: The palaeogeographic evolution of the Netherlands during the Quaternary. *Geol. Mijnbouw 53*, 369–385.

Chapter III

The North Sea in Early and Middle Pleistocene times

Introduction

Present data on Lower and Middle Pleistocene deposits are unequally distributed in the North Sea area. There are a number of scattered onshore data in the southern part of the North Sea basin regarding lithology, relative age and depositional environment.In particular deposits of Early Pleistocene age are of wide-spread occurrence. Due to their suitability for palynological investigation a relatively detailed knowledge of climatic evolution during that time is available. Much less known is the climatic history and consequently the exact number of interglacials and glacials during Middle Pleistocene times. Little is known on the nature and extension of inland ice, prior to the Saalian glaciation. This holds likewise true for the glacial regressions and interglacial transgressions during those times.

Offshore data on Early and Middle Pleistocene deposits are scarce. They are essentially restricted to the Southern Bight. It may be expected that the growing interest in offshore activities will considerably augment our knowledge on the Pleistocene geology in the near future.

The picture in the central and northern North Sea is at the moment very bleak indeed. Reliable data on pre-Weichselian events are virtually absent. Traces of prior glacial and interglacial events may have been removed to a large extent by the erosive forces of the subsequent glaciation. However, a more or less complete record of extreme climatic variations and related sea level stands will probably be present in graben-like depressions with a considerable Quaternary infill (compare the Quaternary isopach map in Caston's contribution elsewhere in this volume). Unfortunately, these scientific treasures may be deeply buried by deposits of the subsequent Pleistocene glaciations and transgressions.

The present contribution by W. H. Zagwijn summarizes the existing knowledge on Early and Middle Pleistocene events in the Netherlands and in the Dutch part of the North Sea. Of interest is the attempt to correlate the resulting local stratigraphy with that established in Northern Germany and England.

III-*a*

Early and Middle Pleistocene coastlines in the southern North Sea basin

WALDO H. ZAGWIJN

Zagwijn, W. H., 1979: Early and Middle Pleistocene coastlines in the southern North Sea basin. In E. Oele, R. T. E. Schüttenhelm & A. J. Wiggers (editors), The Quaternary History of the North Sea, 31–42. *Acta Univ. Ups. Symp. Univ. Ups. Annum Quingentesimum Celebrantis: 2*, Uppsala. ISBN 91-554-0495-2.

A review is presented on the shoreline displacements in the southern North Sea area during the Early and Middle Pleistocene. The building of large deltaic fans by the River Rhine and North German rivers has strongly influenced the shape of the coastline of The Netherlands. Generally speaking the paleogeographic reconstruction of former coastlines is only feasible for interglacial times.

No relevant data from the present North Sea floor are available permitting reconstruction of the shorelines of glacial episodes, when sea level was lowered.

Six paleogeographic maps, illustrating the coastline evolution of the area during interglacials are included, as well as a tentative reconstruction of the maximum glacier advance during the Elsterian glacial stage.

Dr. W. H. Zagwijn, Rijks Geologische Dienst, Spaarne 17, Haarlem, The Netherlands.

Introduction

In this paper a short review is presented on the shoreline displacements in the southern North Sea area during Early and Middle Pleistocene time. In a number of maps the assumed shorelines are drawn. It should be noted, however, that much of the evidence available is circumstantial, and that the patterns pictured should be considered as attempts only of

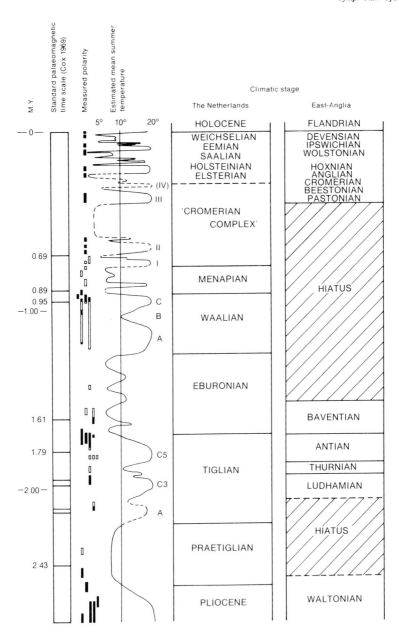

Fig. III-1. Paleomagnetic measurements (mainly from van Montfrans 1971), climatic curve and stratigraphic subdivision of the Quaternary in The Netherlands and East Anglia.

paleogeographic reconstruction.

Except for the coastlines the main directions of deposition of the North German rivers and the River Rhine are indicated. These rivers have

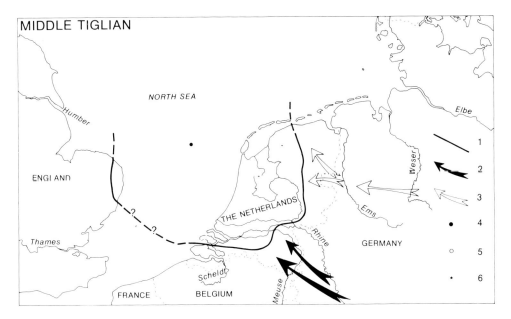

Fig. III-2. Paleogeography of the southern North Sea area during the Middle Tiglian.
Legend to Figs. III-2 to 6 and III-8. 1) coastline. 2) flow direction of River Rhine. 3) flow direction of North German rivers. 4) boring with marine deposits. 5) boring without marine deposits. 6) boring with near-coastal deposits.

built large deltaic fans in the present area of The Netherlands during the period considered. The shape of the coastlines in these areas may be understood in relation to the mentioned process of fluviatile sedimentation.

For further information on the paleogeography of the fluviatile Pleistocene sediments in The Netherlands see Zagwijn (1974).

In order to facilitate comparison between stratigraphical schemes in use on both sides of the southern North Sea, reference is made to Fig. III-1. which is based on the views expressed in some more detail in Zagwijn (1975).

Discussion of the maps

During the Late Tertiary a general regression took place in the areas surrounding the North Sea basin. At the onset of the Pleistocene only two of these areas, namely a part of East Anglia and the greater part of The Netherlands, were covered by sea.

Middle Tiglian (pollen-zone C3) (Fig. III-2)

This part of the Tiglian of the stratigraphic scale in The Netherlands may be correlated with the Ludhamian stage in East Anglia according to the views expressed by Zagwijn (1975). It may roughly be dated around 2 million years ago (van Montfrans 1971).

In The Netherlands a considerable part of the present land surface was covered by sea. The configuration of the coastline was as shown on the map and is called concave; this configuration persisted for several million years during the Late Tertiary. The shape of the coastline in this area was evidently related to tectonic activity in

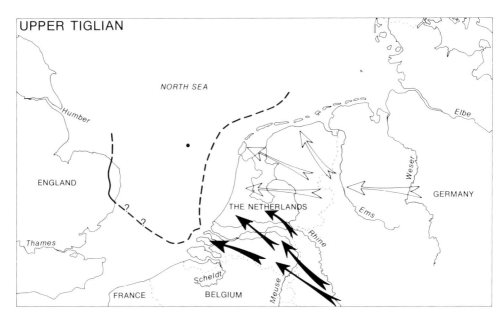

Fig. III-3. Paleogeography of the southern North Sea area during the Late Tiglian. Legend see Fig. III-2.

the area of the Central graben Rhine embayment in the southeast.

In East Anglia the deposits found in the deepest part of the Crag Basin are to present knowledge of Ludhamian to Thurnian age (West 1968, 1971). These deposits are correlative with the Middle and Upper Tiglian beds of The Netherlands. In the maps presented in Figs. III-2 and III-3 Crag deposits found in East Anglia below −50 ft O. D. (West 1968:fig. 12. 6) have been attributed to these stages in the reconstruction of Tiglian coastlines.

Late Tiglian (C5)(Fig. III-3)

In this phase, dating only some 200.000 years later than the previous one, the configuration of the coastline in The Netherlands was completely different. In this part of the southern North Sea the coastline had shifted such a distance to the west, that the sea did not occupy the present territory of The Netherlands anymore. The coastline, as drawn on the map, is, however, largely conjectural, as the data in the area covered by the present North Sea are still very scarce. A more detailed account of the data used has been published by Zagwijn (1974).

The shape of the coastline, as shown on the map, is convex, reminiscent of the shape of the present-day coastline of The Netherlands. Zagwijn (1974) assumes this shape to be related to the formation of an enormous delta of the River Rhine and North German rivers combined, protruding into the North Sea.

In East Anglia the area covered by sea was probably of approximately similar size as during the previous phase.

From two wells on the Danish part of the continental shelf respectively at 55°24'20" N, 5°03'45" E and 55°24'21" N, 5°03'34" E, marine beds containing *Azolla tegeliensis* have been described (Bertelsen 1972). These marine deposits can be dated within the Tiglian.

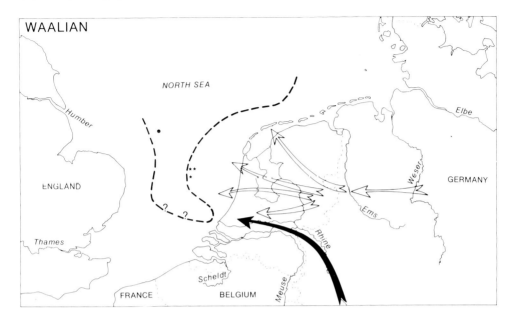

Fig. III-4. Paleogeography of the southern North Sea area during the Waalian. Legend see Fig. III-2.

Eburonian and later glacial stages

After the Tiglian a long phase of cold followed. namely the Eburonian glacial. which according to paleomagnetic studies (van Montfrans 1971) may be dated to start around 1.6 million years ago.

During this time the large Nebraskan inland ice covered North America. This may be assumed to have resulted in a considerable eustatic lowering of sea level as early as 1.5 million years ago.

Similarily during at least a number of subsequent glacial stages, noticeably during the Menapian (correlative with part of the Kansan glaciation of North America) the sea level has also dropped.

As at these times the North Sea was a shallow shelf sea, parts of varying extent of the present North Sea floor may have been dry land during these cold phases. However, at present pertinent geological information, which could permit a reconstruction of the coastline during these Early and Middle Pleistocene glacial episodes is lacking.

Waalian (Fig. III-4)

During this interglacial stage the sea level must have risen, as compared to that in the preceding Eburonian glacial stage. True marine beds presumably dating from this interglacial have been recently encountered in a boring in block J14, where a marine deposit was found at a depth of about 54 m below sea-floor (the waterdepth is about 33 m). Pollen analysis points to an Early Pleistocene interglacial age. Inferred from the stratigraphic position a Waalian age for the marine deposit seems probable. In a number of other localities clear indications have been found in non-marine beds of Waalian age, that the sea was at a short distance from these localities (according to Zagwijn 1974). These observations have led to the construction of the assumed position of the coastline as presented on the map.

In East Anglia no marine beds dating from the Waalian interglacial have been found, if the stratigraphic correlation presented by Zagwijn (1975), see Fig. III-1, is accepted.

In two wells on the Danish part of the con-

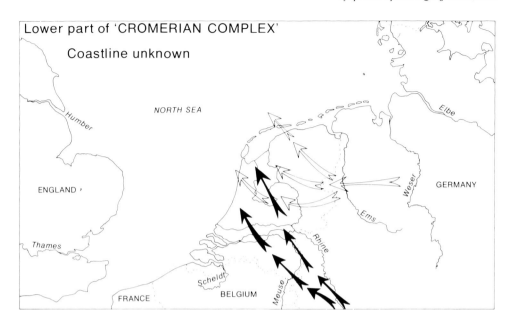

Fig. III-5. Paleogeography of the southern North Sea area during the earlier part of the "Cromerian complex". Legend see Fig. III-2.

tinental shelf the presence of marine beds of Waalian age has been assumed by Bertelsen (1972), on the basis of finds of *Azolla filiculoides*. However, additional pollenanalytic data to confirm this assumption are required.

Lower part of the "Cromerian complex" of The Netherlands (Fig. III-5)

No data are available at present, to permit a reconstruction of the coastline during the first two interglacials of this Middle Pleistocene complex, which unfortunately has been indicated with the incorrect term of "Cromerian" in The Netherlands. At any rate, during these times the coastline has been somewhere northwest of the present territory of The Netherlands.

Pastonian to Cromerian (Upper part of the "Cromerian complex" of The Netherlands (Fig. III-6).

Marine beds are known from the upper interglacials of the earlier mentioned Middle Pleistocene complex in East Anglia as well as in The Netherlands. Detailed correlation of the East Anglian stratigraphic sequence of Pastonian-Beestonian-Cromerian with the sequence in The Netherlands is awaiting additional data for confirmation. In the reconstruction of the coastline presented in Fig. III-6 no attempt has been made to separate the configuration during the Pastonian from that in the Cromerian. Therefore, the map probably shows a composite configuration.

In East Anglia, part of the Icenian Crag and part of the Cromer Forest Bed Series show marine and estuarine facies. West (1971) pointed out that the Norwich Crag, Chillesford Crag and Weybourne Crag (Harmer 1902) at a number of sites have been deposited during the temperate Pastonian stage as indicated by the results of pollen analysis. It probably means that most of the Younger Crag deposits, which in East Anglia overlap Older Crag deposits (Red Crag and Coralline Crag) in the south and Chalk in the north, were laid down during one

Early and Middle Pleistocene coastlines 37

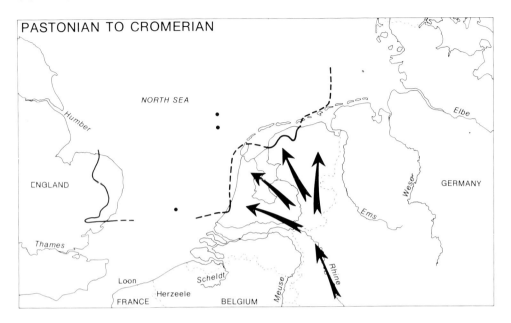

Fig. III-6. Paleogeography of the southern North Sea area during the later part of the "Cromerian complex" (Pastonian to Cromerian). Legend see Fig. III-2.

transgression which took place during the Pastonian. The generalized coastline as represented for East Anglia in Fig. III-6 has been drawn on this assumption (cf. West 1968: figs. 12.5 and 12.6). According to Funnell (1972) the deep Stradbroke trough NE of Showmarket is filled with Red Crag and therefore has been omitted from the reconstruction of the Pastonian to Cromerian coastline. According to West (1972) the Cromerian transgression plane is near present sea level in northern Norfolk and marine beds of Cromerian age are not found above 4 m O.D.

In the northern Netherlands marine beds, previously regarded to be of Holsteinian (or Mindel-Riss) age, are frequently found at depths of 50 to 60 metres below present sea level. They have appeared to be definitely older than Holsteinian (Zagwijn 1974, Doppert et al. 1975) and are according to pollenanalytic data probably correlative with the deposits of Cromerian age in East Anglia. Marine beds of Pastonian to Cromerian age occur in the western Netherlands near The Hague. However, here it cannot be ascertained whether they should be correlated with the Pastonian or with the Cromerian beds of East Anglia.

The coastline of The Netherlands during the Pastonian to Cromerian was similar in shape and position as the present coastline and shows the convex shape, which remained since Late Tiglian time.

For the reconstruction the assumption has been made that no direct connection existed between the North Sea and the sea south of the Strait of Dover. However, the presence of marine beds in SW Belgium and NW France (Sommé and Paepe & Baeteman, this volume), which may be of the same age, may contradict this assumption (see below).

Elsterian glacial stage (Fig. III-7)

During the Elsterian glacial stage, which is considered to be correlative with the Anglian of the British stratigraphic scale, inland ice invaded the

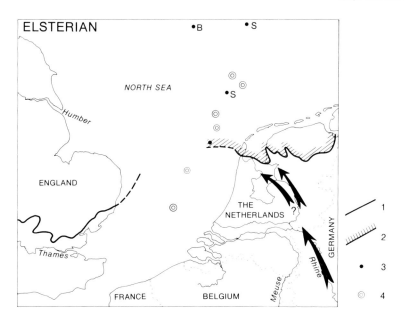

Fig. III-7. Paleogeography of the southern North Sea during the maximum advance of the Elsterian inland ice. Legend: 1) maximum glacial advance in eastern Britain. 2) southern limit of "potklei" in The Netherlands. 3) boring with Elsterian glacial deposits. 4) boring without Elsterian glacial deposits. S=pollen assemblage of Scandinavian inland ice. B=pollen assemblage of British inland ice.

southern part of the North Sea basin. Additional data permit a tentative reconstruction of the ice front during the maximum glacial advance of the Elsterian. Bristow and Cox (1973) have argued that East Anglia has been invaded only once by inland ice, namely during the Anglian. These authors assume this advance, which was the maximal in this part of Britain, to be correlative with the Saalian maximum advance in Central Europe. This assumption is, however, in contradiction with paleobotanical and other paleontological evidence, which indicate that the Hoxnian interglacial deposits overlying the Anglian glacial deposits in East Anglia are indisputably correlative with the Holsteinian beds on the continent (see discussion remarks by Francis and Turner in Bristow and Cox 1973).

Therefore only one conclusion may be drawn from the above mentioned observations of Bristow and Cox (1973), namely that the maximum glacial advance in eastern Britain was during Elsterian time in terms of continental stratigraphy.

In the northern part of The Netherlands a series of deposits of presumably glacial origin underlying the Saalian till is known since a long time. Whether or not this part of the country has been subjected to more than one invasion of inland ice has been a much disputed topic in the first two decades of the century. After settlement of the dispute in favour of one glaciation (during Saalian or Riss) only, agreement was reached on the assumption that the deposits concerned (among which the so-called "potklei", i.e. pottery clay) were laid during the advance of the Saalian inland ice (e.g. Zonneveld 1958). However, since it has been demonstrated that these deposits have to be of Elsterian age (Zagwijn 1973) their mode of origin has again become a matter of dispute. Various observations indicate that part of these deposits may be of glacial origin (warved clays, Scandinavian pebbles). Glacial till is extremely scarce, however, and only very recently the presence of Elsterian till was demonstrated in a boring in the Wadden Sea area (Doppert et al. 1975)

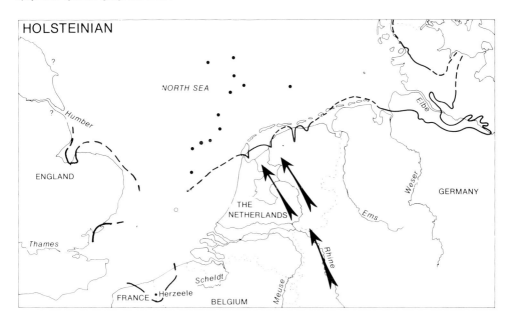

Fig. III-8. Paleogeography of the southern North Sea area during the Holsteinian. Legend see Fig. III-2.

Moreover, it has become evident that the "potklei" deposits are the very counterpart of the "Lauenburger Ton" of NW Germany, which has been attributed to the Elsterian glacial advance since the beginning of the century. However, details on the character of the Elsterian glacial advance into the northern Netherlands are indistinct. Evidently, the effect of this advance has been very different in various aspects from that of the Saalian advance, among others with regard to the absence of push-moraines and large erratic boulders and the scarcity of tills. Most of the deposits involved are of lacustro-glacial origin, which have partly filled deep tunnel-valley-like depressions.

In the map on Fig. III-7 the assumption has been made that the area showing "potklei" in the northern Netherlands was at least during a short time glaciated during the Elsterian. It is interesting to note that also in East Anglia the Anglian glaciation is associated with tunnel-valleys (Woodland 1970), a feature similar to that in The Netherlands and NW Germany. However, no pertinent data are available at present from the North Sea floor to trace the Elsterian ice front across the North Sea from Holland to Britain, though the presence of Elsterian glacial deposits of Scandinavian provenance below marine beds, dated as Holsteinian, has been demonstrated in a few borings in the area east of Dogger Bank (Fig. III-7).

In another drilled hole, on the eastern flank of Dogger Bank, two glacial sequences have been found separated by a peat bed containing a very cold, tundra-like pollen flora. Each of the glacial sequences is over 20 m thick. Pollen analysis has shown that the upper sequence contains reworked assemblages of both Scandinavian and British provenance indicating intermixing of inland ice from the two mentioned areas. The lower glacial sequence contains exclusively assemblages indicating a British provenance.

As the upper sequence appeared to be of Saalian glacial age, the lower one, below the arctic peat bed, may be assumed to be of Elsterian age. This means that during Elsterian times British inland ice invaded the Dogger Bank area.

In this connection it should be noted that the fluvioglacial deposits of Scandinavian provenance in the area east of Dogger Bank, assumed by Oele (1969) to be of Elsterian age, have since appeared to be of Saalian age (Zagwijn, unpubl.; Oele, this volume).

Holsteinian (Fig. III-8)

During this interglacial the North Sea invaded for the first time since the Miocene parts of the northern German lowlands.

As indicated on the map (according to Woldstedt 1950) the sea penetrated deeply into the area of the recent Elbe estuary in the Hamburg region. Also further to the north in Holstein marine deposits occur far inland. In The Netherlands the coastline was near and parallel to the present northern coastline of this country. No Holsteinian deposits are known from the western coastal part of Holland. It suggests non-sedimentation in the adjacent area of the southern North Sea. As pointed out by Sommé and Paepe & Baeteman (this vol.) the upper part at least of the marine beds found at Herzeele in the Franco-Belgian border region are of Holsteinian age. This may indicate that at this time the sea could penetrate the Strait of Dover (see below). In the coastal areas of southeastern Britain marine beds of Holsteinian (Hoxnian) age are known from Clacton (Pike and Godwin 1953) and from the Nar Valley in northern Norfolk (Stevens 1960). The presence of these marine beds at Clacton also indicates that the southernmost part of the North Sea was possibly through the Strait of Dover inundated by the sea during Holsteinian time. In Lincolnshire and in the Holderness region of Yorkshire marine beds of Hoxnian age have been reported from Kirmington and Speeton respectively (Watts 1959, Catt and Penny 1966, Catt 1977). The age assignment is, however, open to discussion (Gaunt et al. 1974).

The altitude of the marine Holsteinian beds at Clacton is from about 3 to 9 metres above present mean sea level, which is compatible with the altitude of the upper marine beds at Herzeele. The marine beds in the Nar Valley area, however, range from 5 to 20 metres above present sea level, and the corresponding maximum sea level during deposition has been estimated at 23 metres above present sea level as a minimum by West (1972).

Recently the presence of marine Holsteinian beds has been demonstrated by pollen-analytical and stratigraphical investigations in a number of borings from the North Sea floor to the north of The Netherlands. In general these beds occur between about 30 to 50 metres below sea-floor, which means that the top of these marine deposits is at 50 to 60 metres below present sea level.

The above mentioned data suggest that the east-west running coastline in the northern Netherlands extended westward for some distance. Based on these data the present configuration of the German bight may be assumed to have developed during the Holsteinian. It can not be ascertained yet whether the southern occurrences of marine Holsteinian beds at Clacton and at Herzeele indicate an extension of the North Sea to the south or a northward extension of the Channel through the Strait of Dover or perhaps an open connection of the North Sea with the Channel through the aforementioned Strait.

The opening of the Strait of Dover

It is assumed to date that a connection of the North Sea with the Channel through the Strait of Dover came into existence in Late Pleistocene time (Destombes et al. 1975).

As pointed out by Funnell (1972) the composition of the marine faunas found in Pliocene deposits of East Anglia and the presence of some buried throughs eroded in the surface of the Chalk, are in favour of a connection of the southern North Sea with the ocean through the Channel in Late Tertiary time. The decrease in the number of species in Lower Pleistocene marine faunas in East Anglia as compared to Pliocene marine faunas makes it unlikely, that a connection with the open ocean persisted during the early part of the Pleistocene.

The new data presented by Sommé, Paepe and Baeteman (this volume) may contribute to a better understanding of the entrance of the sea through the Strait of Dover. As pointed out by these authors marine deposits of Middle Pleistocene age have been found in the coastal area of northernmost France and southwestern Belgium (Herzeele Formation, Loon Formation). From their geographical distribution it is believed that these beds have been laid down in a sea which may have penetrated from the Southwest through an open Strait of Dover.

For at least part of these deposits an Holsteinian age has been advocated (see also Vanhoorne 1962).

However, as deposits of more than one transgression appear to be present, an older age of at least some of these should not be excluded. In connection with the paleogeographic data discussed before, it may be suggested that the deposits at Herzeele and Loon are partly of Holsteinian age, and possibly also represent a counterpart of the Younger Crag deposits of East Anglia, which are related to a transgression during Pastonian time, as evidenced by West (1971).

With regard to the possibility of the Strait of Dover being open in pre-Saalian times attention should be drawn to the presence of deeply incised buried channels just south of the submerged Chalk outcrop in the Strait of Dover (Destombes et al. 1975). The inferred subglacial origin of the buried channels during the Warthe stadial of the Saalian glaciation is incompatible with our present knowledge on the paleogeography more to the North. Moreover the alleged Eemian age of the sedimentary infill is very questionable as the pollen spectra are too featureless to assist in arguments on the age of these deposits (see discussion remark by C. Turner in Destombes et al., loc. cit.).

In our opinion the above mentioned buried channels may be tentatively considered to be caused by tidal scour during an early stage in the development of an open Strait of Dover.

REFERENCES

Bertelsen, D., 1972: Azolla species from the Pleistocene of the Central North Sea area. *Grana Palynologica 12*, 131–145.

Bristow, C. R. and Cox, F. C., 1973: The Gipping Till: a reappraisal of East Anglian stratigraphy. *J. Geol. Soc. 129*, 1–37.

Catt, J. A., 1977: Guidebook for Excursion C7 (Yorkshire and Lincolnshire), Xth INQUA Congress, Birmingham 1977, 56 pp.

Destombes, J.-P., Shephard-Thorn, E. R., Redding, J. H. and Morzadec-Kerfourn, M. T., 1975: A buried valley system in the Strait of Dover. *Philos. Trans. R. Soc. London A 279*, 243–256.

Doppert, J. W. Chr., Ruegg, G. H. J., Staalduinen, C. J. van, Zagwijn, W. H., and Zandstra, J. G., 1975: Formaties van het Kwartair en Boven-Tertiair in Nederland. In: W. H. Zagwijn & C.J. van Staalduinen (editors), Toelichting bij Geologische Overzichtskaarten van Nederland, 11–56, Rijks Geologische Dienst, Haarlem.

Funnel, B. M., 1972: The History of the North Sea. *Bull. Geol. Soc. Norfolk 22*, 2–10.

Gaunt, G. D., Bartley, D. D., Harland, R., 1974: Two interglacial deposits proved in boreholes in the southern part of the Vale of York and their bearing on contemporaneous sea levels. *Bull. Geol. Survey G. B. 48*, 1–23.

Harmer, F. W., 1902: A Sketch of the Later Tertiary History of East Anglia. *Proc. Geol. Assoc. London 17*, 416–479.

Montfrans, H. M. van, 1971: Palaeomagnetic dating in the North Sea Basin. Thesis, Amsterdam, 113 pp.

Oele, E., 1969: The Quaternary Geology of the Dutch part of the North Sea, North of the Frisian Isles. *Geol. Mijnbouw 48*, 467–480.

Pike, K. & Godwin, H., 1953: The Interglacial at Clacton-on-Sea, Essex, *Quat. J. geol. Soc. Lond. 108*, 261–272.

Stevens, L. A., 1960: The interglacial of the Nar Valley, Norfolk, *Quat. J. geol. Soc. Lond. 115*, 291–315.

Vanhoorne, R., 1962: Het interglaciale veen te Loo (België). *Natuurwet. Tijdschr. 44*, 58–64.

Watts, W. A., 1959: Pollen spectra from the interglacial deposits at Kirmington, Lincolnshire, *Proc. Yorks. geol. Soc. 32*, 145–152.

West, R. G., 1968: Pleistocene Geology and Biology with special reference to the British Isles. Longmans, London, 377 pp.

West, R. G., 1971: The Stratigraphical position of the Norwich Crag in relation to the Cromer Forest Bed Series. *Bull. Geol. Soc. Norfolk 21*, 17–23.

West, R. G., 1972: Relative land-sea-level changes in southeastern England during the Pleistocene.

Philos. Trans. R. Soc. London A, 272, 87–98.

Woldstedt, P., 1950: Norddeutschland und angrenzende Gebiete im Eiszeitalter. Stuttgart, 464 pp.

Woodland, A. W., 1970: The buried tunnel-valleys of East Anglia. *Proc. Yorksh. Geol. Soc. 37,* 521–578.

Zagwijn, W. H., 1973: Pollenanalytic studies of Holsteinian and Saalian Beds in the Northern Netherlands. *Meded. Rijks Geol. Dienst Nieuwe Ser. 24,* 139–156.

Zagwijn, W. H., 1974: The Palaeogeographic evolution of the Netherlands during the Quaternary. *Geol. Mijnbouw 53,* 369–385.

Zagwijn, W. H., 1975: Variations in climate as shown by pollen analysis, especially in the Lower Pleistocene of Europe. In A. E. Wright and F. Moseley (editors), Ice Ages: ancient and modern, 137–152, *Geol. J. Spec. Issue 6.*

Zonneveld, J. I. S., 1958: Lithostratigrafische eenheden in het Nederlandse Pleistoceen. *Meded. Geol. Sticht. Nieuwe Ser. 12,* 31–64.

Chapter IV

The development of ancient shorelines since the Eemian

Introduction

In this chapter the present knowledge of the development of ancient shorelines since the beginning of the Eemian is compiled. In eight contributions geologists, physical geographers and palynologists from Norway, Sweden, Denmark, the Federal Republic of Germany, the Netherlands, Belgium, France and the United Kingdom have described the Late Pleistocene and Holocene history of their coastal area and, as far as possible, of the adjacent part of the North Sea. A ninth contribution deals with the Late Quaternary sedimentation in the North Sea itself. In some cases the authors had to start with a review of the older Pleistocene development.

All authors and, in addition, Dr. M. J. Tooley, Durham, contributed to the compilation of the colour chart by providing data pertaining to the Eemian and Holocene shorelines. This chart is described in more detail in the following paragraphs.

The Eemian shoreline depicts the limit of marine sedimentation during the maximum of the Eemian transgression, most probably a metachronous line. As regards the Holocene, the configuration of the shorelines at 8700 B.P., 7500 B.P., 5000 B.P. and 2000 B.P. is shown. Moreover, in order to present a combined picture of the Holocene ingression and to enable inclusion of data from isostatically uplifted areas, a so-called upper marine limit — a metachronous line — was constructed.

The shoreline towards the end of the Boreal is assumed to be represented by the shoreline at 8700 B.P. At that time the sea level in the southern North Sea was about 36 m below the present sea level. Consequently, an extensive part of the southern North Sea area was still dry land.

At 7500 B.P. the shoreline in this area locally reached the present shoreline. Rapid rise of sea level during Early Holocene time resulted in the interpreted shoreline configuration.

The transition Atlantic — Subboreal is generally dated at 5000 B.P. The shoreline at this date is more or less representative of the shoreline during the Calais III transgressive phase (5300—4700 B.P.).

The shoreline at 2000 B.P. is that of the earlier part of the Subatlantic stage and the regression phase after the deposition of the Dunkerque I sediments.

IV-*a*

Late and Post-Weichselian shore level changes in South Norway

ULF HAFSTEN

Hafsten, U., 1979: Late and Post-Weichselian shore level changes in South Norway. In E. Oele, R. T. E. Schüttenhelm & A. J. Wiggers (editors), The Quaternary History of the North Sea, 45—59. *Acta Univ. Ups. Symp. Univ. Ups. Annum Quingentesimum Celebrantis: 2*, Uppsala. ISBN 91-554-0495-2.

The results obtained so far show clearly the great differences in Late and post-Weichselian (Flandrian) shore level changes from one region to another along the coast of South Norway.

In the eastern part (viz. the Oslofjord and Trondheim regions), formerly heavily depressed by glaciation, the early post-Weichselian drop in shore level was so sudden that no less than two thirds of the total post-Weichselian shore displacement had taken place before the transition to the Atlantic period (i.e. prior to about 8200 B.P.); this means a rate of 7—8 m per century on the average for the Preboreal (about 10300—9400 B.P.) and the Boreal (about 9400—8200 B.P.).

As a result of differential uplift, the marine isobases generally become situated lower from the central part towards the coast. In western Norway the shoreline isobases during the deglaciation period (Younger Dryas/Preboreal) drop from 120—130 m at the head of the fjord systems to 20 m or less in the outer coast regions. Within the Oslofjord region the isobases for Early Preboreal time show a lowering from north (max. height 221 m) to south of about 0.8 m/km.

Submergence with a shoreline rise, as a result of post-Weichselian (Tapes) transgressions, occurred only in the peripheral, western and southernmost coast regions. The rise in shoreline since the Boreal transgression minimum on the extreme southwest coast (Jæren—Fonnes) may be in the order of 6—7 m. At Fonnes, the Middle Atlantic transgression maximum reached a height above present sea level of 11 m, whereas the Boreal regression minimum is 4 m.

Submarine peat, probably of Boreal age, collected by chance off the coast of Lista near the southernmost point of Norway, indicates that the Boreal regression minimum was well below present sea level in that area, probably at a depth of about 4 m. Beach ridges and stratigraphic evidence of the post-Weichselian transgression up to a height of 7 m above present sea-level suggest a maximum rise of the shoreline of about 11 m.

The southwest coast, which became deglaciated already during early Late Weichselian time (prior to 13000 B.P.), shows also evidence of a Late Weichselian transgression.

Dr U. Hafsten, Universitetet i Trondheim, Norges Lærerhøgskole, Botanisk Institutt, N-7000 Trondheim, Norway.

Introduction

Norway's long coastline bordering the Atlantic Ocean and its location close to the Weichselian glaciation centre offers ample opportunities for studying ancient shoreline changes, i.e. primarily the relative shoreline changes resulting from the interaction of eustatic sea-level changes and isostatic crustal movements. Because the last glaciation obliterated or disordered nearly all pre-Weichselian evidence, the possibilities of ancient shoreline studies in Norway are almost exclusively restricted to Late and post-Weichselian indications.

Marine sediments deposited during Late Saalian, Eemian and Early Weichselian time at Fjøsanger in Bergen, western Norway, indicate that the shoreline during that time was situated at least 5 m higher than today (Mangerud, in prep.). Marine sediments and high-lying strand marks probably of pre-Weichselian or Early Weichselian age are also found in Jæren, in the southwest.

The traces of eustatic and isostatic movements for many years have drawn the attention of Norwegian research workers. As early as 1838, the presence was known of subfossil mollusc beds and other traces of the sea far above present sea-level; Keilhau concluded on the basis thereof to a postglacial uplift of as much as 600 feet.

Research was continued in the 1860's by the pioneer of mollusc fauna studies, Michael Sars (e.g. 1865). His studies promoted a long series of shore level investigations, which culminated in W.C. Brøgger's impressive monograph (1900—01) on the raised, subfossil mollusc beds in the Kristiania (now Oslo) region, and stressed their significance for postglacial "niveau changes" within this area. Much attention has also been paid to the numerous glaciofluvial terraces in the fjord district of western Norway and to the raised beaches and strandlines, most of which are found in northern Norway.

Information on shoreline changes obtained from peat bog studies was first mentioned by Blytt (1876). He observed that the thickness (viz. age) of the peat increases and that the sequence of peat and stump layers (viz. wet and dry climatic periods) becomes more complete from present sea-level towards the higher, early postglacial marine level. Blytt realized that isostatic recovery and related shoreline displacements had been continual processes, although at rates that varied, from Late Weichselian to present time. In this respect he was opposed to e.g. the view of Kjerulf (1871, 1879), who considered isostatic uplift as a stepwise process, interrupted by stagnation periods in which the glaciofluvial terraces were formed or modified by the sea.

Methods

More than a hundred years of observations on the highest shoreline ("marine limit", M.L.) during or after the deglaciation period and on the post-Weichselian sea-level transgression maximum (O. Holtedahl 1953, 1960), have been useful for obtaining a preliminary, general in-

Fig. IV-1. The Late and/or post-Weichselian shoreline displacement curves for South Norway, obtained up to 1976. M. L. — highest marine level (marine limit) in the area.

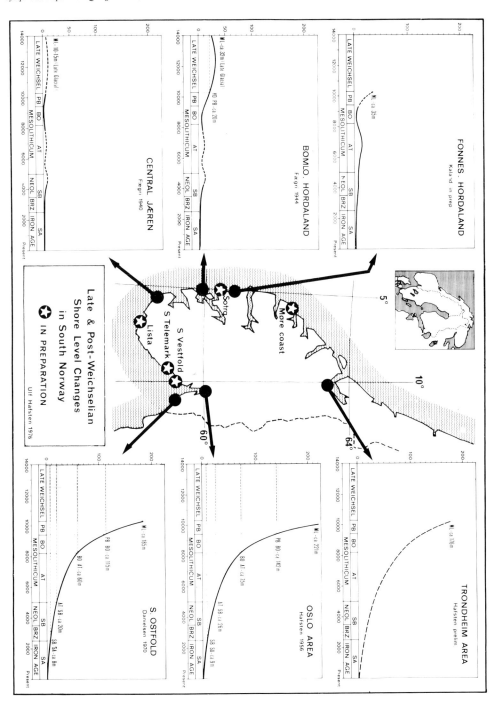

sight in the sea-level changes in Norway. These results are inadequate, however, for constructing continuous curves for the shoreline displacement, in view of the lack of precise age determinations.

The present contribution will be restricted to the results of studies on establishing curves for the Late and/or post-Weichselian shorelevel changes in various regions, based primarily on lithostratigraphical and biostratigraphical investigations (pollen and phytoplankton analyses) of bog and lake deposits (see Fig. IV-7 and Hafsten 1958*a*) or on radiometrically dated material from tills or boulder clay, glaciofluvial accumulations or abrasion terraces (see Fig. IV-2). As several of the investigations were completed or started before the application of the radiocarbon dating method, the dating of most shoreline displacement curves presented in Fig. IV-1 was based only on pollen chronological data. In order to facilitate the comparison between the various displacement curves, the author has incorporated all curves into a standard (Blytt—Sernander) climatic period division, based on the available ^{14}C-datings (Hafsten 1970).

It should be realized that most studies of this kind do not provide more than a general picture of the shoreline displacement. The main reason is that small areas very seldom have a sufficient number of suitable basins between the Late Weichselian marine limit and the present day sea-level for a detailed reconstruction of the actual shore displacement, particularly of the smaller oscillations that may have taken place.

The six displacement curves shown in Fig. IV-1 have been converted to an equal time scale (referred to above) and to an equal scale of depth, regardless of the variations in highest marine level from one region to another, and regardless of the fact that some of the regions were covered either by ice or by sea in Late Weichselian time (see shading). The displacement curve for Jæren has been constructed by the present author on the basis of the results of studies by Fægri (1940), which were described but not graphically depicted. The curve for the Trondheim area is based on investigations of only four (orographically well-distributed) sites, and therefore must be considered as tentative.

The shorelines indicated for various period transitions do not imply that these levels always refer to sites situated at precisely those heights. The asterisks for areas from which similar curves are in preparation refer to investigations by Henningsmoen (S. Vestfold), Stabell (S. Telemark), Hafsten and collaborators (Lista, Trøndelag), Kaland and collaborators (Sotra) and Mangerud and collaborators (Møre coast).

Discussion of differential movements and the resulting shorelines

The shoreline displacement diagrams in Fig. IV-1 distinctly show the variation from one region on the coast of South Norway to another. One obvious difference between the studied areas is that the highest marine level (M.L.), i.e. the highest recorded relative sea level, is located considerably higher in the central and eastern areas (viz. the Oslofjord and Trondheim regions) than in the peripheral, western and extreme southern areas. The position of glaciofluvial terraces in the fjord district of western Norway indicates that the position of the highest marine level becomes considerably higher from the coastal regions towards the inland part of the fjord systems. In the Hardangerfjord and Sognefjord marine deposits are known from the head of the eastern tributary fjords, and reach heights of 120—130 m above present sea-level (O. Holtedahl 1953, 1960). The different positions of the highest marine level in the eastern and western fjord areas must be attributed to the proximity of the eastern areas to the north-European glaciation centre.

The highest marine levels in Norway are in most cases related to the local event of deglaciation and are consequently not always synchronous. This applies also to the Oslofjord region, where Danielsen's Østfold diagram refers to the Younger Dryas Ra moraine district, whereas the diagram for the Oslo area, 60—70 km further north, refers to a somewhat later stage in the Preboreal ice retreat, viz. the Aker substage (Holtedahl 1924). The shoreline changes outside the Younger Dryas Ra moraine in the Oslofjord prior to Preboreal time cannot be traced as this part of the country does not contain areas situated above sea-level during Late Weichselian time. This is in contrast to western Norway, where a rather broad strip of the outer coast region became deglaciated and rose above sea-level already in early Late Weichselian time (Fig. IV-2).

The highest marine level in Norway is recorded in the Oslo area, situated at 60° N latitude, where the Early Preboreal shoreline reached a height of 221 m above present sea-level (Hafsten 1956). From Oslo to the south, the early post-Weichselian marine level drops to about 185 m in Østfold (Danielsen 1970) and to about 155 m in south Vestfold (Henningsmoen, personal comm.). Further southward, on the coast of Telemark county, it drops to about 140 m in the Langesundfjord district and to 120 m in the Kragerøfjord district (Stabell 1976, unpubl.). These data imply a lowering of the early Preboreal marine level from north to south in the Oslofjord region of about 0.8 m/km.

In the Trondheimsfjord region, located approximately at the same longitude as the Oslofjord region but about 3.5° further north (at 63°30' N), the early post-Weichselian shoreline can be observed at an altitude of c. 175 m in the Trondheim area and at c. 90 m on the island of Hitra, 80—90 km west of Trondheim (Undås 1942).

On the farthest southwest and west coast, which was free of ice and covered with vegetation already in early Late Weichselian time, the highest marine level nowhere exceeds a height of 35—40 m above present sea-level. The Late Weichselian or early post-Weichselian marine limit is found higher, going from the extreme southern part (Lista-Lindesnes area) to the north, i.e. 6—8 m in Lista (at lat. 58° N), around 10 m in the southern part, 15 m in the central and 24 m in the northern part of Jæren, between 27 and 37 m in the Bømlo archipelago in the outer part of the Hardangerfjord (at lat. 59°45' N), and at 35 m (Early Preboreal) at Fonnes (at lat. 60°45' N). In Lista an almost complete Late Weichselian sequence, deposited entirely above sea-level, has been found in a basin at 7 m above present sea-level, indicating that the marine limit at that time was below the present 7 m level (Hafsten 1963).

The differences in upheaval of the land causes the marine limit to rise from the peripheral coast districts towards the central and eastern parts of South Norway. Studies on the deglaciation history of various parts of western Norway, carried out fairly recently by Follestad (1972), H. Holtedahl (1964, 1975), Mangerud (1970), Undås (1963), Aarseth (1971), Aarseth & Mangerud (1974), and others, have produced detailed information about the shore levels, particularly in the Younger Dryas time (Fig. IV-2). The shoreline data and the isobase map presented by Aarseth & Mangerud (1974) for the Younger Dryas are primarily based on the observed heights of glaciofluvial terraces related to moraines. These terraces are either abrasion terraces formed on the distal side of the moraines, or accumulation terraces built up to sea-level, at the glacier front. The dated samples are almost exclusively from shell material embedded in till.

The post-Weichselian transgressions can be found only in the most southern and western coast regions. The absence of transgressions in the eastern part (viz. the Oslofjord and Trondheim regions) must be attributed to the fact that the strong postglacial isostatic uplift here prevented a eustatic submergence. The reduced rate of shoreline displacement is represented by a bend in the displacement curves approximately at the Boreal/Atlantic boundary. In the Oslofjord diagrams, this bend may be traced at c. 75 m in Oslo, at c. 60 m in Østfold and at c. 50 m in South Telemark.

The Late Weichselian shoreline displacement

Curves for the Late Weichselian shoreline displacement in South Norway have been constructed so far only for the southwest coast, viz. the lowlying coastal plain of Jæren in Rogaland county, and for the Bømle archipelago at the mouth of the Hardangerfjord in South Hordaland (Fig. IV-1). Isobase maps and equidistant shoreline diagrams for various Late Weichselian shorelines in South Norway have been presented, however, e.g. for the Older and Younger Dryas in the extreme southern part (Fevik-Kristiansand-Lista) by Andersen (1960), and for the Younger Dryas in the Hardangerfjord-Sognefjord-Dalsfjord region by Aarseth & Mangerud (1974) and others.

The curve for the Late Weichselian shoreline displacement in Jæren (drawn on the basis of Fægri's results) is based primarily on generally well-preserved old beach ridges. In the northern part Fægri (1940) distinguished in the system of beach ridges an upper level ("oberste Stufe"), which probably corresponds to the highest Late Weichselian shore level in Jæren, viz. about 24 m at Randaberg and 19 m at Tananger. In the central part (Fig. IV-1) this beach ridge is supposed to be present at 15 m, but it has not yet been definitely identified. Further north, in Bømlo there are no raised beach ridges or other morphological features on the basis of which the highest Late Weichselian shore level can be determined, Fægri (1944) judged this level somewhere between 27 and 37 m from the position of erratic boulders down to a height of 37 m.

Of special interest is the evidence of a Late Weichselian transgression, postulated also for the outlying Bømlo archipelago, although the Late Weichselian stratigraphy is here not well known and raised beach ridges are practically absent. In Jæren the maximum level of this socalled Alvevatn transgression has been determined at about 9 m (mean tide) in the central part (Figgjo valley), and at 14 m (Tananger) and 19 m (Randaberg) in the northern part. This indicates an actual tilting or a difference in submergence in Jæren at that time of as much as 0.5 m/km from south to north. In Bømlo, Fægri found stratigraphic evidence of a Late Weichselian transgression from below 27 m to about 32 m. He suggested that this transgression must be parallel to the Alvevatn transgression in Jæren; he emphasized, however, that at that time he was unable to prove the synchronism. The beach ridge damming up the small lake Alvevatn (in central Jæren), after which the

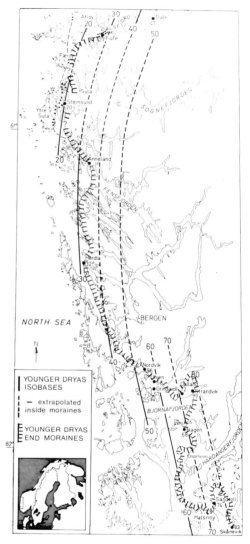

Fig. IV-2. Palaeogeographical map of the coastal region between Hardangerfjord and Dalsfjord in western Norway, with the position of the Herdla moraines, formed during the advancing Younger Dryas ice, and the corresponding shoreline isobases, according to Aarseth & Mangerud (1974). Localities and elevations (in m) of Younger Dryas marine terraces are marked with dots and triangles respectively.

transgression was named, was considered of pre-Allerød (Bølling?) age.

A preliminary shoreline displacement curve for the island of Sotra has recently been constructed showing a Late Weichselian transgression starting in Bølling, when the shoreline regression had reached the present 25 m level, and culminating during early Allerød, just above the present 35 m level. The main regression began in the Allerød, and by mid-Boreal the shoreline had fallen to a minimum height of c. 5 m above present sea-level.

For northern Jæren, Fægri also mentioned a system of beach ridges and abrasion terraces ("mittlere Stufen"), located at a level somewhat lower (16 m) than the ridges belonging to the Alvevatn transgression. The results, however, which may be based on these indistinct features, are too inconclusive to be used in the context of this paper. It is obvious that the whole complex of Late Weichselian shoreline changes in Jæren (which is a most puzzling area in many respects) requires additional studies, preferably in combination with radiometric dating methods. Such studies are now in progress.

In the Yrkje region, c. 40 km SE of Bømlo, stratigraphic and radiometric evidence has been reported recently for a Late Weichselian, marine transgression which exceeded the present 38 m level between c. 10.500 and 9.500 B.P. (Anundsen 1978).

The Post-Weichselian (Flandrian) shoreline displacement

As mentioned above, the shoreline displacement curves are distinctly different in the western coast regions and the central and eastern part of the country. In the Oslofjord and inner Trondheimsfjord regions, which were deeply submerged at the time of deglaciation (Fig. IV-3), the post-Weichselian isostatic uplift was followed by an uninterrupted recession of the shoreline, which reached its maximum during the early post-Weichselian periods of rapid ice melting (Preboreal and Boreal). Assuming 800 years for the part of the Preboreal represented in the diagrams of the Oslofjord region, 1200 years for the Boreal, 3200 years for the Atlantic, and 2500 years for the Subboreal and Subatlantic, respectivley, one obtains the following average values for displacement per century for each climatic period in Oslo and the Ra district of Østfold (Fig. IV-4 and IV-5).

It appears that the early post-Weichselian rise in shore level was so rapid in the Oslo area and in Østfold, that as much as two thirds of the total post-Weichselian shoreline displacement had taken place at the transition to the Atlantic period. Even then, during the rapid melting in the polar regions that caused a considerable rise of the sea-level, the eustatic sea-level rise here never exceeded the isostatic uplift. This applies also to south Vestfold and Telemark, as well as to the region east of the Skagerrak and as far south as Bohuslän in Sweden, about 230 km SSE of the Oslo area. On the western Skagerrak coast (viz. the coast region of Aust-Agder), evidence of post-Weichselian shoreline displacement is not yet available, but somewhere in this region the Tapes transgression is expected to be represented by submergence of areas that had already been raised above sea-level. Palaeogeographic maps with shorelines in the Oslofjord region in the Early Preboreal and at the Atlantic/Subboreal transition are shown in Fig. IV-5.

In the western coastal area of Norway the glacial depression was moderate and the subsequent isostatic uplift insignificant. Deposits at low altitudes provide here evidence of a eustatic rise in sea-level, for a long time exceeding the isostatic land uplift; these marine deposits rest on terrestrial peat or other material of freshwater origin and are covered again with deposits of lacustrine or terrestrial origin (Fig. IV-7). Resting on peat or other terrestrial surface material, many distinct beach ridges appear to consist of regular thick layers of pebbles, extending for kilometres along the coast. In many places, such beach ridges may be damming up bogs or lakes, e.g. in Lista and Jæren, or they are bordered landwards by more or less dried up or artificially drained depressions, which represent earlier lagoonal areas, e.g. on the coast of Møre and Romsdal (Hafsten & Tallantire 1978).

The post-Weichselian transgression phase can be correlated to the younger zones in the

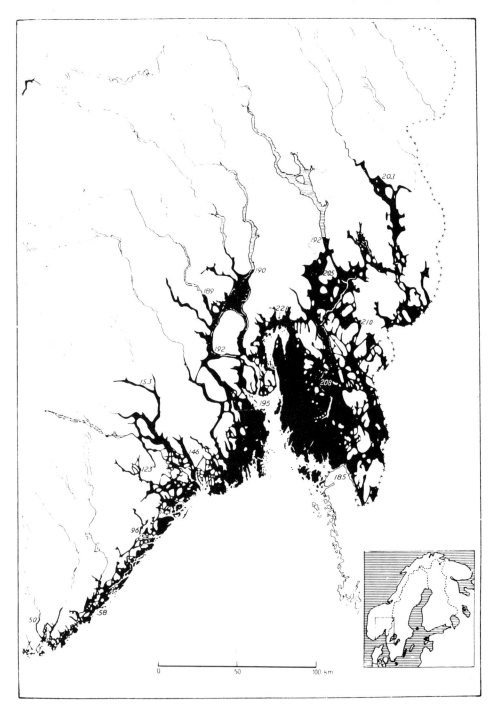

PERIOD REGION	PB	BO	AT	SB	SA
Oslo area	10 m/c	5.5 m/c	1.5 m/c	0.7 m/c	0.36 m/c
Østfold	9 m/c	4.5 m/c	1.3 m/c	0.5 m/c	0.32 m/c

Fig. IV-4. Approximate average values in metres (m) for shoreline displacement per century (c) for each post-Weichselian period, for the Oslo area and the Ra district of Østfold, according to Hafsten (1956) and Danielsen (1970).

system of subfossil mollusc beds that were described long ago, mainly for the Oslofjord and Trondheimsfjord regions; the *Tapes, Trivia* and part of the *Ostrea* zones, in the Oslofjord region, may be found up to the present 75 m level (Hafsten 1958a: 86—87). For this reason the post-Weichselian transgression(s) in the Skagerrak—Kattegat region is widely known as the Tapes transgression which is a synonym for the Baltic Litorina transgression.

In addition to the transgression areas presented in Fig. IV-1, a number of other localities is known where typical transgression sequences are present, for which continuous shoreline displacement curves have not yet been constructed. Many of these localities are on the extreme south coast (Kristiansand-Lista region), where investigations have been made (Andersen 1960:80) and, more recently, the results of radiocarbon datings are being applied (see later). On the west coast, beach ridges, abrasion platforms and sedimentary sequences with evidence of a Tapes subsidence may be found up to 10—15 m above present sea-level.

The Boreal regression minimum

The regression minimum, i.e. the moment when the eustatic rise in sea-level caught up with and began to exceed the isostatic recovery, belongs in all diagrams to the Boreal time, although the actual height of the shoreline at that time varies from one region to another. In Lista, where peat and lacustrine gyttja of Boreal age was found offshore at 4 m below present sea level (Holmboe 1909, Sollesnes & Fægri 1951), the shoreline at that time must have been lower than present sea-level. This may be true also for Jæren, although the most probable shoreline here corresponds to the present one (Fægri 1940). Futher north, at Bømlo, Sotra and Fonnes, the Boreal regression minimum was definitely higher than present sea-level, viz. at c. 5 m in Bømlo (Fægri 1944) and Sotra, and close to 4 m at Fonnes (Kaland in prep.).

In the southernmost part of Norway, the Lista-Kristiansand region, terrestrial peat at varying altitudes underlies marine sediments deposited during the Tapes transgression. Radiocarbon dating was applied to a series of these sequences. On the basis of the data reported in Nydal (1959, 1962) and Nydal et al. (1970), the actual Tapes inundation can be traced at fairly close intervals (see Fig. IV-6).

Considering the Boreal regression minimum at Lista probably at 4 m below present sea-level and the two lower groups listed to refer to higher isobase levels than the first group, the figures indicate a fairly logical sequence for the gradual transgression and subsequent regression of the Tapes sea. The oldest (mid-Boreal) date obtained for the Tapes transgression, viz. 8770 ± 100 B.P. (T-80), applies to the lower layer of terrestrial peat (and lacustrine gyttja) in the Dostad sequence, situated 20 km WSW of Kristiansand (Gabrielsen 1959). The ^{14}C-dating is confirmed with pollen analysis (viz. samples from base and top of the terrestrial peat layer providing 15—17 % hazel and almost no alder pollen) and by a supplementary ^{14}C-dating from the base of the peat layer, indicating an age of 9410 ± 220 B.P. (T-81). From the presence of marine diatoms in the essentially terrestrial peat layer, Gabrielsen (l.c.) concluded that the sea level at the time when this peat layer was formed cannot have been much lower than the basin

Fig. IV-3. Palaeogoegraphical map of SE Norway, showing (black) areas which were below sea level during Late Weichselian and early post-Weichselian time, according to O. Holtedahl (1953). Figures indicate marine limit in m.

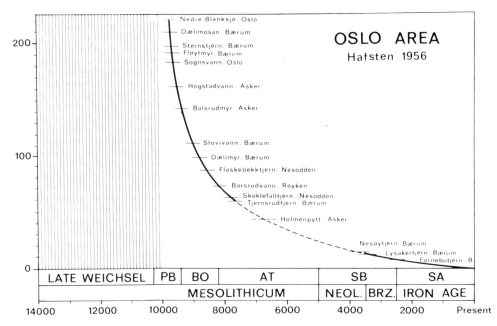

Fig. IV-5. Post-Weichselian shoreline displacement curve for the Oslo area, constructed on the basis of biostratigraphical analysis of sediments and peat deposits from 17 lakes and bogs situated between present sea-level and the early Preboreal marine limit (221 m). Vertical axis: Height scale in metres above present sea-level. Horizontal axis: Time scale for Late and Post-Weichselian time, comprising Blytt-Sernander's climatic periods as well as main cultural epochs. As the investigation was carried out prior to the application of the radiocarbon method, the age determination of the marine/lacustrine ("isolation") contact in each basin is based only on relative pollen chronology. However, as the major features in the vegetational development within the area is today determined also by radiocarbon measurements, the shore displacement curve presented here rests, at least indirectly, on an absolute chronology. Hatched: Period deprived of investigation, because area at that time was glaciated. Revised version of Hafsten's (1956) curve.

threshold, which is 2 m above present sea-level. The distinct separation between basal pine wood peat (1.2 m thick) and overlying marine clay (2.15 m thick) at Oddernes (lowermost group) was not radiocarbon dated but represents, on the basis of pollen analysis, the time of rapid expansion of alder and warmth-demanding tree species (QM), i.e. the Zone V/VI transition (Hafsten 1958b).

The post-Weichselian transgression maximum

In south Scandinavia the Tapes, or Litorina, transgression does not represent one, prolonged, phase of flooding but it includes intervals of temporary recessions, viz. the four transgression phases (Tapes I-IV) in Denmark (Iversen 1937, 1967, Jessen 1937, Krog 1968) and the many post-Weichselian shoreline oscillations in South Sweden (Florin 1944, Berglund 1964, Mörner 1969).

The west coast. — In Jæren and Bømlo, long before the invention of the radiocarbon dating method, Fægri collected evidence of a Late Atlantic lowering of the shoreline, separating an "older Tapes transgression" and a "younger" one (Fægri 1940, 1944). The two transgression phases were dated with pollen analysis at mid-Atlantic and Early Subboreal time respectively. The younger transgression phase was reported to represent the highest transgression level or

Locality	Threshold level	Marine sequence Lower boundary	Marine sequence Upper boundary
Vaage, Spangereid	c. 0.5 m	7950 ± 170 (T-208)	not dated
Røyrtjønn, Lista	c. 7 m	6770 ± 150 (T-635)	4850 ± 100 (T-635)
Dostad, Harkmark	c. 2 m	8770 ± 100 (T-80)	not dated
Ravnemyr, Søgne	c. 4 m	7400 ± 250 (T-209)	3050 ± 100 (T-281)
Oftenes, Søgne	c. 5 m	8300 ± 250 (T-273)	not dated
Oddernes, Kr. sand	c. 9.5 m	Zone V/VI bound.	not dated

Fig. IV-6. Radiocarbon years B.P. for the Tapes transgression and the subsequent separation from the sea of various basins in Vest-Agder, grouped geographically from west to east and arranged according to height of basin threshold above present sea-level. According to Nydal (1959, 1962), Nydal et al. (1970), Gabrielsen (1959) and Hafsten (1958b and unpublished).

post-Weichselian transgression maximum. In Jæren this maximum was mentioned to reach a height of 6 m in the south (Ogna) and 8 m in the north (Randaberg), whereas the older phase (at Figgjo valley, in the central part) might have reached a height of only 3 m above present sea-level. The corresponding heights for Bømlo, which were based primarily on heights of raised pumice layers, are 11 m and 8 m, respectively. The height of the intermediate regression minimum was determined (on stratigraphic and archaeological evidence) between 5 and 6.5 m, indicating a temporary lowering of the shoreline in Late Atlantic time of 2–3 m.

A re-examination of the stratigraphic section on which the evidence of the two Tapes transgression phases in Bømlo was based, indicates that the idea of a transient, Late Atlantic lowering of the shoreline is the result of a stratigraphic misinterpretation. The revised Bømlo curve is, in fact, very similar to the one presented for Sotra, in showing only one, very protracted, transgression peak in excess of the present 11 m level. The transgression maximum of c. 11 m in Bømle and Sotra thus respresents a total shore displacement since Boreal time in excess of 6 m.

In contrast to the Late Weichselian Alvevatn transgression, which indicates a tilting of Jæren at that time of as much as 0.5 m/km from south to north (see above), the Tapes shorelines reflect practically no tilting. Based primarily on heights of beach ridges formed during the transgression maximum, Fægri arrived at a rise in "the Tapes line" from south (Figgjo valley) to north (Randaberg) of not more than 0.1 m/km, whereas the gradient from east to west (at right angles to the isobases, which are practically parallel to the coast) was estimated at 0.3 m/km. A corresponding gradient was estimated by Undås (1955, 1964) for the Hardangerfjord and Sognefjord regions.

The shoreline displacement diagram for Fonnes, in northern Hordaland, 130 km north of Bømlo, differs from the post-Weichselian development described for Jæren and Bømlo, both with regard to the moment of the transgression maximum and to the general course of the displacement curve. Detailed stratigraphic studies, combined with many radiocarbon datings, lead here to only one, Atlantic, transgression phase, followed by a prolonged and slow regression during Late Atlantic and Subboreal time. The transgression maximum is placed at c.11 m, and the Boreal regression minimum at 4 m above present sea-level. This indicates a maximum post-Weichselian rise in height of the shoreline at Fonnes of 7 m, which is in good accordance with the results obtained for Bømlo and Sotra.

The northwest coast. — North of Fonnes, in the northwestern corner of South Norway, no detailed investigations on the shoreline changes have been made so far. However, in Møre and Romsdal county, viz. in Borgund, east of Ålesund, at 62°30′N (Fig. IV-7), and on Haramsøy in the Nordøyane islands off the Møre coast, at 62°40′ N (Hafsten & Tallantire 1978), stratigraphic sequences have been found, which according to pollen analysis and ^{14}C-datings belong to the Tapes transgression. The mentioned sequences have a very sharp boundary between basal terrestrial peat or lacustrine gyttja and overlying marine clay or beach ridge

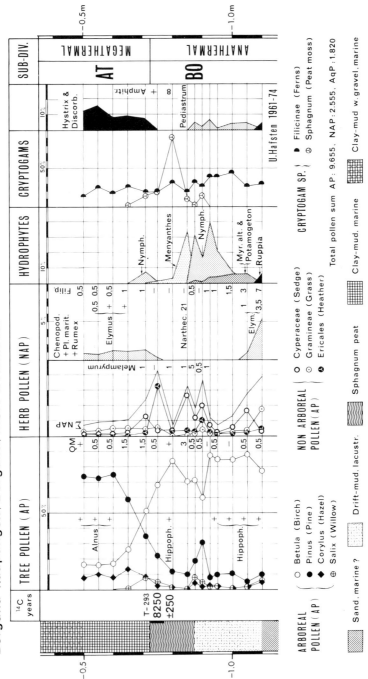

pebbles. A sample from the top layer of the lagoonal gyttja underlying the beach ridge at Haramsøy indicates an age for the Tapes inundation of 7280 ± 100 B.P. (T-831), whereas the peat layer just below the marine clay in the Borgund sequence may be a thousand years older, viz. 8250 ± 250 B.P. (T-293). The difference in age is attributed to the position of the Borgund site closer to present sea-level, whereas the base of the beach ridge at Haramsøy is located at 7.5 m above present sea level. The beach ridge proper, extending kilometres along the coast, has a level top at a height of 10 m, while the lagoon floor is 2.3 m lower. The shoreline positions at Haramsøy are similar to those described for Fonnes, both with regard to the Tapes transgression and to the highest marine limit, reported to be located at approximately the same height in this archipelago as at Fonnes (O. Holtedahl 1953). In both places the sea must have been higher in Early Atlantic time than the present 7.5–8 m level.

The south coast. — The investigations in the southernmost coast region have not yet revealed more than one post-Weichselian transgressive sequence. In the Lista area the investigations are hampered by the fact that the Boreal regression minimum and probably also some of the critical levels for tracing possible regressive shoreline phases during the Tapes transgression are situated below present sea-level.

As to the marine limit, the Lista peninsula seems to equal the areas further south in Scandinavia; the highest marine level coincides with the Tapes transgression maximum. The accurate height of this maximum has not yet been determined by means of stratigraphical studies. Judging from the Røyrtjønn sequence (Fig. IV-6), however, the assumption by O. Holtedahl (1953, 1960) and Andersen (1960) that the top level of 8 m of the post-Weichselian beach ridge system in Lista represents the marine limit, may be correct. This is based on the very moderate thickness of the marine clay that overlies the lacustrine gyttja, and covers a period of less than 2000 years. Therefore, the threshold level of the Røyrtjønn basin, 7 m above present sea level, may not be situated far below the maximum transgression level in this area. A corresponding very thin marine deposit at 5.4 m below the present bog surface, dating from "the age of maximum transgression" (Gabrielsen, in Nydal 1962), has also been found in Launesmyr in Harkmark, 60–70 km east of Lista, the basinal threshold level of which reaches a height of 8 m above sea-level. The age of this marine deposit (shell fragments, mainly of *Ostrea edulis*), 6800 ± 170 B.P. (T-292), also coincides with the transgression maximum in Kaland's diagram for Fonnes (Fig. IV-1).

It remains to be mentioned that central Jæren, as well as the Boknfjord islands immediately to the north, contains marine deposits (not yet radiocarbon dated), which are too low to be related to the Tapes transgression levels (compilation by Fægeri 1940). It may be assumed that these beds ("niedrigere Stufen") were formed in a next transgressive or stagnating regressive stage following the period of the Tapes transgression, e.g. during Late Subboreal (Bronze Age?) or Early Subatlantic time (Roman Iron Age?).

The objective of the Norwegian working programmes launched under the I.G.C.P. sea level project is to obtain additional information on the Late and post-Weichselian shoreline changes in Norway, particularly in regions in South Norway where little or no information is available.

Acknowledgements. — The author wishes to express his gratitude to Mr. A. Kjemperud for his assistance in preparing the shoreline maps, to Mrs. R. Waadeland for drawing the illustrations, and to Drs. A. Danielsen, K. Fægri, K. E. Henningsmoen, P. E. Kaland and J. Mangerud, for critical examination of the manuscript. The editors are responsible for the final revision of the English text.

Fig. IV-7. The result of the analyses of the Tapes transgression sequence underlying the Medieval trade centre at Borgund on the northwest coast of South Norway. The diagram illustrates the investigation method which forms the basis for the shoreline displacement curves being discussed (see Fig. IV-1).

REFERENCES

Andersen, B., 1960: Sørlandet i sen- og postglacial tid (The late- and postglacial history of Southern Norway between Fevik and Åna-Sira). *Nor. Geol. Unders. 210*, 142 pp.

Anundsen, K. 1978: Marine transgression in Younger Dryas in Norway. *Boreas 7*, 49–60.

Berglund, B. E., 1964: The Post-Glacial shore displacement in Eastern Blekinge, southeastern Sweden. *Sver. Geol. Unders. C, 599*, 47 pp.

Blytt, A., 1976: Essay on the immigration of the Norwegian Flora during alternating rainy and dry periods. 89 pp., Cammermeyer, Kristiania (Oslo).

Brøgger, W. C., 1900–01: Om de senglaciale og postglaciale nivåforandringer i Kristianiafeltet (Molluskfaunaen). *Nor. Geol. Unders. 31*, 731 pp.

Danielsen, A., 1970: Pollen-analytical Late Quaternary studies in the Ra district of Østfold, Southeast Norway. *Univ. Bergen Årb. Mat.-naturvitensk. Ser. 1969, 14*, 146 pp.

Florin, S., 1944: Havsstrandens förskjutningar och bebyggelseutvecklingen i östra Mellansverige under senkvartär tid. *Geol. Fören. Stockh. Förh. 66*, 551–634.

Follestad, B. A., 1972: The deglaciation of the southwestern part of the Folgefonn Peninsula, Hordaland. *Norg. Geol. Unders. 280*, 31–64.

Fægri, K., 1940: Quartärgeologische Untersuchungen im westlichen Norwegen. II. Zur spätquartären Geschichte Jærens. *Bergens Mus. Årb. 1939–40 Naturvitensk. R. 7*, 201 pp.

Fægri, K., 1944: Studies on the Pleistocene of Western Norway III. Bømlo. *Bergens Mus. Årb. 1943 Naturvitensk. R. 8*, 100 pp.

Gabrielsen, G., 1959: A marine transgression of Boreal age in the southernmost part of Norway. *Nature 183*, 1616–17.

Hafsten, U., 1956: Pollen-analytic investigations on the late Quaternary development in the inner Oslofjord area. *Univ. Bergen Årb. 1956, Naturvitensk. R. 8*, 161 pp.

Hafsten, U., 1958a: De senkvarære strandlinjeforskyvningene i Oslotrakten belyst ved pollenanalytiske undersøkelser (Application of pollen analysis in tracing the Late Quaternary displacement of shorelines in the inner Oslofjord area). *Nor. Geogr. Tidsskr. 16*, 74–99.

Hafsten, U., 1958b: Funn av Boreale furustammer fra Oddernes i Vest-Agder. Påvisning av Tapes-transgresjonen (Boreal pine trunks from Oddernes (Vest-Agder, Norway). Demonstration of the Tapes transgression). *Nor. Geol. Tidsskr. 38*, 313–25.

Hafsten, U., 1962: Hva myrer og tjern kan fortelle. Oslo-trakten gjennom 10 000 år. *Naturen 86*, 450–512.

Hafsetn, U., 1963: A Late-Glacial pollen profile from Lista, South Norway. *Grana Palyn. 4*, 326–337.

Hafsten, U., 1970: A sub-division of the Late Pleistocene period on a synchronous basis, intended for global and universal use. *Palaeogeogr. Palaeoclim. Palaeoecol. 7*, 279–96.

Hafsten, U. & Tallantire, P. A., 1978: Palaeoecology and Post-Weichselian shore-level changes on the coast of Møre, western Norway. *Boreas 7*, 109–122.

Holmboe, J., 1909: En undersjøisk torv ved Nordhassel paa Lister. *Naturen 33*, 235–243.

Holtedahl, H., 1964: An Allerød fauna at Os, near Bergen, Norway. *Nor. Geol. Tidsskr. 44*, 315–22.

Holtedahl, H., 1975: The geology of the Hardangerfjord, West Norway. *Nor. Geol. Unders. 323*, 87 pp.

Holtedahl, O., 1924: Studier over israND-terrassene syd for de store østlandske sjøer. *Skr. Vidensk. Selsk. Kristiania, Mat.-naturvidensk. kl. 1924, no. 14*.

Holtedahl, O., 1953: Norges geologi. *Nor. Geol. Unders. 164*, 1118 pp.

Holtedahl, O., 1960: Geology of Norway. *Nor. Geol. Unders. 208*, 540 pp.

Iversen, J., 1937: Undersøgelser over Litorinatransgressioner i Danmark. *Medd. Dansk Geol. Foren. 9*, 223–32.

Iversen, J., 1967: Naturens udvikling siden sidste istid. In "Danmarks natur" I, 345–448, Politikens Forl., København.

Jessen, K., 1937: Litorinasænkningen ved Klintesø i pollenfloristisk belysning. *Medd. Dansk Geol. Foren. 9*, 232–36.

Keilhau, B. M., 1838: Om Landjordens Stigning. Undersøgelser om hvorvidt i Norge, saaledes som i Sverig, findes Tegn til en Fremstigning af Landjorden i den nyere og nyeste geologiske Tid. *Nyt Mag. Naturvidensk. 1*, 105–254.

Kjerulf, Th., 1871: Om Skuringsmærker, Glacialformationen og Terrasser, samt Grundfjeldets og Sparagmitfjeldets Mæktighed i Norge I. Univ. Progr. 1870, 1ste Sem. 101 pp. Kristiania (Oslo).

Kjerulf, Th., 1879: Udsigt over det sydlige Norges Geologi. 262 pp. Christiania (Oslo).

Krog, H., 1968: Late-glacial and postglacial shoreline displacement in Denmark. In Means of correlation of Quaternary successions 8, 421–435, *Proc. VII. Congr. Int. Assoc. Quaternary Res.*, University of Utah press.

Mangerud, J., 1970: Late Weichselian vegetation and ice-front oscillations in the Bergen district, Western Norway. *Nor. Geogr. Tidsskr. 24*, 121–48.

Mörner, N.-A., 1969: The Late Quaternary history of the Kattegatt Sea and the Swedish West

Coast. Deglaciation, shorelevel displacement, chronology and eustacy. *Sver. Geol. Unders. C, 640,* 487 pp.

Nydal, R., 1959: Trondheim natural radiocarbon measurements I. *Radiocarbon 1,* 76–80.

Nydal, R., 1962: Trondheim natural radiocarbon measurements III. *Radiocarbon 4,* 160–181.

Nydal, C., Løvseth, K. & Syrstad, O., 1970: Trondheim natural radiocarbon measurements V. *Radiocarbon 12,* 205–237.

Sars, M., 1865: Om de i Norge forekommende fossile Dyrelevninger fra Qvartærperioden. 134 pp. Christiania (Oslo).

Sollesnes, A. F. & Fægri, K., 1951: Pollenanalytisk undersøkelse av to avleiringer fra Norges sydvestlige kyst. *Blyttia 9,* 41–58.

Stabell, B., 1976: Den senglaciale strandforskyvningen i Telemark, undersøkt ved hjelp av diatoméanalyse. Unpubl. degree thesis, Oslo Univ.

Undås, I. 1942: On the Late-Quaternary history of Møre and Trøndelag (Norway). *K. Norske Vid. Selsk. Skr. 1942 Nr. 2,* 92 pp.

Undås, I., 1955: Nivåer og boplasser i Sør-Norge. *Nor. Geol. Tidsskr. 35,* 169–78.

Undås, I., 1963: Ra-morenen i Vest-Norge. 78 pp, J. W. Eide, Bergen.

Undås, I., 1964: When were the heads of the Hardangerfjord and Sognefjord ice-free? *Nor. Geogr. Tidsskr. 19,* 291–96.

Aarseth, I., 1971: Deglaciasjonsforløpet i Bjørnafjordsområdet, Hordaland. Unpubl. degree thesis, Bergen Univ.

Aarseth, I. & Mangerud, J., 1974: Younger Dryas end moraines between Hardangerfjorden and Sognefjorden, Western Norway. *Boreas 3,* 3–22.

IV-b

Late Pleistocene and Holocene shorelines on the Swedish West Coast

K. GÖSTA ERIKSSON

Eriksson. K. Gösta, 1979: Late Pleistocene and Holocene shorelines on the Swedish West Coast. In E. Oele, R. T. E. Schüttenhelm & A. J. Wiggers (editors), The Quaternary History of the North Sea, 61—74. *Acta Univ. Ups. Symp. Univ. Ups. Annum Quingentesimum Celebrantis: 2*, Uppsala. ISBN 91-554-0495-2.

The Swedish West Coast is situated within the area once covered by the Weichselian land ice. The ice load depressed the earth's crust, which since the deglaciation has risen with decreasing intensity and will continue to do so until reaching equilibrium. The shoreline displacements along the Swedish West Coast are caused by a complex interaction between the isostatic land upheaval and the eustatic sea level movements. The northernmost part of the coastal area has risen nearly twice as much as the southernmost part. As a consequence the sequences of small transgressions and regressions are usually not synchronous along the coast. The main stages of the shore level displacements are: (a) Immediately after the deglaciation the isostacy exceeded the eustacy; this resulted in the Late Weichselian regression, the maximum of which was reached around 7500 B.C.; (b) The postglacial climatic amelioration caused the regression to turn into the early Flandrian transgression, which culminated around 5000 B.C.; and (c) The irregular regression during Middle and Late Flandrian times still continues.

Prof. K. Gösta Eriksson, Geologiska institutionen, Chalmers tekniska högskola och Göteborgs universitet, Fack, S;402 20 Göteborg 5, Sweden.

Introduction

The Swedish West Coast

The Kattegat and Skagerrak Seas, separating Sweden and Norway from Denmark and the North Sea, are, from an oceanographic point of view, a continuation of the North Sea proper.

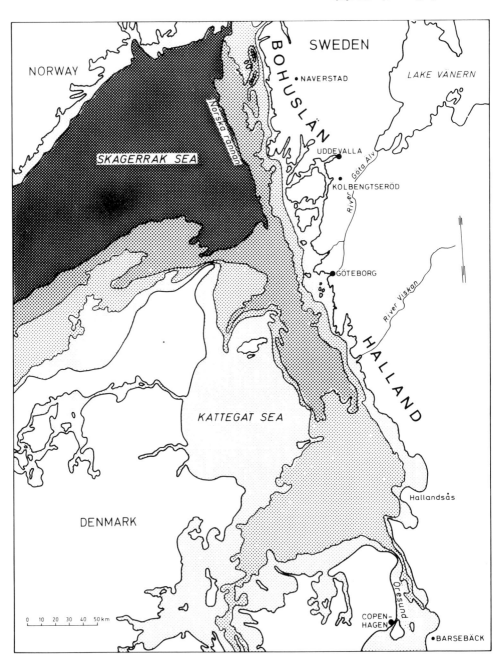

Fig. IV-8. Map of the Swedish West Coast and adjacent areas.

They occupy a depression, the deepest part of which is located along the Norwegian/Swedish coast and is referred to as the Norska Rännan (Fig. IV-8). This deep part shallows towards the South. The channel coincides approximately with an old fault zone which separates the two NW European blocks, the Baltic Shield consisting of Precambrian rocks to the North and East, and the Russian Platform containing principally Paleozoic and Mesozoic rocks (Holmes 1969) to the West.

The Swedish West Coast can be divided into two fairly well-defined topographic regions, roughly coinciding with the two provinces of Bohuslän and Halland. The two regions saw rather different shoreline displacements, mainly caused by a higher rate of land upheaval in the North than in the South. The northern province topographically consists of many small granite plateaus, separated by a network of fissures, most of which with a N-S, NW-SE or NE-SW direction. There is no evidence of the different granite plateaus having moved at different rates during the Late Weichselian and Flandrian land upheaval. This rough landscape continues westward into the sea, forming a large archipelago. Bohuslän has a very thin or no cover of loose sediment, except for some river valleys and smaller basins, and traces of ancient shorelines are scarce.

Bohuslän is separated from the southern province, Halland, by the Göta älv valley, passing through Göteborg. In the lower and wider part of this river valley the clays and sands may have a thickness exceeding 100 metres.

The topography of the central and southern part of Halland is dominated by a coastal plain surrounded to the East and South by hills up to 100 metres high; offshore there is no archipelago. The area is covered with a great variation in deposits of clay, sand, gravel and till. Several valleys, from the coast to the hills in the East, contain deposits from different stages of the Late Weichselian and Flandrian. This part of Halland is a classic area for studies on shoreline displacements in western Sweden (Halden 1929, von Post 1938, Mörner 1969).

Earlier investigations

In the part of western Sweden that is bounded by the Kattegat and Skagerrak Seas, unequivocal evidence of both the Saalian Glaciation and the Eemian Interglaciation is missing. Deposits of Weichselian age, older than the Late Weichselian, were found at a few places (Alin & Sandegren 1947, Hillefors 1969). Only deposits from the Late Weichselian deglaciation in this area provide some information on the history of the North Sea.

The botanist Carl Linnaeus was one of the first scientists to observe shoreline displacements on the Swedish West Coast, when, in 1747, he visited some large mollusc beds in Uddevalla, Bohuslän. As from 1876 (Erdmann), geologists began to pay more attention to elevated marine deposits, and some years later the main features of the process of shoreline displacement in this area became known (De Geer 1890). Reliable evidence of the Postglacial transgression was published around the turn of the century (Sernander 1902). First attempts to construct isobases for the shoreline displacements (Munthe et al. 1923) were made during the geological mapping of the provinces of Halland and Bohuslän at the beginning of this century.

The Late Weichselian glaciation covered the entire Swedish territory with ice and, ieast west of the province of Bohuslän, the ice reached as far as the North Sea proper (Andersen in prep.). Owing to the ice load the earth's crust was depressed and contemporaneously the sea level was lowered about 80—100 metres. As the depression of the earth's crust is related to the thickness of the ice cover, the northernmost part of Bohuslän was depressed nearly twice as much as the southernmost part of Halland — a distance of about 300 kilometres. When later the ice melted, the isostatic land upheaval began to restore the earth's crust to its former equilibrium. At the same time the sea level began to rise eustatically. In the beginning the isostatic effect was stronger than the eustatic effect, causing a regression, i.e. the Late Weichselian regression.

In early Postglacial time, however, the conditions changed, and a considerable climatic amelioration caused an accelerated eustatic rise of the sea level. In northern Bohuslän, land upheaval continued to exceed the eustatic rise of the sea, but in southern Bohuslän and Halland a transgression took place — the Postglacial (Flandrian) transgression — which during the later part of the Atlantic period turned gradually into a regression. The declining isostatic land upheaval exceeded the eustatic water rise up to the present time, except in the southernmost part of Sweden (Fig. IV-15).

During this century a large number of papers have been published on different aspects of shoreline displacements along the Swedish West Coast (reviews by Sandegren 1921, 1946, Lundqvist 1971). This report contains only a brief compilation of some of the papers, a few of which contributed to an understanding of the history of the North Sea proper.

Methods used in marine shoreline displacement studies

The shorelines

Numerous shell beds and locally developed shorelines indicate that the Swedish West Coast was once submerged. The shorelines are usually preserved on the western slopes of deltas, eskers and till deposits and in areas with a fairly thick cover of loose sediment. As mentioned previously, this cover is comparatively thin in the northern half of the area. As a result, there are only few localities where the highest shoreline can be determined.

The highest level to which the sea once reached is of special interest. Investigators continue to disagree on the interpretation and the dating of the highest shoreline in certain localities, i.e. on the height of the sea level when the shoreline was formed. The "highest shoreline" around Bohuslän in recent time is, according to the upper boundary of the yellow encrusting lichen *Caloplaca marina*, 6—7 m in the outer archipelago and 3—4 m in the inner; locally the lichen level is about 10 m above sea level (Rudberg 1970).

The exact sea level forming the raised beaches, therefore, cannot be determined, except in a very few cases. An investigation has been made by Påsse (1976) to establish a uniform interpretation of the earlier published highest shoreline measurements in the Göteborg area in accordance with the criteria recommended by Hörnsten (1964) and Wedel (1967).

Elevated marine shell beds

During earlier investigations on shoreline displacement along the Swedish West Coast and around the whole of Scandinavia, the main interest was focused on the large accumulations of subfossil mollusc shells and their relation to raised beaches. Asklund (1936) and Hessland (1943) presented summaries of the investigations made before and Fredén (in preparation) after the start of C-14 dating.

On the basis of his own investigations, also Hessland criticized the conclusions based on shell beds in shoreline analysis, and he proved that they have no accurate value for dating raised beaches. The shell beds provide essential information, however, on the ecological and hydrological conditions during the lifetime of the mollusc population.

The species in different shell beds indicate the changes from arctic to boreal and temporate conditions. Also a few lusitanic species (e.g. *Tapes* (*Venerupis*) *decussatus* have been found in some beds dating from the Postglacial climatic optimum (Odhner 1927, Antevs 1928). It has been proved in Bohuslän that favourable conditions for several mollusc populations prevailed during the Alleröd Interstadial and at the beginning of the Younger Dryas Stadial, mainly because bathymetric conditions allowed Atlantic water to enter the Vänern basin as bottom currents (Hessland 1949, Björsjö 1953, Fredén in prep.).

Pollen and diatom analyses

Several geologists have studied shoreline displacements by using pollen and diatom analyses of mires and lakes isolated from the sea. Such analyses, which are a classical method for determining transgressions and regressions, have a high degree of accuracy and were used by Munthe et al. (1923), Halden (1929), Sandegren (1931), Thomasson (1934), von Post (1938), Fries (1951), Mörner (1969), Persson (1973) and others. The pollen zone system used in most of the publications is that proposed by Jessen (1935) and modified by Fries (1951), Nilsson (1961) and Berglund (1971). Boundaries between zones of special interest for the interpretation of shoreline displacements can briefly be defined as follows:

PB/BO-1	rational *Corylus* limit (C°)	7600–7750 B.C.
BO-1/BO-2	rational *Alnus* limit (A°)	6500–6600 B.C.
BO-2/AT-1	rational *Tilia* limit (T°)	5200 B.C.
AT-2/SB-1	decreasing *Ulmus* and *Tilia* frequences	3300 B.C.

The rational limits of *Ulmus* (U°) and *Quercus* (Q°) are situated in zone BO-2.

Shoreline displacements

The deglaciation and the highest shoreline

The deglaciation of the Swedish West Coast area took place mainly between the Ågård and the Alleröd Interstadials (Mörner 1969). The direction of the icefront was NNW to SSE; this is largely parallel to the present coastline, except during the Ågård Interstadial, when the ice from Norway still covered most of the Skagerrak Sea. The icefront began to recede from the southernmost part of Halland as from about 11000 B.C., and the northernmost part of Bohuslän became free of ice about thousand years later. The deglaciation — an irregular and interrupted recession of the icefront — caused extensive terminal moraines, deltas or ice-marginal deposits. Of special interest for the interpretation of shoreline displacements in the area are two major ice-marginal deposits, the Göteborg Moraine (= the Fjärås line according to Mörner 1969) formed during Oldest Dryas time, and the Berghem line of Older Dryas age (Mörner op. cit.).

The main part of the west coast area is a fissure valley landscape, locally with a rather rough topography. During the deglaciation, the lower parts of the valleys were successively covered by the sea. During the final part of the ice-recession, valley glaciers or dead-ice remained for some time in many of the valleys, thus delaying the inlet of sea-water (Fig. IV-9). This is a complication for interpreting raised shorelines. In general, the highest shoreline is not a synchronous line in so far as the northwards receding ice successively exposed new ground, where still younger shorelines were formed. The highest shorelines are, therefore, younger towards the Northeast. Another question is whether or not the highest shoreline is lower towards the East in connection with a lengthy standstill (or as a result of oscillations) of the ice. This subject was studied before by De Geer (1909), Gillberg (1952) and Strömberg (1969), and is at present investigated at Ödskölts moar in Bohuslän by B. Johansson (in preparation). The colour chart shows a compilation of highest shoreline measurements published by Munthe et al. (1923), Sandegren (1931), Asklund (1936), Björsjö (1949), Gillberg (1952), Wedel (1967), Hillefors (1969), Fredén (1975) and Påsse (1976).

Next to the problems in obtaining unequivocal evidence for the highest shoreline, it has proved difficult to compare measurements made during earlier periods of investigations with more recent ones, mainly because different distinctive features have been used to determine a shoreline. At the present time, therefore, it is not possible to present reliable elevations of the

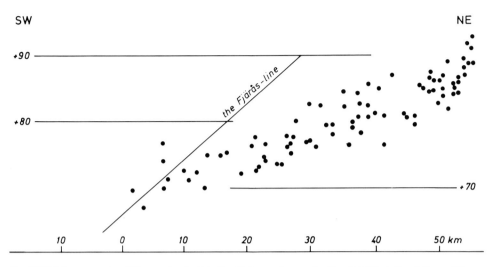

Fig. IV-9. Measurements of the marine limit in the Viskan valley on the Swedish West Coast (from Mörner 1969). The deglaciation from the southwestern part of the valley to the northeastern part took about 1000 years. During that time the land in the eastern part had been elevated about 25 metres more than in the western part.

highest shorelines. The map shows the maximum extent of the Late Weichselian sea during the retreat of the land ice.

Roughly calculated, the gradient of a suggested synchronous shoreline between Hallandsås in the South and Naverstad in the North is about 4.7 m/10 km. This might be an approximate gradient for land upheaval along the coastline, but it need not be valid a few kilometres further east.

The Late Weichselian regression (about 11000−7500 B.C.)

Isostatic land upheaval probably was a rather regular process, which immediately after the deglaciation was comparatively strong, and thereafter gradually decreased. The eustatic rise of the sea level, on the other hand, was a more irregular process with several minor variations due to changing rates of melting of the ice. Naturally, isostatic land upheaval started long before the final deglaciation of the Swedish West Coast. The most rapid crustal movements after the disappearance of the ice might have amounted to 100 mm/year (the present shoreline displacement is 1−2 mm/year). During the deglaciation the isostatic effect was stronger than the eustatic effect; as a result, the marine transgression which followed the retreating icefront, turned rather soon into a regression, i.e. the Late Weichselian regression.

This regression included a number of larger or smaller stagnations or transgressions due to variations in the eustatic movements. The differences in the regression between Bohuslän and Halland must be attributed to differences in isostatic land upheaval (Fig. IV-10).

More exact information on shoreline displacements from the regression period is rather scarce. In Bohuslän, shoreline displacements are mainly indicated by shell beds and in Halland by fairly large accumulations of subfossil mollusc shells. Asklund (1936) distinguished three malacological levels within this regression period, representing the changing environment from the time soon after the deglaciation up to

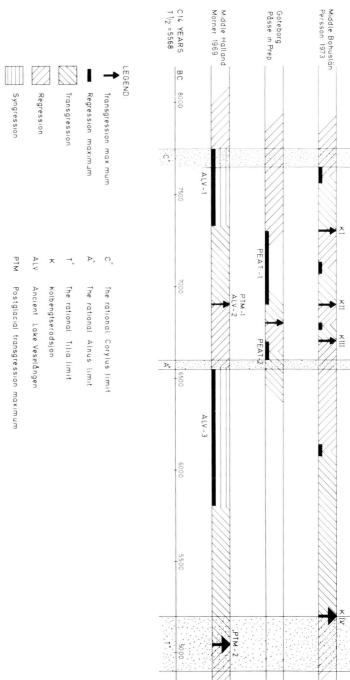

Fig. IV-10. A schematic correlation of the shoreline displacements in Bohuslän, Göteborg and Halland (from Påsse 1976). Due to differences in the magnitude of the isostatic land upheaval, the altitude of a transgression is different in the three regions.

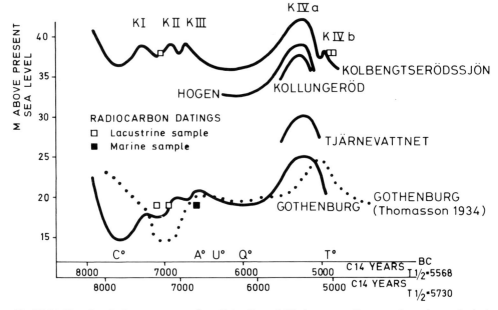

Fig. IV-11. Shoreline displacement curves from Bohuslän and Göteborg according to a micropalaeontological investigation of mires and lakes by Persson (1973).

the maximum of the regression:
— A high-glacial zone with predominant *Portlandia arctica*.
— A middle- to low-glacial zone with *Pecten islandicus* and *Buccinum undatum*
— A boreal zone with *Zirphaea crispata* and *Cyprina islandica* (*Arctica islandicus*) as the most abundant fossils.

An attempt was made also by Asklund to correlate these zones with similar zones in Denmark and Norway.

In his work "The Late Quaternary History of the Kattegatt Sea" Mörner (1969), referring to von Post (1968) and to his own investigations of the ancient lake Veselången in the Viskan valley and adjacent areas, described some distinct shore marks related to this regression.

The end of the regression — the Late Weichselian regression maximum — is placed in southern Bohuslän at "the time before C°", according to Persson (1973:35), whereas in Halland it reached its maximum between 7750 and 7330 B.C. (Mörner 1969:241). The regression maximum apparently has not been synchronous along the whole coastal area.

In Bohuslän the isostatic factor seems to have dominated (Persson 1973), while at least in the middle of Halland "the sea level was constant" (Mörner 1969:301).

In Bohuslän and northern Halland the regression did not reach below the present sea level as it did south of the Viskan valley in Halland. Knowledge of the bed marking the regression maximum in northern Bohuslän is incomplete, but in the center of the province, at Kolbengtseröd (Fig. IV-8), it is found at 35 m above present sea level, in Göteborg at 15 m, close to the Viskan valley it intersects the present sea level but south of Halmstad the regression maximum shoreline is situated about 20 m below present sea level.

The Early Flandrian transgression (about 7500—5000 B.C.)

The Late Weichselian regression maximum

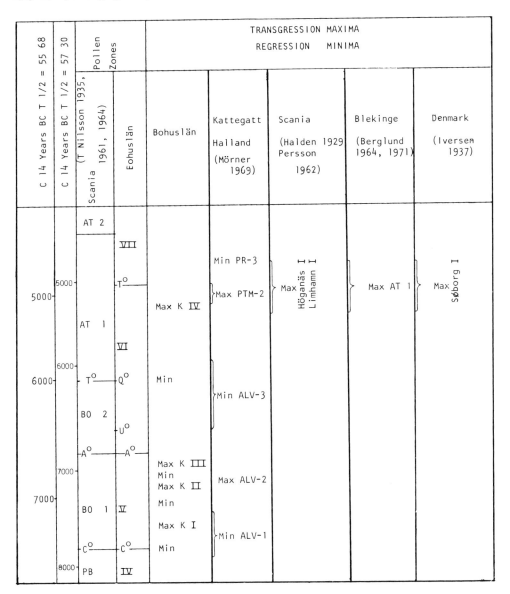

Fig. IV-12. Chronology and changes of the sea level in south Scandinavia during early Postglacial time (from Persson 1973).

changed gradually into the Early Flandrian transgression. The many shorelines, mires and littoral sediment sequences facilitated obtaining more exact information on this period than on the earlier periods. The transgression is registered as a sequence of minor and irregular rises and falls of sea level due to variations in the climate and indirectly to the rate of melting of

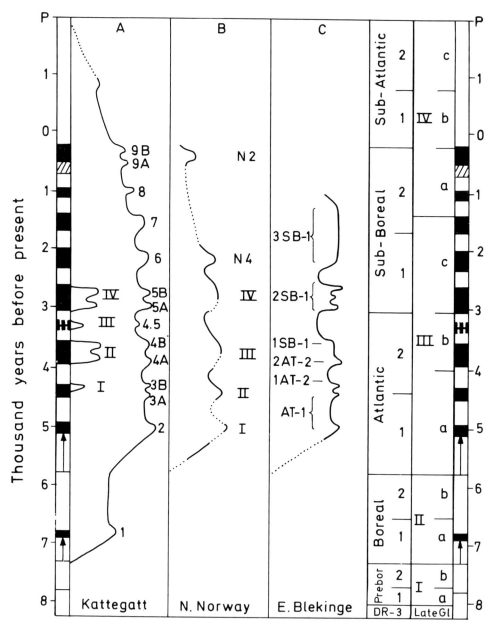

Fig. IV-13. A compilation of Postglacial transgressions, Postglacial transgression maximum (= black areas), and regressions from different parts of Scandinavia. Scale does not represent the amplitude of the fluctuations (from Mörner 1969).

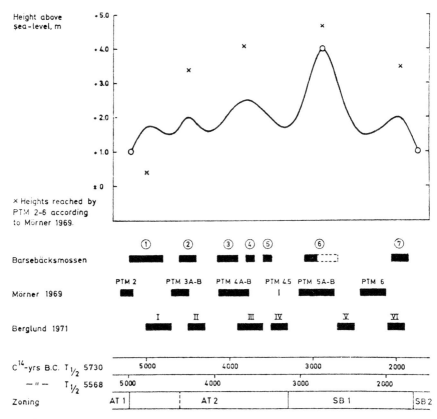

Fig. IV-14. Correlation between the Holocene transgressions and the preliminary shore displacement curve in the Barsebäck area, Skåne, southernmost Sweden. The transgressions Barsebäcksmossen 3—5 are not indicated separately, but placed together in a "complex" transgression (from Digerfeldt 1975).

the ice. Persson (1973) proved that south of the Kolbengtseröd area, in Bohuslän, the eustatic sea level rises exceeded isostatic crustal movements, while north of the area isostacy was stronger than eustacy; therefore, shore marks from this period appear to be absent.

The dominant isostatic land upheaval in Bohuslän can be convincingly concluded from the observation that the Vänern basin became isolated from the Skagerrak Sea at the very beginning of the Flandrian transgression. This event was of great importance for the hydrography in western Sweden and the North Sea. Approximately at the same time the Vänern basin was isolated from the Baltic Sea also (Ancylus Lake stage). Prior to this event the Vänern basin was also connected with the Oslo fjord area for some hundreds of years.

The rise of sea level south of Kolbengtseröd had a slightly increasing magnitude southward. Moreover, the beginning as well as the culmination of the transgression seems to have occurred at somewhat different times in the North and in the South.

The earliest Flandrian transgression in southern Bohuslän coincides with the boundary between the pollen zones PB and BO-1 (C°), i.e. about 7650 B.C. according to Persson (1973). In the middle of Halland (Viskan valley) it is placed about 300 years later or 7330 B.C. (Mörner 1969).

In the Göteborg-Kolbengtseröd area, the

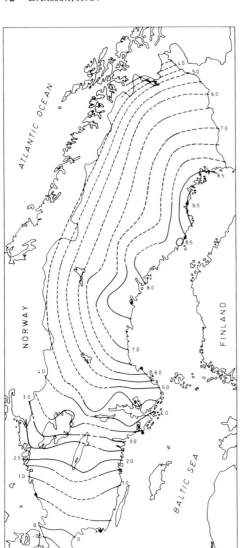

or 5300—5100 B.C. (Persson 1973:39). In Halland the corresponding maximum can be delimited between 5270 ± 150 and 4905 ± 90 B.C. (Mörner 1969:333) as shown in Figs. IV-11 and 12.

A comparison between Persson's and Mörner's curves shows that Persson investigated five localities in southern Bohuslän and that he had evidence of two additional transgressions between C° and T° (Fig. IV-11). In Halland, however, Mörner found evidence of only two of the transgressions indicated by Persson, the numbers KII and KIV in Fig. IV-12. This might indicate that the transgressions cannot have been synchronous.

The Middle and Late Flandrian regression (about 5000 B.C. — present time)

Since the culmination of the transgression during the first part of the Atlantic time, generally speaking there has been a slow regression interrupted by some eustatic transgressions of smaller extent and magnitude.

The regression curve is irregular and shows many stagnations and smaller transgressions. Mörner described and C-14 dated thirteen transgression maxima from the time of T° until the present (Fig. IV-13), with an average duration of about 400 years:

2	= 500 B.C.	5A	= 3000 B.C.
3A	= 4500 "	5B	= 2600 "
3B	= 4300 "	6	= 2100 "
4A	= 3900 "	7	= 1600 "
4B	= 3600 "	8	= 1000 "
4, 5	= 3250 "	9A	= 500 "
		9B	= 250 "

Fig. IV-15. Recent shore displacement in Sweden, measured in mm per year. 1 = Selfregistering tide gauge used in the computation of shore displacement. 1 = Newly established tide gauge. 3 = Isobases of shore displacement in mm per year.

culmination of the classical Postglacial transgression maximum has been dated in the upper part of the pollen zone BO-2, close to T°,

Most of the older above-mentioned transgressions were also described by Digerfeldt (1975) in his study on an ancient lagoon, the Barsebäcks mire, close to Öresund in western Scania (Fig. IV-14); the transgressions must have been synchronous and of a considerable magnitude. Many of them can be recognized also around the North Sea proper. During the last few hundred years isostatic crustal movements have been strongly diminishing. In Bohuslän, shoreline displacement is about 1–2 mm/year and in southern Halland half of that or less (Fig. IV-15).

The Late Weichselian and Flandrian shoreline displacements are rather well known from the province of Halland, which was investigated by Mörner (1969). He also attempted to distinguish isostatic and eustatic components in the shoreline displacements. In addition, he placed the shoreline displacements on the Swedish West Coast in a more global perspective. Persson (1973) made a similar attempt in southern Bohuslän, but only for the Postglacial period. The main features in relation to the Flandrian are fairly clear, but information is rather scarce on the Late Weichselian in Bohuslän and Skagerrak, and also on the deglaciation of the whole West Coast. In spite of very detailed investigations of the shoreline displacements on the Swedish West Coast in the last decade, there are still many problems to be solved for a better understanding of the history of the North Sea proper.

Acknowledgements. – The author wants to express his thanks to Dr. Per Wedel and Dr. Curt Fredén for reading the manuscript and to Mr. Tore Påsse and Mr. Lars-Magnus Fält for preparing the maps.

REFERENCES

Alin, J. & Sandegren, R., 1947: Dösebackaplatån. *Sver. Geol. Unders. C 482*, 44 pp.
Andersen, B. B., (in prep.): Report with map of Weichselian ice borders in Europe. *Geol. Inst.*, Bergen.
Antevs, E., 1928: Shell beds on the Skagerrack. *Geol. Fören. Stockh. Förh. 50*, 479–750.
Asklund, B., 1936: Den marina skalbärande faunan och de senglaciala nivåförändringarna. Med särskild hänsyn till den gotiglaciala avsmältningszonen i Halland. *Sver. Geol. Unders. C 393*, 103 pp.
Berglund, B., 1971: Late-Glacial stratigraphy and chronology in South Sweden in the light of biostratigraphic studies on Mt. Kullen, Scania. *Geol. Fören Stockh. Förh. 93*, 11–45.

Björsjö, N., 1949: Israndstudier i södra Bohuslän. *Sver. Geol. Unders. C 504*, 231 pp.
Björsjö, N., 1953: Bohuslinjen och den senglaciala landhöjningen. *Göteborgs Kungl. Vet. Samh. Handl. Sjätte följden, B, 6*, 8, 23 pp.
De Geer, G., 1890: Om Skandinaviens nivåförändringar under quartärperioden. *Geol. Fören. Stockh. Förh. 10*, 366–379 and *12*, 61–110.
De Geer, G., 1909: Dal's Ed. Some stationary Iceborders of the last Glaciation. *Geol. Fören. Stockh. Förh. 31*, 511–556.
Digerfeldt, G., 1975: Investigation of Littorina transgression in the ancient lagoon Barsebäcksmossen, western Skåne. *Univ. Lund Dept. Quaternary Geol.*, Report 7.
Erdmann, E., 1876: Mötet den 4 Maj 1876. *Geol. Fören. Stockh. Förh., III*, p. 150.
Fredén, C., 1975: Subfossil finds of arctic whales and seals in Sweden. *Sver. Geol. Unders. C 710*, 62 pp.
Fredén, C., (in prep.): Molluscan shell deposits in the Vänern region.
Fries, M., 1951: Pollenanalytiska vittnesbörd om senkvartär vegetationsutveckling, särskilt skogshistoria, i nordvästra Götaland. *Acta Phytogeogr. Suec.*, 220 pp., Uppsala.
Gillberg, G., 1952: Marina gränsen i västra Sverige. *Geol. Fören. Stockh. Förh. 74*, 71–103.
Halden, B. E., 1929: Kvartärgeologiska diatomacéstudier belysande den postglaciala transgressionen å svenska Västkusten. *Geol. Fören. Stockh. Förh. 51*, 311–366.
Hillefors, Å., 1969: Västsveriges glaciala historia och morfologi. *Medd. Lunds Univ. Geogr. Inst. 60*, 319 pp.
Hessland, I., 1943: Marine Schalenablagerungen Nord-Bohusläns. *Bull. Geol. Inst. Ups. 31*, 348 pp.
Hessland, I., 1949: Note on *Balanus hammeri* (Askanius) as a hydrological indicator. *Bull. Geol. Inst. Ups. 34*, 24 pp.
Hörnsten, Å., 1964: Ångermanlands kustland under isavsmältningsskedet. *Geol. Fören. Stockh. Förh. 86*, 181–205.
Holmes, A., 1969: Principals of Physical Geology, 1288 pp., London.
Jessen, K., 1935: The Composition of the Forests in Northern Europe in Epipaleolithic time. *Kungl. Dansk Vid. Selsk. Biol. Medd. XII*, 1, Copenhagen.
Johansson, B., (in prep.): Ödskölts moar. A Complex ice margin deposit in the Fennoscandian moraine, western Sweden.
Lundqvist, J., 1971: Kvartärgeologisk forskning i Sverige 1946–1970. *Geol. Fören. Stockh. Förh. 93*, 303–334.
Munthe, H., Johansson, H. E. & Sandegren, R., 1923: Göteborgstraktens geologi. In Göteborgstraktens natur – Göteborgs jubileumspublikation, Göteborg.
Mörner, N.-A., 1969: The Late Quaternary History of the Kattegatt Sea and the Swedish West

Coast. Deglaciation, shorelevel displacement, chronology and eustacy. *Sver. Geol. Unders. C, 640.* 487 pp.

Nilsson, T., 1961: Ein neues Standardpollendiagramm aus Bjärsjöholmssjön in Schonen. *Lunds Univ. Årsskr. N.F. 56,* 18, 34 pp.

Odhner, N., 1927: Några fakta till belysning av skalbanksproblemet. *Geol. Fören. Stockh. Förh. 49,* 77—111.

Persson, G., 1973: Postglacial transgressions in Bohuslän, southwestern Sweden. *Sver. Geol. Unders. C 684,* 47 pp.

von Post, L., 1938: Isobasytor i den senkvartära Viskafjorden. *Geol. Fören. Stockh. Förh. 60,* 434—456.

von Post, L., 1968: The Ancient Sea Fiord of the Viskan Valley, Chapter X. Stages of Ancient Lake Veselången. *Geol. Fören. Stockh. Förh. 90,* 37—110.

Påsse, T., 1976: Kontroll av marina gränsen i Götaborgstrakten. Geol. Inst., Chalmers Tekn. Högsk./Göteborgs Univ., B 32, Stencil 56 pp.

Rudberg, S., 1970: Naturgeografiska seminarieuppsatser vid Göteborgs universitet höstterminen 1959 — vårterminen 1969. *Gothia 10,* 164 pp., Göteborg.

Sandegren, R., 1923: Göteborgstrakten geologi. In Göteborgstraktens natur — Göteborgs jubileumspublikationer, Göteborg.

Sandegren, R., 1931: Beskrivning till kartbladet Göteborg. *Sver. Geol. Unders. Aa 173,* 141 pp.

Sandegren, R., 1946: De senkvartära nivåförändringarnas problem. *Geol. Fören. Stockh. Förh. 68,* 303—318.

Sernander, R., 1902: Bidrag till den västskandinaviska vegetationens historia i relation till nivåförändringarna. *Geol. Fören. Stockh. Förh. 24,* 125—144 and 415—466.

Strömberg, O., 1969: Den mellansvenska israndzonen. *Stockh. Univ. Naturgeogr. Inst.,* Forskningsrapport 8.

Thomasson, H., 1934: Boplatsens tidsställning. In J. Alin, N. Niklasson and H. Thomasson, 1934: Stenåldersboplatsen på Sandarna vid Göteborg. *Göteborgs Kungl. Vet. Samh. Handl. Femte följden. A, 3,* 6.

Wedel, P., 1967: Några MG-bestämningar i norra Halland. In Teknik och Natur. Studier tillägnade Gunnar Beskow, 419—433, Göteborg.

IV-c

Late Pleistocene and Holocene shorelines in Western Denmark

HARALD KROG

Krog, H., 1979: Late Pleistocene and Holocene shorelines in Western Denmark. In E. Oele, R. T. E. Schüttenhelm & A. J. Wiggers (editors), The Quaternary History of the North Sea, 75–83. *Acta Univ. Ups. Symp. Univ. Ups. Annum Quingentesimum Celebrantis: 2*, Uppsala. ISBN 91-554-0495-2.

In the Danish North Sea area, the position of the shorelines indicated on the colour chart is rather uncertain, as little modern research has been done in this area. The Eemian sea covered only the coastal areas of southwestern Jylland, and the distribution there is similar to that of the present salt marsh area which is an extension of the Dutch and German tidal flat and salt marsh areas. The Holocene shoreline displacement in this area was apparently similar to that in the German and Dutch areas. The area north of Nissum Fjord differs from the southern area; the Holocene shorelines are located above present sea level, and the height of the shorelines increases gradually towards the NE owing to a gradually increasing isostatic upheaval. The maximum sea level was not synchronous in this region but was reached successively later from north to south. In the area from Hanstholm to Nissum Fjord, the maximum level apparently was reached gradually between the middle of the Subboreal and the middle of the Late Subatlantic.

Dr. H. Krog, Danmarks Geologiske Undersøgelse, Thoravej 31, DK-2400 København NV, Danmark.

Introduction

Denmark, bordered by the North Sea in the West, by the Skagerrak and Kattegat in the North and Northeast, and by the Baltic Sea in the East, offers ample opportunities for studying shoreline displacements. The Danish North Sea area extends from the Danish—German border in the South to the Skagerrak in the North; the separation line runs from Hanstholm, the NW point of Jylland just north of 57° N, to the southern tip of Norway.

Modern research on shorelines in Denmark proper has been carried out on a limited scale only, and Danish studies on submerged ancient shorelines in the North Sea practically have not been made. The information on the position of shore marks, salt marshes, beach ridges and cliffs permits an approximate interpretation of former sea level position only.

The Saalian

Horns Rev, west of Blåvands Huk, is a complex of submarine ridges, the configuration of which based on isobaths was described by A. Jessen (1925). The outer part of Horns Rev is an E-W, 40 km long narrow hilly ridge, generally at 4—7 m depth. It rises 10—20 m above the sea bottom, and it has relatively steep sides. The ridge is covered with sand; the lithological composition of the subsurface of the ridge is unknown. Jessen (l.c.) considered the ridge as a moraine on the extreme southside of a former ice cap. As the Weichselian ice has not reached this area, the ridge may represent a stage of the Saalian glaciation. This supposition may be supported by investigations by Ödum (1968); from the distribution of indicative boulders, he reconstructed the extreme limits of the last phase of the Saalian glaciation in Jylland, represented by a line that runs NE-SW towards Blåvands Huk, that can be extended towards Horns Rev. According to Reinhard (1974), Horns Rev belongs to the Fläming moraine, in his opinion the last stage of the (Saalian) Warthe phase. He extended the line in the North Sea towards the NW, and on land mainly towards the SE and S. The latter interpretation is not in agreement with the position of Ödum's borderline just mentioned.

According to Gripp (1944), Horns Rev consists of sands deposited by opposite (confluent) bottom currents meeting in the Horns Rev, in his opinion a still active process.

The Eemian

The Eemian in Denmark is represented by fresh water deposits and marine deposits. Most of the fresh water deposits are found in western and central Jylland, the marine deposits in southwestern Jylland and the western Baltic area, mainly in the coastal areas of eastern South Jylland and southern Fyn. The distribution was mapped by S. T. Andersen (1967), Hansen (1965) and Konradi (1976). In the Baltic area the marine deposits are disturbed to various degrees by the Weichselian glaciation. Two sites in the Kattegat area (Ejby and Skärumhede, marked on the colour chart) contain marine Eemian deposits which are supposed to be in situ.

In southwestern Jylland, which was not glaciated during the Weichselian, the marine deposits are undisturbed by ice. They are known from borings between Blåvands Huk and the Danish-German frontier, overlying Saalian deposits and covered with Weichselian meltwater deposits. The top of the Eemian deposits in the area is 7—12 m below present sea level, but in several places the top layers have been removed by erosion. It is unknown whether the Eemian sea ever covered the area of the Danish Frisian islands (Römö, Manö, Fanö), but it is now certain that it did not extend across southern Jylland to the Baltic area, a possibility which was held open by Hansen (1965:75). The extension of the Eemian sea as indicated in this paper is only slightly different from that by Konradi (1976). It may be noted that the area of the Eemian sea in southwestern Jylland roughly coincides with the present salt marsh area.

The Holocene

Fig. IV-16 shows the emerged or reclaimed Late Weichselian and Holocene former marine areas in Denmark. Late Weichselian marine deposits are known only from northernmost Jylland. Information on Late Weichselian shorelines in the Danish part of the North Sea area is not available.

Holocene marine deposits are found along nearly the entire Danish North Sea coast (Fig.

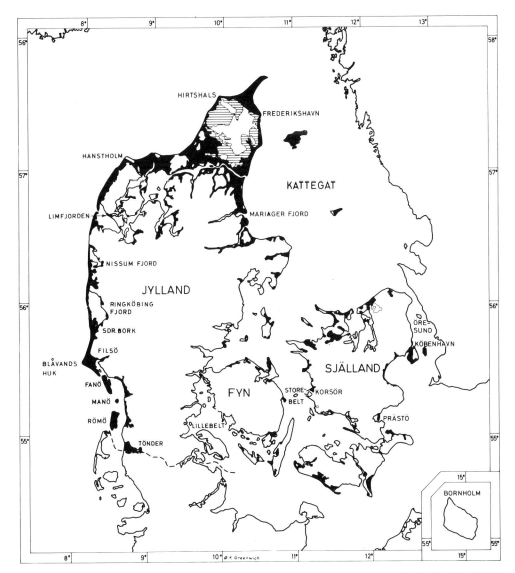

Fig. IV-16. Map of Denmark showing emerged Late Weichselian former marine areas (hatched) and emerged or reclaimed Holocene former marine areas (black). Mainly after Hansen (1965). The locations of discussed areas are indicated.

IV-16). They will be briefly described, from south to north.

The Danish Frisian islands and salt marshes. — The southernmost coastal area, S of Blåvands Huk and bordered by the Danish Frisian islands (Römö, Manö, Fanö), is a continuation of the tidal flat and salt marsh area along the Dutch and German North Sea coast. The salt marsh formation takes place above present sea level,

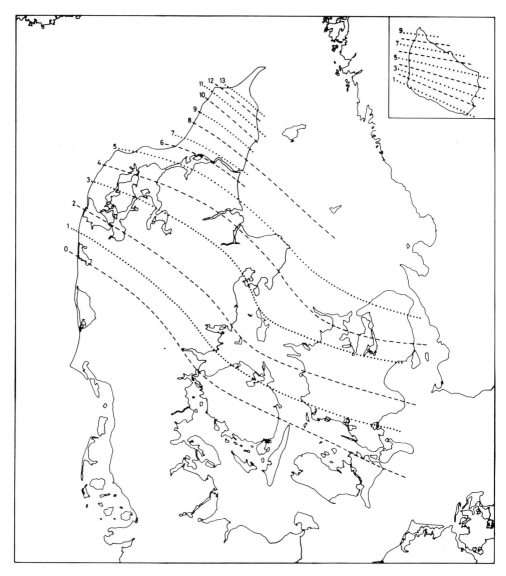

Fig. IV-17. Isobases of the maximum of the Holocene sea (Litorina or Tapes sea). Interval 1 m. After Mertz (1924).

and the surface of the salt marshes is 0.7—1.7 m above mean high tide or, on an average, about 2 m above present sea level (B. Jakobsen 1954). Several authors have dealt with the recent processes of the salt marsh formation in the area, but only few have discussed the older development. N. K. Jacobsen (1964), however, reviewed all the essential previous work and thoroughly investigated the Tönder salt marsh area. He found variations in the Holocene sediments and correlated these to a series of transgression phases and intermediate stationary or regressive phases. His transgression phases were marked as follows (figures in brackets indicate the altitude of shore marks in relation to present sea level): A_1 (−6.00 m), A_2 (−4.00 m), A_3 (−3.00 m), C_1 (−2.00 m), C_2 (−1.20 m), S (−0.75 m), I, V, M, N.

Jacobsen assumed that the sea level was about 2 m lower than the shore marks during the phases A_1–C_1. The phases A_1–A_3 were dated as Middle and Late Atlantic. A regression phase followed during the Subboreal (ca. 2300 to 1800 B.C.). C_1 and C_2 indicate the "Cardium" transgressions and were correlated with the Late Subboreal. The transgression phase S was correlated with the Early Subatlantic, I with the Iron Age, V with the Viking Period, M with the Medieval Age, and N with post-Medieval time. The age was estimated for each transgression phase, but Jacobsen emphasized that the datings are hypothetical, as they are based on correlations with corresponding German and Dutch data. Only two of the shorelines required for the colour chart are included in Jacobsen's results, namely the 5000 B.P. and 2000 B.P. lines, and both of them are questionable. From these data it seems most probable that 5000 B.P. corresponds to a sea level at −5 m, and 2000 B.P. to −0.5 m.

Recent investigations in the area just south of Denmark indicate sea levels at −3 m for 5000 B.P., and at −0 m for 2000 B.P. (Menke 1976, in press).

N. K. Jacobsen's results summarized above may be roughly applied to the northern part of the Danish salt marsh area, which was investigated by A. Jessen (1916, 1925). His assumption of an upheaval of the salt marsh area after the Bronze Age was based on the idea that the salt marsh formation at +2 m may only have been possible if the sea level was higher than at present, which was later proved incorrect (B. Jakobsen 1954).

Blåvands Huk. − From Blåvands Huk northwards, the smooth course of the present shoreline must be attributed to the activity of longshore currents. The sharp corner at Blåvands Huk is a rather recent feature, modulated by the currents. According to A. Jessen (1925), the inner coastal area consists of young Holocene gravel and sand deposits, mainly beach ridges. The maximum height of these ridges is 4.7–4.8 m above present sea level; this surpasses the height of the recent beach ridges in the area. Jessen, assuming a later minor upheaval of the area, tentatively correlated it to the upheaval that he suggested for the salt marsh mentioned above. The Holocene deposits extend to at least −20 m in the area (Jessen 1925). Accordingly, the water depth and the wave action may have been much greater during the formation of the beach ridges than at present. The beach ridges thus may have been formed at a sea level stand at or below the present one.

In the same area Jonassen (1957) carried out pollen analytical investigations of sediments from lake Filsö and from a basin just south of the lake. He proved that Filsö during its early stages was a fjord connected with the North Sea, and he dated the fjord period from the end of pollen zone VI to the end of pollen zone VII. In my opinion pollen zone VI is not represented in his diagram. The lowermost part belongs rather to pollen zone VII, and the zone border VII/VIII should be placed at a lower level than done by Jonassen. In that case, the beginning of marine sedimentation in the fjord coincides with pollen zone VII and the end with pollen zone VIII; sedimentation lasted from Middle Atlantic to Middle Subboreal. The dated marine sediments are situated between −6.5 m and −3 m, and these levels indicate minimum sea levels at the start and the end of the fjord period. The connection between Filsö and the sea was subsequently interrupted, possibly by beach ridges, and since then Filsö has been a fresh-water lake.

Jonassen suggested a continuous rise of sea level in the area until the Early Subboreal with a maximum level of +5 m, a regression during the Late Subboreal down to +1.5 m, a rise to +2.5 m during the Early Subatlantic, and a slow regression thereafter. His +5 m level was based on the assumption that shore marks at the foot of the inland Grärup cliff, which borders the basin south of Filsö to the east, were formed by

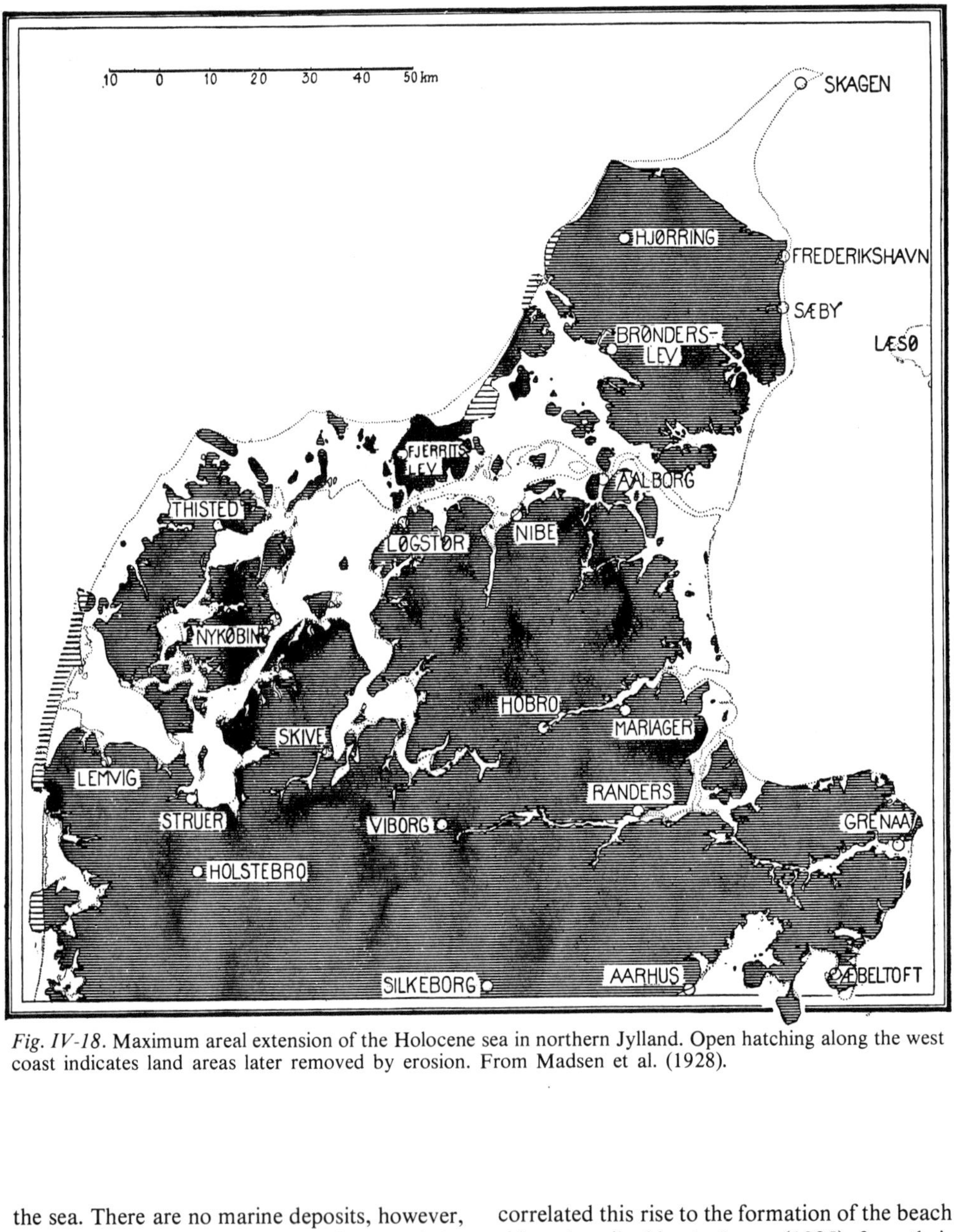

Fig. IV-18. Maximum areal extension of the Holocene sea in northern Jylland. Open hatching along the west coast indicates land areas later removed by erosion. From Madsen et al. (1928).

the sea. There are no marine deposits, however, in the basin, hence there is no evidence of the Grärup cliff having been formed by marine erosion. The presumed sea level rise in the Early Subatlantic was based on an assumed water level rise of Filsö to a maximum of +7.2 m. Jonassen correlated this rise to the formation of the beach ridges described by A. Jessen (1925); from their maximum height, +4.7 to +4.8 m, he deduced a sea level stand of +2.5 m. As mentioned above, in my opinion the height of the beach ridges does not indicate a sea level higher than the present one, and the water level rise in Filsö to +7.2 m may be attributed to a local damming-up by beach ridges.

The Ringköbing Fjord. — Hatt (1942) investigated an Iron Age settlement at Sönder Bork, situated in a salt marsh area just south of Ringköbing Fjord, about 15 km north of the Filsö area. The geology at the site was described by K. Jessen (in Hatt 1942). His profile may indicate two small transgressions, represented by brackish water gyttja and marine clay, separated by a peat layer. The first layer, situated at about present sea level, was dated as beginning Subatlantic and contained Roman Iron Age artifacts. The marine clay was situated 43—72 cm above present sea level, and was assigned to a later part of the Subatlantic. The two marine layers may be correlated to sea level stands below but very close to the present one.

Ringköbing Fjord is the southernmost and largest in a series of present and former lagoons. According to A. Jessen (1920, 1925), Ringköbing Fjord has always been sheltered from the North Sea during the Holocene, as it is today, by the present beach barrier. At that time strong marine erosion or North Sea beach ridges were unknown for the shores of Ringköbing Fjord; moreover, most of the marine sediments are salt marsh clays, which are deposited only in sheltered areas. Jessen considered the height of a sea level like the present one adequate for the formation of local beach ridges up to +2 m.

S. A. Andersen (1963) described the Ringköbing Fjord area in detail. He found two beach ridge systems of apparently different origin and age, an older one at about present sea level, and a younger one reaching up to +2 m. The first one, at the northern end of the fjord, is composed of coarse material of North Sea origin, and Andersen supposed an Early Subboreal origin; it suggests a rather open fjord stage at that time with a shoreline at −1 m. The second beach ridge consists of less coarse material, and it must have been formed by tidal movements during more sheltered fjord conditions. To explain its position he suggested a sea level stand of at least +1 m.

In the salt marsh area at the northern end of Ringköbing Fjord, Andersen found a lithological development, which he correlated to the one described above for Sönder Bork, and to the formation of the highest beach ridge which he assigned to 400 B.C. He suggested a period of regression between the formation of the two beach ridges. In part he based his conclusions on Jonassen's investigations in Filsö. Andersen (l.c.) illustrated his ideas of the Holocene shoreline movements on the west coast of Jylland in a diagram, showing selected shorelines as straight lines sloping from north to south as a result of the proved increasing isostatic land rise from south to north. However, the datings are uncertain and in some cases even in disagreement with the text.

Contrary to the opinions expressed above by various authors of a former sea level stand above the present one, the present author considers the beach ridges and the marsh clay to have been formed at a sea level stand equal to or slightly below the present one.

In the smaller lagoonal areas north of Ringköbing Fjord the marine sediments mainly consist of thin layers of marsh clay upon peat. Datings are not available, but it may be supposed that most of the marsh clay is of Subatlantic age, as in Ringköbing Fjord.

Nissum Fjord to Hanstholm. — Isobases of the maximum levels of the Holocene sea (Litorina or Tapes Sea), constructed by Mertz (1924), are shown in Fig. IV-17. The area N and NE of Nissum Fjord is different from the southern area just discussed; the Holocene shorelines are above present sea level, and the height of the shorelines increases gradually towards the NE. A maximum height of 5 m is indicated at Hanstholm, the northernmost point in the North Sea area under discussion. The maximum area occupied by the Holocene sea in northern Jylland (Madsen et al. 1928) is shown in Fig. IV-18. It should be noted that the western part of the present Limfjord, north of Lemvig, is supposed to have been sheltered from the North Sea by a beach barrier, indicated in Fig. IV-18. The reason for this assumption is the absence of

North Sea beach ridges in the area and the presence of Holocene sediments indicating fjord conditions (A. Jessen 1920). The shoreline indicated in Fig. IV-18 north of the beach barrier and west of Thisted is probably placed too far inland (Petersen 1976). Due to the differential isostatic rise, increasing to the north, neither the maximum sea level nor the shorelines in Fig. IV-18 are synchronous. Apparently, the transgression maximum was attained successively later from north to south (Krog 1968). In northern Jylland, from NE to SW, the following datings of the transgression maximum are available: at Dybvad, SW of Säby, the middle of the Atlantic (Iversen 1943), at Brovst, NW of Nibe, apparently Late Atlantic (S. H. Andersen 1970), and at Bjergegård, north of Lögstör, probably Early Subboreal (Petersen 1976). This sequence, correlated to that in eastern Denmark (Krog 1968), may indicate that from Hanstholm to Nissum Fjord the maximum level was attained successively from the middle of the Subboreal to Middle or Late Subatlantic.

Reconstruction of Holocene shorelines

From the above it is evident that Holocene North Sea shorelines can hardly be reconstructed, as only scarce data are available and former coastlines did not follow the present depth contour lines of the North Sea owing to a differential isostatic rise, at least in the northern part of the area. A reconstruction must be based on the assumption that the shorelines followed the present depth lines south of the O isobase in Fig. IV-17, but are gradually higher from there towards the north. This is a rough generalization, and it should be mentioned that in the area south of Blåvands Huk the bottom relief changed to a great extent during the younger part of the Holocene, as the present Frisian islands and the Blåvands Huk area consist of Holocene marine sediments down to -20 m (A. Jessen 1925, Jacobsen 1964). No datings of these deposits are available and the coastline positions must be roughly estimated.

The oldest dating of a Holocene shoreline above present sea level in Denmark is from northernmost Jylland E of Hirtshals (Fig. IV-16), where a height of $+5.7$ m was reached about 8500 B.P. (Krog, unpubl.). At Grenå, the easternmost point of Jylland (Fig. IV-18), the shoreline at that time was probably about -16 m (Krog, unpubl.), and by correlating the isobases in Fig. IV-17, the same shoreline may be supposed just SW of Hanstholm. These are the only points which may serve as a basis for reconstructing the 8700 B.P. shoreline.

A shoreline for 7500 B.P. is not available. As already stated, the dating from Filsö by Jonassen (1957), according to which a level of -6.5 m had been passed already at that time, must be rejected. Hanstholm, however, is situated on the same isobaseline in Fig. IV-17 as a place SW of Fjerritslev, where a shore level of $+3$ to $+4$ m was dated at about 7000 B.P. (Petersen 1976). A lower level, possibly -5 m, can be expected 500 years earlier. The -5 m level at Hanstholm may be the starting point of the 7500 B.P. coastline that can be connected with the -17 m level south of the Danish area.

It was suggested above that the maximum sea level at Hanstholm, $+5$ m (Fig. IV-17), was attained during the middle of the Subboreal. This leads to the conclusion that the 5000 B.P. level, the beginning of the Subboreal, must have been somewhat lower, probably $+4$ m. At Filsö (Jonassen 1957), the level at that time was at least -4 m or higher, but not higher than present sea level as stated by Jonassen. In the discussion of the Tönder area, I suggested -5 m for the 5000 B.P. level, which seems too low compared to Filsö and to the area just south of Denmark with a sea level at -3 m at that time (Menke 1976, in press). Considering these data, -3 m in the southern Danish area seems acceptable.

The 2000 B.P. coastline in the southern area from Tönder to Nissum Fjord may be estimated at -0.5 m, according to the mentioned opinions. If the above conclusions on the maximum Holocene sea level in northern Jylland can be accepted, a maximum level at 2000 B.P. may be expected at $+2$ m in the western part of the Limfjord (Fig. IV-17), and a slightly higher level, possibly $+2.5$ m, at Hanstholm.

REFERENCES

Andersen, S. A., 1963: Geologisk förer over Holmsland og dens klit. Issued by Historisk Samfund for Ringköbing Amt, 134 pp.

Andersen, S. H., 1970: Brovst, en kystboplads fra ældre stenalder. Summary in English. *KUML, Årb. for Jysk Arkæologisk Selsk. 1969*, 67–90.

Andersen, S. T., 1967: Istider og mellemistider. In Danmarks Natur I, 199–250, Politikens Forl., København.

Gripp, K., 1944: Entstehung und künftige Entwicklung der Deutschen Bucht. *Archiv der Deutschen Seewarte und des Marineobservatoriums 63*, 2, 45 pp.

Hansen, S., 1965: The Quaternary of Denmark. In K. Rankama (editor), The geologic systems. The Quaternary 1, 1–90.

Hatt, G., 1942: En sænket og hævet Jernaldersboplads ved Ringkøbing Fjord. Summary in English. *Sven. Geogr. Årsb.*, 314–329.

Iversen, J., 1943: Et Litorinaprofil ved Dybvad i Vendsyssel. *Meddr Dan. Geol. Foren. 10*, 324–328.

Jacobsen, N. K., 1964: Træk af Töndermarskens naturgeografi. Summary in English. *Folia Geogr. Danica 7*, 1, 350 pp.

Jakobsen, B., 1954: The tidal area in southwestern Jutland and the process of the salt marsh formation. *Geogr. Tidsskr. 53*, 49–61.

Jessen, A., 1916: Marsken ved Ribe. Résumé en français. *Dan. Geol. Unders. II, 27*, 66 pp.

Jessen, A., 1920: Kortbladet Blaavandshuk. Résumé en français. *Dan. Geol. Unders. I, 16*, 76 pp.

Jessen, A., 1925: Stenalderhavets Udbredelse i det nordlige Jylland. Summary in English. *Dan. Geol. Unders. II, 35*, 112 pp.

Jonassen, H., 1957: Bidrag til Filsöegnens naturhistorie. Abstract in English. *Meddr Dan. Geol. Foren. 13*, 192–205.

Konradi, P. B., 1976: Foraminifera in Eemian deposits at Stensigmose, southern Jutland. *Dan. Geol. Unders. II, 105*, 57 pp.

Krog, H., 1968: Late-glacial and postglacial shoreline displacement in Denmark. In Means of correlation of Quaternary successions 8, 421–435. *Proc. VII. Congr. Int. Assoc. Quaternary Res.*, University of Utah Press.

Madsen, V., Nordmann, V., Andersen, J., Böggild, O. B., Callisen, K., Jessen, A., Jessen, K., Mertz, E. L., Milthers, V., Ravn, J. P. J. & Ödum, H., 1928: Summary of the geology of Denmark. *Dan. Geol. Unders. V, 4*, 219 pp.

Menke, B., 1976: Befunde und Überlegungen zum nacheiszeitlichen Meeresspiegelanstieg (Dithmarschen und Eiderstedt, Schleswig-Holstein). *Probleme der Küstenforschung im südlichen Nordseegebiet 11* (in press).

Mertz, E. L., 1924: Oversigt over de sen- og postglaciale Niveauforandringer i Danmark. Résumé en français. *Dan. Geol. Unders. II, 41*, 50 pp.

Petersen, K. S., 1976: Om Limfjordens postglaciale marine udvikling og niveauforhold, belyst ved mollusk-faunaen og C-14 dateringer. Abstract in English. *Dan. Geol. Unders., Årb. 1975*, 75–103.

Reinhard, H., 1974: Genese des Nordseeraumes im Quartär. *Fennia 129*, 96 pp., Helsinki.

Ödum, H., 1968: Flintkonglomeratet i Jylland. Abstract in English. *Meddr Dan. Geol. Foren. 18*, 1–32.

IV-*d*

The Quaternary geological development of the German part of the North Sea

KARL-ERNST BEHRE, BURCHARD MENKE & HANSJÖRG STREIF

Behre, K.-E., Menke, B. & Streif, H., 1979: The Quaternary geological development of the German part of the North Sea. In E. Oele, R. T. E. Schüttenhelm & A. J. Wiggers (editors), The Quaternary History of the North Sea, 85—113. *Acta Univ. Ups. Symp. Univ. Ups. Annum Quingentesimum Celebrantis: 2*, Uppsala. ISBN 91-554-0495-2.

The Pleistocene development in the German part of the North Sea basin is discussed with special emphasis on the ingressions of the Holsteinian sea and the Eemian sea, and also the effects of the Weichselian glaciation. The marine environment reached its maximum extension in the Holsteinian. The marine sediments of this phase permit a reconstruction of the paleogeography and the coastline, part of which is far inland of the present shore. During the transgression of the Eemian sea, the sea level reached 5 to 9 m below the present level. A coastline was formed, which resembles the present one.
  The Holocene transgression on top of the basal peats can be traced back to a sea level of —46 m, at about 7500 B.C. The course of the rising sea level is represented in a sea-level curve. The marine sediments will be described, as well as their regional distribution, and the changes in the paleogeographic pattern. Finally, the interrelations between variations in sea level and human occupation of the coastal zone will be discussed. For a long time, man was able to react only passively to natural changes. After A. D. 1000, this passive resistance was replaced by active construction of dikes for protection against the sea.

Dr. K.-E. Behre, Niedersächsisches Landesinstitut für Marschen- und Wurtenforschung, Viktoriastrasse 26/28, D-294 Wilhelmshaven, Federal Republic of Germany. Dr. B. Menke, Geologisches Landesamt Schleswig-Holstein, P. O. Box 5029, D-23 Kiel 21, Federal Republic of Germany. Dr. H. Streif, Niedersächsisches Landesamt für Bodenforschung, P.O. Box 510153, D-3 Hannover 51, Federal Republic of Germany.

Introduction

The German part of the North Sea shelf is an irregular area. In the north, the boundary is a line from the island of Sylt to the WNW. The western boundary is formed by a line connecting the island of Borkum with the Nordschillgrund bank, from where a narrow strip reaches as far as the northeastern point of the Dogger Bank.

The German Bight — the southeastern corner of the North Sea — has an average depth of 45 m. The sea may be very choppy due to the predominant westerly winds. By submarine photographs Figge (1974) proved that the bottom of the sea is affected by waves and currents to a depth of at least 35 m. As a result of the prevailing winds and the Gulf Stream, the currents at the surface of the North Sea have an eastern direction in front of the East Frisian islands and turn to the north along the coast of Schleswig-Holstein.

The tidal movements in the German Bight are influenced by two tidal currents. A minor one runs through the English Channel, a major one enters the North Sea between Scotland and Norway and moves in a counter-clockwise direction. The average tidal range on the open sea is 2.4 m (Reineck 1970), at Borkum 2.2 m, at Cuxhaven 2.85 m, and at Sylt 1.7 m. The range is larger in the river estuaries.

The rocky island of Helgoland in the German Bight (Fig. IV-19) is an unusual morphological element in an area largely covered with unconsolidated sandy and clayey sediments. The island has been uplifted by halokinetic movements, for the greater part during the Tertiary (Wurster 1962). These movements continued at least during the Pleistocene. The exposed part of Helgoland consists of Middle Bunter (Triassic) sediments (Boigk 1962). An abrasion platform on the north side and east side of the island consists of Upper Bunter to Lower Cretaceous rocks.

The German Bight is bordered in the south by the west-east orientated chain of East Frisian barrier islands and the islands Scharhörn and Neuwerk. The Wadden Sea with the Ems, Weser and Elbe estuaries, the Dollert and the Jade bay is situated between these islands and the mainland. The eastern limit is formed by the coast of Dithmarschen, the Eiderstedt peninsula and the North Frisian islands with adjoining tidal flat ("Wadden") areas.

The present knowledge of the Quaternary development of the German North Sea region is mainly based on the results of investigations in the coastal zone, to a lesser degree on observations in the shelf area.

The geology has been studied predominantly by the former "Preussische Geologische Landesanstalt" and its successors, the "Geologische Landesämter" of Niedersachsen, Schleswig-Holstein and Hamburg. Studies have been made by the regional "Landesinstitute für Marschen- und Wurtenforschung" at Wilhelmshaven and Schleswig and the institute for marine research "Senckenberg am Meer" at Wilhelmshaven as well as by Kiel university. The results of these studies have been gratefully used for the present compilation.

Methods

Geological methods

A wide range of actuogeological and actuopaleontological methods can be used for a detailed litho- and biofacies analysis of the various stages of coastal development. Starting from modern environments a reconstruction of their fossil counterparts can be achieved. From many investigations, only the most important ones will be mentioned in this paper.

Schäfer (1962) discussed the animal communities of the North Sea and their influence on deposition. Usually, each environment is characterized by a specific fauna, a biocoenosis. Diatoms are a valuable tool for analysing the transitional zone between marine and freshwater conditions (Brockmann 1950). Ziegelmeyer (1957, 1966) wrote a compilation on the North Sea molluscs and their ecology. The mollusc ecology in the Jade Bay was studied by Dörjes et al. (1969). Bartenstein (1938), Rottgard (1952), Van Voorthuysen (1960), Haake (1962) and Richter (1964, 1965) studied the foraminiferal assemblages in the intertidal region. Wagner (1960) dealt with the os-

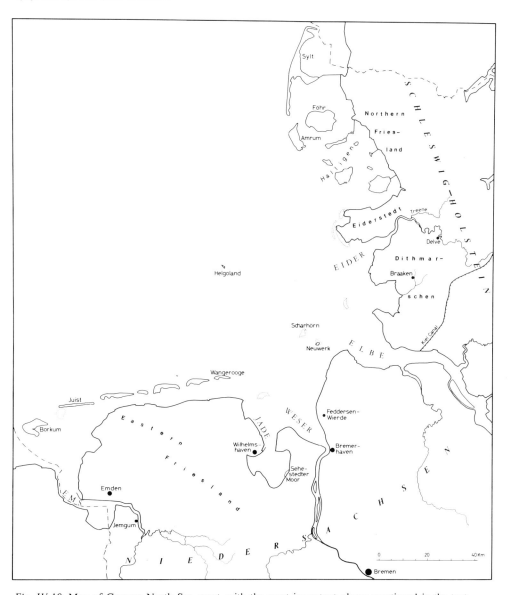

Fig. IV-19. Map of German North Sea coast, with the most important places mentioned in the text.

tracod biocoenosis and thanatocoenosis. Scheer (1953) studied the ecology of *Phragmites communis* in the brackish water environment. A summary of geological and paleontological observations in the shallow marine and tidal flat areas of the southern North Sea was presented by Reineck (1970) and Reineck and Singh (1975).

Based on these actuogeological studies, Streif (1971) investigated Holocene sedimentary se-

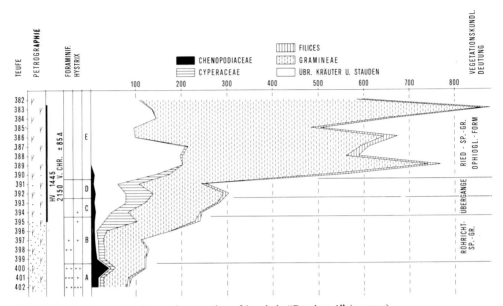

Fig. IV-20. Environmental changes in a section of borehole "Braaken 1" (see text).
Teufe = depth below surface in cm.
Petrographie = petrography; in the lower part: clay with remains of reed; in the upper part: reed-sedge peat.
Local zonation (A—E) see text.
Calculation: sum of arboreal pollen = 100 %

quences. Ludwig, Barckhausen and Streif (1973) carried out geochemical analyses on the paleosalinity of coastal deposits. Streif (1975) and Schmidt (1976) made attempts to determine the sedimentation rates during different stages of the Holocene.

Peat layers which are generally interfingering with the clastic sediments are important not only for dating, but also for understanding the transgressive and regressive tendencies in the course of the Holocene. The evidence of intercalated peat layers as indicators of a temporarily dropping sea level has been discussed since the very beginning of the coastal research, but no agreement has been obtained so far. In order to solve this problem paleobotanical investigations have been carried out.

Several papers dealt with the reconstruction of fossil plant communities and presented conclusions on environmental conditions and their changes in marsh areas during the Holocene (Behre 1970a, 1972, 1976, Grohne 1957, Grosse-Brauckmann 1962, 1963, Körber-Grohne 1967, Menke 1968b, 1969a, 1969b). Investigations of macro-remains (e.g. seeds, fruits, leaves, rhizoms, wood) and micro-remains (e.g. pollen, spores), lead to the same conclusion: forests were scarce in the hinterland. *Alnus* and *Betula* swamp vegetation was found in the back swamp area, hardwood forests on the river banks (Behre 1970a, 1977). Shrub communities were common, but large areas were covered with various reed and sedge communities, the first mainly on wetter, often inundated grounds, the latter generally in drier and partly not inundated areas. Communities similar to those of raised bogs (with Ericales and *Sphagnum*) were at least from time to time common, too. Real halotrophic communities were probably limited to relatively small shore and bank areas.

Of special interest are plant communities indicating different halotrophic and hydrographic conditions, alternating with scarce vegetation or clastic sediments. Based on local trends "allogenic series" have been distinguished from

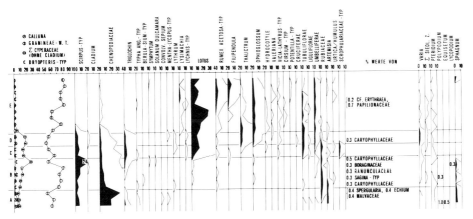

Fig. IV-21. The same section as Fig. IV-20, non-arboreal pollen diagram. Calculation: sum of non-arboreal pollen + Filices = 100 %.

"autogenic series" (Menke 1968, 1969a). The "allogenic series" are mainly characterized by increasing humidity and inundation, eutrophy and eventually halotrophy. The "autogenic series" are chiefly characterized by decreasing humidity and inundation, freshening and often by oligotrophy and dystrophy.

Figs. IV-20 and 21 for example show the beginning of an "autogenic series" on marginal sediments of a tidal creek, which was active from 4600 to 4300 B.P. Then the sediments were covered with peat-forming plant communities. During the time of the local zone A, the surface was lower than the mean high water level, and vegetation was scarce. During the time of the zone C/D, inundation stopped, and freshening started. The brackish reed community was replaced by a meadow-like sedge and grass vegetation, and the surface was then distinctly above the mean high water level. A higher level (zone E) does not show any signs of inundation. Above the section, shown in Fig. IV-20, dystrophying started; at 4000 B.P. this caused a raised bog-like vegetation, with predominant *Calluna* and *Sphagnum*. The "autogenic series" started at least at the border C/D (Figs. IV-20, 21). About 3900 B.P. and "allogenic series" started again.

The following questions remain to be solved: are certain trends synchronous for larger areas, or are they local phenomena only; are the "autogenic series" caused e.g. by sedimentation and peat-forming (rising surface), or is it possible that cycles indicate a real lowering of the mean high water and ground water level?

In principle all datings are based on radiocarbon dates. Menke (1969a) recommended the dating not only of base and top of a peat layer, but also of well-defined and comparable paleoecological markers e.g. base and top of "autogenic series". A problem is to obtain and select suitable samples for radiocarbon dating (Benzler & Geyh 1969, Menke 1969a, Streif 1971), preferably in combination with pollenanalytical studies.

Archaeological methods

The distribution of datable archaeological finds provides useful clues for the extension of settlement areas in different prehistoric periods. From the bottom of the North Sea such finds are already known for Palaeo- and Mesolithic times.

The distribution of remains of former cultures in coastal areas provides strong indications that the North Sea floor was visited and presumably also inhabited by man in earlier periods. Schwabedissen (1951) mentioned the distribu-

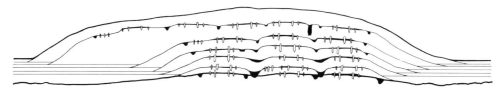

Fig. IV-22. Schematic representation of a wurt of Roman time. At the bottom a natural levee with a flachsiedlung, followed by several clay layers deposited by man, each with a habitation layer containing the foundations of houses etc.

tion of the core- and flake-axe culture during middle Mesolithic (Boreal) times; finds were made in Denmark and Northern Germany, about as far as the river Elbe, and also in England. In the intermediate continental area of Northern Germany and Holland, finds of this kind are known only from a small district on the coast of Dutch Friesland. It may be concluded that this culture must have spread across the area of the present southern North Sea, which was still dry in Boreal time. The archaeologists naturally paid special attention to the coastal areas, where finds from the Neolithic and younger periods are present in the Wadden Sea and in the sand bars along the coast (see e.g. Bantelmann 1949 and Kersten & La Baume 1958 on Schleswig-Holstein).

The mapping of the horizontal and vertical distribution of archaeological finds provides additional information. Archaeological finds can generally be well dated, and within limits may provide indications of the position of the former sea level at certain times. Finds from graves have only a limited use; the position was generally below ground level, and not always above storm flood level.

The settlement layers in Holocene sediments are much more important. They can be identified by pottery and may contain the foundations of houses and other remains, indicating that the marsh area was above mean high water (MHW) and also above the storm flood level at that time. An example is the flachsiedlung ("Flachsiedlung": this is the name for a settlement on the flat marsh) in the Late La-Tène period (around the birth of Christ), at + 0.30 m N.N. under the present wurt Feddersen Wierde, north of Bremerhaven ("Wurt" is a mound, built to raise the settlement above the storm flood level, Fig. IV-22). The area was later covered with marine sediments up to + 1.30 m N.N. (Haarnagel 1973).

Details as to the position of the settlement in relation to the sea level can be obtained in many places from botanical information. The excellent preservation of plant remains in the settlement layers facilitates a reconstruction of the vegetation in the vicinity of the settlement. Thus, not only information becomes available on marine, brackish or fresh-water environments, but also on local hydrographical conditions and their development (see Körber-Grohne 1967, Behre 1970a, 1972, 1976).

The archaeological sources of information about the North Sea are supplemented in more recent times by written records. The oldest of these, appertaining to Niedersachsen, are Roman (compiled by Schmid 1969). Many of the written sources of the Roman period and the Middle Ages are inaccurate, only those later than about 1500 are more reliable (compare e.g. the old maps, compiled by Lang 1969). These are generally concerned with the devastating storm floods and the dyking history. Useful indications of flood levels are also found in plaques on houses and height markers in the field (Prange 1965).

The Pleistocene

The distribution of Pleistocene sediments gives an idea about the paleogeographical development of the North Sea basin. Conclusions can be drawn as to tectonical movements, the morphological influence of glaciation, and isostatic movements during periods of deglaciation. Systematic mapping, as in the Dutch part of the North Sea (Oele 1969, 1971), was not carried out in the German part.

Sindowski (1970) studied the sedimentary cores of 23 borings, which penetrated as much as 30 m of Pleistocene deposits and reached a maximum depth of 75 m below S.K.N. (datum of hydrographic charts, corresponding to the local mean low-water spring). Sindowski (1970:table 6) subdivided the sedimentary sections into nine units and proposed the following stratigraphical classification:

— Upper marine layers; 1—15 m fine sand with molluscs (Holocene),
— Limnic fluvial layers; up to 17 m fine and middle sand (Weichselian),
— Middle marine layers; 4—20 m fine to coarse sand with molluscs (Eemian),
— Basin deposits; 0—3 m dark clays (Saalian),
— Upper glacio-fluvial layer; fine-coarse sand > 2 m (Saalian),
— Lower marine layer; fine to middle sand with molluscs (Holsteinian),
— Basin deposits; 30 m dark, partly laminated clay ("Lauenburg clay", Elster),
— Lower glacio-fluvial layer; < 2 m fine to middle sand, partly gravel (Elster).

In addition to the borings studied by Sindowski, about 600 shallow holes with an average depth of 4 m were drilled in the German part of the North Sea; many of these holes did not reach the Pleistocene. So far, Pleistocene marine deposits are only known from a few places on the North Sea shelf. Their relation to the corresponding marine deposits in the coastal zone will be discussed (for locations, see Fig. IV-19).

The Holsteinian interglacial

Marine sediments of the Holsteinian interglacial are known from two places in the North Sea. Sindowski (1970) described fine to middle sand with a mollusc fauna containing *Astarte montagui*, a species characteristic of the marine Holsteinian (Grahle 1936). In the adjacent onshore region, marine deposits of the Holsteinian are well known. These interglacial deposits are a useful index horizon for the stratigraphical subdivision of the glacial deposits in Northwest Germany. Unfortunately, in many places the layers were disturbed by the Saalian ice.

The coastline of the Holsteinian sea is rather irregular and partly located far inland. According to Woldstedt & Duphorn (1974:fig. 25), the coastline in front of the East Frisian islands had a west-east direction and a bay in the Jade-Weser area. Linke (1970) described marine Holsteinian from the islands Scharhörn and Neuwerk northwest of Cuxhaven. The transgression of the Holsteinian sea (*Mytilus* clay, *Cardium* sands) followed the deposition of the lacustrine deposits of the early warm period and also the deposition of the Lauenburg clay. The Niederelbe Bay of the Holsteinian sea extended in southeastern direction far inland, and had bights to the northeast, towards western Mecklenburg and southern Schleswig-Holstein. In western Schleswig-Holstein, the coastline followed many small and shallow bays. Marine deposits of the Holsteinian near Esbjerg indicate a continuation of the coast in northwestern direction. From this pattern Gripp (1964) concluded that the present-day configuration of Schleswig-Holstein had already originated before the Holsteinian interglacial.

An ecological interpretation of the mollusc fauna (Grahle 1936) and of the microfauna (Woszidlo 1962, Lange 1962), and pollen analysis (Menke 1970) indicates a rise in temperature in the Holsteinian sea from arctic to boreal conditions. The temperature reached maximum values that resemble the present values in the southern North Sea. Palynological investigations by Menke (1968a) confirmed that the marine transgression was locally noticeable already while the ice gradually melted during the Late Elsterian period. In western Schleswig-Holstein the sea obviously advanced directly into the meltwater basins of the Elster glaciation, in which the Lauenburg clay had been deposited. In nearly all complete marine Holsteinian sections, the transition from marine to estuarine and fluvial terrestrial depositions, i.e. the process of a regressive overlap, can be observed.

The Saalian glaciation

With the exception of highland ridges in the southernmost part of the country, the whole North Sea basin, all Niedersachsen, Westfalen, and the Rhine district as far as the Ruhr area and beyond the Rhine were covered with ice during the Saalian period. The ice-pushed moraine ridges of Krefeld-Kleve west of the Rhine represent the farthest advance of the Saalian ice in West Germany (Fig. V-1). Various hypotheses about the course of the glacial boundaries and the stages of deglaciation in the North Sea region were developed by Gripp (1937), Richter (1937) and Woldstedt (1955). Recently, Reinhard (1974) attempted an interpretation of the morphology that was formed during the Saalian glaciation.

The Eemian interglacial

Deposits of the Eemian interglacial, described by Sindowski (1970), are widely distributed on the North Sea shelf, at depths between −62 and −37 m S.K.N. The sediments consist of coarse sand and contain a mollusc fauna; in deeper water deposits, *Arctica islandica* is predominant. In the coastal zone Sindowski distinguished a fine sandy shallow water facies with *Venerupis senescens* and *Bittium reticulatum*, and a sandy to clayey tidal flat facies with basal peats and intercalations of coastal bogs and fluviatile deposits. A paleogeographic map of the marine Eemian was published by Woldstedt & Duphorn (1974: fig. 24). A more detailed representation of the configuration of the coastline of the Eemian sea is presented in the colour chart. The sediments of the Eemian sea are less widely distributed than the marine Holsteinian sediments. The Eemian coastline had a certain similarity with the present coast.

The representation of the Eemian coastline in the colour chart is based on previous investigations. In the region between the Ems and Elbe rivers, the fundamental work was carried out by Dechend (1950, 1954, 1958) and Sindowski (1958, 1961, 1965). According to these authors the Eemian transgression in the coastal zone can be traced to a depth of −35 m N.N. (plane of reference of German topographic maps). Deposition of sandy to clayey material without macrofossils started in deep gullies. Next, these gullies were filled up with sands containing molluscs, e.g. *Venerupis senescens* and *Bittium reticulatum*. These deposits are found at depths of −26 to −17 m N.N. They are covered with sediments of a muddy tidal flat sequence, subdivided lithologically by Dechend (1954:fig. 5) and Sindowski (1958:figs. 7−9). A comprehensive representation of the observations between the Ems and the Weser rivers by Sindowski (1965:fig. 2) is based on an evaluation of 120 drilling logs and shows the complicated distribution of different facies in this region.

On the basis of palynological investigations, Sindowski (1965) placed the beginning of the Eemian transgression in the present coastal region in pollen zones f and g (Jessen & Milthers 1928). Lithological evidence indicates a single-phase Eemian transgression, possibly interrupted by two phases of standstill. Already in the upper Eemian h zone the sea had with-drawn to such an extent that the former tidal flat area was covered by bogs.

Along the west coast of Schleswig-Holstein, the Eemian sea advanced over a geomorphological configuration formed during the Saalian glaciation. The general morphology of existing bays and fjords is known from the investigations of von der Brelie (1951, 1954), Dittmer (1941, 1951, 1954), Gripp (1952, 1964) and Heck (1932).

Results of recent borings are available from the Eider-Sorge area (Lange, Schlüter, pers. communication). Averdieck (1967) described a boring in the inner Miele area, which penetrated brackish Eemian. It can be concluded that the coast between the Eider and the Elbe rivers was morphologically more differentiated than generally assumed. The marginal interfingering with brackish and fresh-water deposits has not been sufficiently investigated in Schleswig-Holstein. The marine Eemian deposits in the "Nordfriesische Rinne" and in the more southward "Nordmann-Rinne" can be petrographically and faunistically distinguished (Dittmer 1951). The Eemian deposits may be thick and situated rather deep. In a boring in the

Treene valley they have been found at depths of 18.6 to 40.8 m N.N. (Dittmer 1951). Even deeper are the limnic sediments (diatomite) in the Brocklandsau area (Menke 1976b, boring Rederstall I), where the Eemian-Weichselian boundary is placed at −35 m N.N.

In Schleswig-Holstein the maximum sea level was probably reached in the Late Eemian, as the "upper brackish clay", which can be correlated with pollen zone h (Jessen & Milthers 1928), shows a transgressive sequence (von der Brelie 1951, Gripp 1964).

From the Eemian faunas, the following molluscs were described: *Venerupis senescens, Gastrana fragilis, Divaricella divaricata, Mytilus lineatus, Abra ovata* and others. Malacological studies were made by Madsen, Nordmann & Hartz (1908), Heck (1932), Dittmer (1941, 1951, 1954) and von der Brelie (1951). The diatoms of the marine Eemian were investigated by Brockmann (1932), König (1953) and von der Brelie (1959), the foraminifera by Lafrenz (1963). From these studies, subarctic to boreal climatic conditions can be concluded for the beginning of the Eemian transgression. Then the temperature rose to values that correspond to the present conditions in the southern North Sea.

With regard to the height of the surface of the marine Eemian, the following observations can be made. In the western Netherlands the surface is located at an average of −10 m N.A.P. (Dutch plane of reference; this corresponds approximately to the German plane of reference N.N.). In the Groningen area, Roeleveld (1974) described the surface of the marine Eemian from −13 m N.A.P. In Niedersachsen, Sindowski (1965) found an average depth of −9 m N.N. The highest known marine Eemian with *Venerupis senescens* in that region has been observed south of the island of Juist (boring 64, sheet Juist-Ost topographic map 1:25,000), at −6.45 m N.N. For Schleswig-Holstein Dittmer (1954:77) and Gripp (1951:99) assumed the highest marine level of the Eemian at the middle Eider river at −5 m N.N. Gripp (1952:99) mentioned the highest known Eemian clay at the Nordostseekanal, km 36.003 S, at −4.1 m N.N. These figures indicate the surface of the marine Eemian to be generally rising from the Netherlands towards Denmark. It is unknown whether this is caused by a stronger epirogenetic subsidence in the south or by a stronger isostatic uplift in the north.

The Weichselian glaciation

During the Weichselian glaciation the German North Sea shelf was not covered with ice. On the basis of glaciomorphological investigations in the east of England and a comparison with the situation in Denmark (Valentin 1957:map 10), it may be postulated that the East-English and Norwegian glaciers reached the Dogger Bank area during the maximum of the Weichselian glaciation and dammed up a meltwater basin. The overflow of this lake discharged through the English Channel to the Atlantic; the lake level was −35 m. During the Late Weichselian the glaciers receded and the basin emptied to the north. Consequently, the rivers of Schleswig-Holstein as well as the Elbe, Weser and Ems rivers discharged to the North and lowered the basis of erosion.

Contrary to Valentin's hypothesis, Reinhard (1974:72) assumed that by far the greater part of the middle and northern North Sea was free of ice during the Weichselian glaciation, except a narrow zone in the east and a larger area in the English-Scottish region. Based on morphological observations, Reinhard (1974: 75) postulated that the Elbe valley took its course to the North and discharged in a basin near the Austern-Grund. At a lake level height of −44 m the overflow took a northwestern course via the middle and northern North Sea to the North Atlantic. He regarded this as the major drainage system of the Weichselian glaciation.

The paleogeographic relations are not definitely known, but the ancient Elbe valley must be considered as one of the most significant morphological elements of the German North Sea shelf (Figge 1974:253). It can be traced to depths of −40 to −25 m. According to a sonosurvey, parts of the valley show erosive features in boulder clay, parts consist of sand. Investigations in the coastal zone indicate that after a phase of accumulation, which presumably corresponds to the Weichselian pleniglacial period, the basis of erosion must have been lower. Consequently, the rivers were forced to cut deeper into the older valley bottoms. Similar morphological patterns are known from the Ems, the Weser, and the Elbe rivers (Valentin 1957:80).

A morphological view of the pre-Holocene surface in the coastal region of Schleswig-Holstein was presented by Dittmer (1952). The morphology of the Pleistocene subsurface is represented in the isopach maps of the Holocene, published by the Geological Survey of Schleswig-Holstein. The configuration of the Pleistocene surface must have been of influence on the development of the Holocene. In the coastal zone, several types of landscape can be distinguished, based on different morphogenetical developments since the Eemian interglacial:

— The North Frisian coastal and tidal flat area: the Saalian depression of the "Nordfriesische Rinne" with its tributary valleys was filled up with marine sediments during the Eemian and was covered with "sandur" deposits of the Weichselian glaciation.
— Eider-Sorge-Treene district: Saalian valley system, filled up with marine Eemian deposits, have been exposed to Weichselian meltwaters.
— Western Dithmarschen and South Eiderstedt: the Pleistocene surface is developed as a broad (predominantly Weichselian) plain, bordered in the east by Pleistocene cliffs.
— Bays and valleys east of the above-mentioned cliffs and in the tributary valleys of the Eider: presumably most of them are tongue basins of the Warthe phase (Saalian) and drainage systems of the Eemian period. During the Weichselian they were exposed to periglacial activities.

All these areas, including the Weichselian "sandur" plains, have been subjected to fluvial erosion in Weichselian late-glacial times and in the Early Holocene. The developing erosional cliffs were moulded by eolic erosion and covered with dunes. For the larger rivers (Eider, Sorge, Treene), two phases of erosion can be distinguished (Lange & Menke 1967). During the first phase (most probably late-glacial, but undoubtedly after the sedimentation of the Weichselian meltwater deposits), wide and rather shallow U-shaped valleys were formed (probably above the eternal frost level). The second, essentially Early Holocene phase of erosion is characterized by relatively narrow but deep V-shaped valleys. This phase of erosion lasted until the sea level rise caused a damming up.

The Holocene

During the Holocene, the rise of the sea level in the North Sea must have been a predominantly eustatic rise. Consequently, large and formerly terrestrial areas were drowned by the transgressing sea. The deepest and well-dated find of semiterrestrial deposits on the bottom of the North Sea is a peat, described by Behre & Menke (1969). It is known from −46 m in a boring on the southern Dogger Bank.

In the following parts of this paper, the Holocene sediments and their distribution will be described, and also the rate of sea level rise and its influence on the habitation of the coastal region by man.

Basal peat

During the Early Holocene, peat forming in the coastal area of today was as common as it was outside this area, but in the coastal area the mires, located deeper, were covered with brackish or marine sediments when the sea level rose. It would be misleading to assume that these old mires generally were formed by the sea level rise.

Many of them were real inland mires, which developed under different conditions (Lange & Menke 1967). The following types can be distinguished:

— mires on river terraces (e.g. Eider, Sorge, Arlau: Lange & Menke 1967, Wiermann 1962);

- large mires at the base of slopes, mainly caused by upwelling ground water (Kulczinski 1949, Succow 1971, Grosse-Brauckmann & Dierssen 1973);
- sedimentation of mud in freshwater lakes and peat forming in depressions, most of which were sink holes, formed by compaction of interglacial organic layers in the underground, after the permafrost of the last glacial (Menke 1976b);
- mires on plains, probably caused by high ground water level due to poor drainage (e.g. Haarnagel 1950);
- raised bogs, independent of the local ground water level (e.g. Juister Watt, Grohne 1957).

A great number of the basal mires definitely originated as a result of rising ground water, caused by the rising sea level. Lange & Menke (1967) proposed to restrict the term "Basistorf" (basis peat) to the peat in this type of mires. A causal relation must be proved between the peat forming and the sea level rise. For this reason, the term "Basaltorf" (basal peat) should be maintained for those cases in which a causal relation between peat formation and sea level rise cannot be proved. Only the "Basistorf" can be used for the reconstruction of the sea level rise.

Sediments of the North Sea and the coastal zone

The Holocene sediments of the North Sea are only thin. Towards the present coast they become thicker, and in the barrier islands the maximum thickness is approximately 30 m (without the eolian dune deposits). In landward direction, the sediments become thinner and are bordered by the outcropping Pleistocene deposits. In the following paragraphs the typical depositional areas and their deposits will be described and the geological development of the different regions will be discussed.

Holocene sediments in the present North Sea. — Behre & Menke (1969) described a core of 3.15 m length, recovered by Kolp at a depth of 43 m on the southern Dogger Bank. This core consists of 2.80 m clayey to fine sandy calcareous deposits with molluscs, discontinually overlying basal fen peat and boulder clay. The basal peat has been correlated palynologically with the Preboreal and might have been formed about 7500–7000 B.C. Descriptions of Holocene cores from the central part of the North Sea (southern Dogger Bank, Outer Silver Pit, and Austern-Grund) were published by Kolp (1976). The cores contain a transition of peat and limnic sediments into brackish, brackish-marine and marine sediments.

Sindowski (1970) examined cores which, in contrast to the above mentioned transitional type, indicate a fully marine environment. He described the "Upper marine layers (Holocene)" as a sequence of fine sand with molluscs, 1 to 15 m thick, and he suggested a lithological connection with Oele's finds (1969) in the Dutch North Sea. The cores investigated by Sindowski contain psammo- and pelophile mollusc faunas, on which a certain depth zonation can be based. He distinguished a *"Macoma baltica* assemblage" in the coastal sand areas with a maximum water depth of 15 m, and a *"Venus gallina* assemblage" in muddy and sandy regions at a depth of 15–50 m.

So far, the Holocene sediments have not been systematically mapped in the German part of the North Sea shelf. The distribution of surface sediments of the German Bight was mapped by the "Deutsches Hydrographisches Institut" on a scale of 1:100,000 (Figge 1974). In the coastal zone, the geological map 1:200,000 (GÜK 200) shows the distribution of the sediments. Detailed investigations on grain size distribution, clay minerals and heavy metals were carried out by Little-Gadow & Schäfer (1974). It is not possible, however, to determine the exact course of fossil coastlines and barrier systems. Consequently, the hypothetical coastlines for 8700 B.P. and 7500 B.P., represented in the colour chart, have been constructed according to the bathymetrical map of the recent North Sea. Modifications due to Holocene erosion and accumulation must be taken into consideration.

The islands. — The island of Helgoland, uplifted by halokinetic movements (see above) has a special position among the German North Sea islands. During the Elster and the Saalian glaciation, Helgoland was covered with ice (Hartung 1964). According to Behre (1970b) limnic sediments were deposited in the Helgoland area during the Eemian interglacial, indicating that the island was considerably larger at that time (see the colour chart), or even connected with Schleswig-Holstein. The present configuration of the island is caused by marine abrasion, but also by human activities, as a result of intensive quarrying during the Middle Ages and in modern time (Hartung 1964).

The genesis of the other islands is quite different. A complete coastal barrier system is known from the coast of the western Netherlands between the Rhine-Maas rivermouth and Alkmaar. Northeast of Alkmaar, there is a system of present or former barrier islands and sheltered tidal flats, the so-called Wadden Sea.

This system includes the Frisian islands of the Netherlands and the German East-Frisian islands from Borkum to Wangerooge. According to their stability, stable islands (Borkum, Juist and Langeoog) and unstable islands (Baltrum, Spiekeroog and Wangerooge) can be distinguished in Eastern Friesland.

The island of Langeoog, which was investigated by Barckhausen (1969), can be regarded as an example of a stable island. The oldest sediments of the Holocene North Sea are brackish deposits, restricted to former valleys, and considered as equivalents of the Baltrum Beds (Brand et al. 1965). On top of the tidal flat sediments of the Dornum Beds, a shoal came into being above mean high tide level during deposition of the Midlum Beds. This shoal is to be regarded as the initial stage of the island. It is not clarified yet, whether this accumulation of sand can be related to an ancient barrier system.

The shoal was enlarged by younger sedimentary units, the "Hydrobienbank" (layer with *Hydrobia*), the "Kleibank" (layer of clayey sediments), the "Untere" and "Obere Moorerdebank" (lower and upper peaty layer), which is covered by the island dunes. The lower peaty layer has been radiocarbon dated. Its age is A.D. 1230 to 1340 (Barckhausen 1969). This age corresponds very well to the YD1B of the Dutch stratigraphic subdivision for the coastal dunes (Jelgersma et al. 1970). Consequently, the morphologically outstanding dunes cannot have been formed before A.D. 1340. The youngest changes in the configuration of the islands have been compiled at a scale 1:50,000 from historical maps (Forschungsstelle Norderney 1961, 1962, 1963, Homeier 1962, 1969).

The above-mentioned principle can be accepted for the stable islands Borkum, Juist and Langeoog. Still doubtful are the causes of different development of the unstable islands (Baltrum, Spiekeroog and Wangerooge), which are situated between the stable islands. The unstable islands show distinct tendencies of a southeast movement, so that their original centres have to be looked for in the recent foreshore area (Sindowski 1973). The southeast movement of Wangerooge amounts to 4 to 5 km in 2000 years (Sindowski 1969). This displacement of the islands is related to sand transport along the beaches of the islands in eastward direction. The amounts of transported sand determine the activity of erosion in the western part and the presence of dunes in the eastern part of the islands.

Such a system of barrier islands does not exist east of the Weser river. There, an open tidal flat is found with scattered islands and shoals, which are partly above mean high tide. The origin of these islands is still unknown. An exception is the Scharhörn-Neuwerk area northwest of Cuxhaven, which was investigated by Linke (1969, 1970).

Another open tidal flat area with scattered islands is present north of the peninsula of Eiderstedt. These islands have had another geological history. The islands of Sylt, Amrum and Föhr consist of Pleistocene material in the central parts. During the Holocene, marine littoral sediments accumulated around the Pleistocene centres. The history of this region was studied by Dittmer (1952), Bantelmann (1966) and Hoffmann (1969, 1974). Southward of these islands, the so-called "Halligen" have developed. They are remnants of a former marshland, which was destroyed by young marine erosion (Fig. IV-23). The Halligen, which have no dykes, have been continuously reduced in size. Only locally, aggradation may

Fig. IV-23. Old well in the tidal flats. In the background Hallig Habel (North Frisian Wadden sea). The present tidal flats are former habitation areas. Foto Bantelmann

be found. In this region the Pleistocene subsoil is relatively high. The Holocene marine ingression started about 4500 B.C. by deposition of tidal flat sediments on top of a basal peat (Dittmer 1952). Since 2000 B.C. a filling-up process took place in the former tidal flat region, and archeological finds indicate that as early as 1500 B.C. large areas must have been suitable for human habitation.

The seaward border of this rapidly but partly incompletely filled-up region (see Fig. IV-19) might have moved from west of Eiderstedt in a northward direction to Amrum. Wiermann (1962) and Prange (1963, 1967, 1968) found evidence of local marine ingressions in Northern Friesland between 500 and 100 B.C., but in large parts of the present tidal flat area the peat bogs remained present.

The beach ridges. — In Niedersachsen, beach ridges were discribed from the eastuary of the Weser river only (Haarnagel 1973). In Schleswig-Holstein they are known from Eiderstedt (Menke 1969a, 1976a) and Dithmarschen.

The "Lundener Nehrung" (Hummel & Cordes 1969, Dümmler & Menke 1970) can be considered as the best known beach ridge in this area. During the younger Atlanticum the formation of the sandy bar started with the deposition of coarse sandy material on top of tidal flat sediments. In the Subboreal, the former large sandy shoal, only little higher than mean high tide, became narrower and reached to +1 m, locally to +2.5 m above N.N. At the same time (about 1800 B.C.) the sandy bar was considerably extended to the south. East of the bar, a wide lagoon was formed, which was connected with the North Sea by a few channels. Hummel & Cordes (1969) discussed several older hypotheses about the development of the "Lundener Nehrung" (Dittmer 1938, 1952, Gripp 1964). According to the authors, the beach ridge had been formed in a zone of shallow water only effected by the prevailing westerly winds. There is no connection with an abrasion area.

The tidal flats and marshes. — Landward of the barrier islands, a zone is found that can be geographically subdivided into tidal flat areas ("Wadden") and marsh areas. From a geological point of view such a subdivision cannot be accepted. The whole region is a transitional zone between the open sea and the outcropping Pleistocene. This transitional zone consists of different tidal flat and lagoonal environments with interfingering fresh-water environments including peat bogs. As to their bio- and lithofacies, parts of the marshland areas can be looked upon as "fossil" tidal flats. On the other hand, peats and lagoonal sediments in the subsurface of the actual tidal zone indicate the presence of former semiterrestrial and lagoonal environments.

Many investigations in the coastal region between Ems and Weser were evaluated by Wildvang (1938), Schütte (1939) and Haarnagel (1950). More recent studies were carried out by Dechend (1956), Barckhausen (1969), Behre (1970a) and Streif (1971), and a summary was published by Sindowski (1972). All these investigations prove that the Holocene sedimentation in the initial phase of transgression was restricted to a few deep channels.

During that phase Early Holocene gully fills (peat and fluvial deposits) and Pleistocene sands were eroded, reworked and redeposited, together with sediments supplied by the ingressing sea. Consequently, the deposits consist of fine to medium partly calcareous sand with wood, peat and organic detritus.

The rising sea caused the recent coastal zone to be flooded, and tidal flats, lagoonal systems, fresh-water and semiterrestrial environments were formed. The beginning of tidal flat sedimentation varies from place to place depending on the height of the depositional area and its exposure to the open sea. Generally a transgressive overlap can be observed of marine and brackish deposits on top of semiterrestrial formations. Thus the coastline moved landward. This general development can be repeatedly interrupted by regressive overlaps. In that case peat layers starting from the backswamp area are intercalated in the clastic deposits in seaward direction. In contrast to the basal peats, these intercalated peat layers are called "schwimmende Torfe" (swimming peats). In many places the formation of the intercalated peats starts about 4500 B.C. and ends in the fifth century B.C.

The vertical sequence of the coastal Holocene normally shows a cyclic sedimentation with the above-mentioned transgressive and regressive overlaps. Based on this cyclic sedimentation, different systems of lithological subdivisions have been developed for the coastal Holocene. Brand et al. (1965) published an attempt of a regional classification of the Holocene in the southern North Sea region. This lithostratigraphical system has been combined with radiocarbon data for correlation purposes. Prange (1967), de Jong (1971), Streif (1971), and Roeleveld (1974) pointed out some fundamental methodical and terminological problems and difficulties in applying this system. Another lithological subdivision of the coastal Holocene was presented by Barckhausen et al. (1977).

Geyh (1969, 1971) attempted an evaluation of ^{14}C-data with statistical methods. Based on the simple concept that any transgression causes a retardation of the peat development in the coastal region, while regressions have the opposite effect, Geyh set up frequency diagrams, in which the frequency of ^{14}C-data is plotted against time. In such histograms, transgression phases, i.e. phases of retarded peat formation, appear as minima, whereas stagnation and regression phases, i.e. phases of full peat growth, appear as maxima of sample frequencies. His histograms for the Netherlands, Niedersachsen and Schleswig-Holstein, and the histogram of the total southern North Sea region indicate a distinct rhythmic development of the transgressions.

Very early the ingressing sea flooded the low-lying pre-Holocene surface in the Dithmarschen area, with the valley systems of the Elbe and the Eider rivers (Dittmer 1952). The Holocene sedimentation started with tidal flat deposits, and continued with a 25 m thick layer of sediments, indicating a deeper water environment, with *Corbula gibba, Nucula nitida,* and *Bittium reticulatum.* According to Menke (1976b), tidal flat deposits were formed since 3000 B.C., and barrier systems since 2000 B.C. in the vicinity of Pleistocene outcrops. This shallow marine sedimentation, interrupted by soil formation during the Roman period, still continues. A general seaward movement of the coastline can be observed in the Dithmarschen region.

The perimarine area. — In the sense of Hageman (1969) the perimarine area must be regarded as an environment with sedimentation and peat formation as a result of relative sea level displacements, but without brackish sediments. Paleogeographically, such a perimarine environment is characterized by a complicated system of gullies with sandy fills, covered with clay interbedded in peat.

Behre (1970a) investigated the perimarine area in the lower Ems region. The geological sequence normally starts with a basal peat. On top of the peat, river bank deposits accumulated parallel to the Ems and about 2 km wide. Farther inland, a thick fen peat is found behind the river banks. In the uppermost part, a raised bog peat may be locally present. Thin clay layers and a system of channels with clayey fills are present in the peat. The whole complex is covered with a clay layer, 20 to 50 cm thick. According to these observations, the Ems must have been a relatively small river with low current velocity and a slight tidal range for a long time. Only in the youngest phase of development, the area changed to a brackish water environment with strong tidal movements. This development can be traced back to a transgressive phase during the last century B.C., but was essentially caused by the formation of the Dollart embayment.

Other perimarine areas in the German coastal region are the Bremer Blockland near Bremen (Cordes 1967) and parts of the Eider-Sorge-Treene area in Schleswig-Holstein (Lange & Menke 1967, Menke 1969a, 1967a).

Time depth diagram

Schütte (1933, 1939, cit. by Haarnagel 1950) was the first to study relative sea level changes (in his opinion coastal uplift and subsidence) in relation to time. Later, sea level changes were studied by Nilsson (1948), Haarnagel (1950), Dittmer (1952, 1960), Dechend (1954), Müller (1962) and others.

The time depth diagram (Fig. IV-24) is based on present data from the Schleswig-Holstein area (Menke 1976a), and some from other parts of the German North Sea area.

Between "Nr. 1" and "Nr. 7" the curve depends on dates from basal peat layers. Except for "Nr. 1", the beginning of marine influence has been dated. Special attention was paid to the variations in the pollen flora in combination with the appearance of dinoflagellates, forams, halotrophic diatoms (especially marine, planktonic forms), influx of clay etc. This marker was assumed to be more significant than the start of peat formation. However, the relation between the start of peat formation and the mean sea level (or mean high water level) at that time will remain to be a puzzling problem, and only the upper parts of most of the mentioned peat layers are "Basistorf" in the sense of Lange & Menke (1967).

Coastal area of Schleswig-Holstein:

Nr. 2: Lower Elbe (Marne, borehole Nr. 35, P. H. Ross): 35 cm freshwater sediment, covered with 10 cm clayey

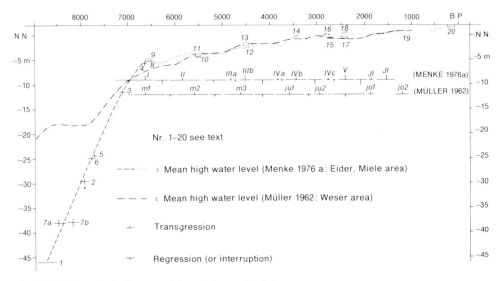

Fig. IV-24. Time-depth diagram of the relative sea level rise.

brackish sediment. Palynological dating: the transition zone is just above the increase in *Alnus* values (border of the zones 7/8, dated in Meldorfermoor by Hv 6189: 8075 ± 60 B.P., 13-C corrected: 8000 ± 60 B.P.; other dates of this border in Overbeck (1975), cit. by Menke (1967a).

Nr. 3: Lower Eider (Delve, borehole Wallen 7a, Lange & Menke, 1967): Boreal peat in a tributary valley of the Early Holocene Eider river (pollen diagram unpubl.): 100 cm Boreal peat. After a phase of desiccation (Lange & Menke 1967) renewed peat forming, 30 cm peat of Early Atlantic age, with indications of marine influence slightly above the base of the "upper" peat layer. Base of the marine influence: Hv 628: 7115 ± 90 B.P.

Nr. 4: Inner Miele Bay (Meldorfermoor, borehole 307a, A. Brande, pollen diagram unpubl.): 80 cm Preboreal limnic sediments, covered with 110 cm Preboreal-Boreal peat; peat forming continued during the earliest Atlanticum. Lowest indications of marine influence (30 cm above the transition Boreal/Atlanticum) dated by Hv 6190: 6705 ± 60 B.P. (13-C corrected: 6645 ± 60 B.P.). The rate of compaction is unknown.

North Sea area:

Nr. 1: Southern Dogger Bank (Behre & Menke 1969).

Nr. 5: Scharhörn (mouth of the Elbe river, borehole 58/67, Linke 1970): thin layer of basal peat, covered with clay. Dated by Hv 2143: 7720 ± 65 B.P.

Nr. 6: Scharhörn (borehole 56/77, Linke 1970): transition from freshwater mud to brackish-water mud. Dated by Hv 2575: 7790 ± 90 B.P.

Nr. 7: German Bight (borehole 235, Ludwig, Müller & Streif 1977, in press): peat with transition from freshwater to brackish water conditions:
a) 37,78–37,80 m below S.K.N.: Hv 7095: 8190 ± 140 B.P.,
b) 37,88–37,90 m below S.K.N.: Hv 7094: 8485 ± 125 B.P.

Within "Nr. 1" and "Nr. 7", the curve (Fig. IV-24) does not give any information on the mean sea level, but indicates a rising high water level. The results point to an average rise of high water level before 7500 B.P. of about 200 cm per century (the rate of isostatic movement is unknown). On the right side of "Nr. 4" (Fig. IV-24) the mode of presentation has been changed: from the younger period no suitable basal peat layers were available. Therefore, the right part of Fig. IV 24 shows some highest tops of dated sediment beds of marine or brackish origin.

The dotted line (Fig. IV-24) is thought to represent the changing mean high water level of the Inner Miele and Eider area (Schleswig-Holstein), estimated by means of vegetational changes and sediments (Menke et al., in prep.). The curve shows the trends only. In the Miele area the sediments were deposited in tidal creeks of a wide bay with probably slight tidal movements. It appears unlikely that these creeks were filled up higher than the mean high water level. The clastic sediments, laterally interfingering with peat, are covered with peat. Datings from the peat are available. The sediments of the Eider area were deposited on river banks; during the main part of the Holocene, the Eider-Sorge valley must have been a large bay, probably with slight tidal movements.

The palynologically investigated and radiocarbon dated sections are generally not suitable for reconstructing time-depth diagrams, as the thick peat and clay layers are more or less compacted. Therefore, the time levels have to be related to non-compacted clastic sediment layers, which can be correlated to the investigated sections. This method can be applied to widely distributed and synchronous peats in the Eider and Miele areas (Menke 1968b, 1969a, 1976a), the formation of which is attributed to temporarily lower high water and ground water levels as a result of climatic changes (Menke 1976a).

Nr. 8: Miele Bay (Meldorfermoor, borehole 307a): thin lower clay layer in the deepest area of a Pleistocene valley. Age estimated on "Nr. 4" (Hv 6190) and "Nr. 10" (Hv 6191).

Nr. 9: Eider (Delve, borehole 92 amongst others, Lange, Menke, pollen diagrams unpubl.): lowest sediment layer, covered with peat. The start of the peat forming (swamp forest peat without clay on brackish sediment with poor vegetation) was estimated on the base of Hv. 629: 6300 ± 85 B.P. (age probably 150–200 years lower due to influx of older material, compare "Nr. 11").

Nr. 10: Miele Bay (Meldorfermoor, borehole 307a, and Braaken, borehole 1, Menke, pollen diagram unpubl.).
a) Sediment layer between Hv 6191: 5635 ± 60 B.P. and Hv 6192: 5145 ± 75 B.P. (Meldorfermoor),
b) Sediment layer somewhat below the border of the zones 8/9 (5000–5100 B.P., Menke 1976a, Braaken).
The age of the sediment layers has been estimated on the base of the mentioned radiocarbon datings and pollen diagrams.

Nr. 11: Eider (Delve, borehole 92 amongst others): diffuse clayey zone in a thick peat layer (sample Hv 629, compare "Nr. 9", was collected in this zone). The top of the clayey zone lies somewhat below the zone 8/9 transition (compare "Nr. 10"; Hv 627: 5030 ± 275 B.P.).

Nr. 12: Miele Bay (Meldorfermoor, borehole 307a, Braaken, borehole 1): sediment layer in Meldorfermoor between the samples Hv 6192: 5145 ± 75 B.P., and Hv 6193: 4105 ± 95 B.P., in Braaken between the zone 8/9 transition (ca. 5100 B.P.) and Hv 1445: 4100 ± 85 B.P. Vegetational development after sedimentation of the clastic material, see Figs. IV-20, 21.

Nr. 13: Eider (Delve, borehole 92 amongst others): sediment layer between the zone 8/9 transition (ca. 5100 B.P.), and Hv 624: 4350 ± 75 B.P. Vegetational development after sedimentation of the clastic material in borehole 92: rapid change from brackish sediments (reed community, distinctly below the mean high water level) to swamp forest (peat without mineral admixture, surface above the high water level) and further changing to a *Sphagnum* and heath covered mire within a 10 cm thick peat layer.

Nr. 14: Miele Bay (Braaken, borehole 1): clastic layer, deposited between Hv 1443: 3720 ± 85 B.P., and the zone 9/10 transition (ca. 3000 B.P., Menke 1976a). The exact dating is made by interpolation (pollen diagram).
In Meldorfermoor (borehole 307a amongst others) a comparable sediment layer has not been found. In the Eider area a clastic layer in this position is absent.

Nr. 15: Miele Bay (Meldorfermoor, borehole 307a): clastic sediment layer between the zones 9 and 10 (ca. 3000 B.P.), and Hv 6196: 2710 ± 60 B.P. The exact dating is made by interpolation (pollen diagram).

Nr. 16: Western Eiderstedt (Leikenhusen, Brande, Elwert, Menke, pollen diagrams unpubl.): in lower positions fresh-water mud and peat, in higher levels fossil soil at the top of marine sediments. The beginning of the peat formation was dated with Hv 6360: 2740 ± 60 B.P.

Nr. 17: Eider (Delve): the correlation of the clastic sediment layer in question cannot be exactly established. A pollen zone in the peat series indicates increasing wetness and eutrophy. The dating was based on interpolation (between the transition of the zones 9/10, ca. 3000 B.P., and 10/11, ca. 2000 B.P., Menke 1976a).

Nr. 18: Miele Bay (Meldorfermoor): clastic sediment layer between Hv 6196: 2710 ± 60 B.P., and the border of the zones 10/11 (ca. 2000 B.P.). Braaken: sediment layer between Hv 1441: 2510 ± 80 B.P., and the zone 10/11 transition.

Nr. 19: Lower Eider (Tönning): mean tidal high water level, estimated by Bantelmann (1966), based on the level of a prehistoric settlement.

Nr. 20: Eider area; highest subrecent surfaces.

The publication of the details is in preparation.

Two dates of the Sahlenburg area (Cuxhaven, mouth of the Elbe river) may be mentioned that have not been used for the diagram (Fig. IV-24): a clay layer gradually passing into a fresh-water peat layer; its base: −2.42 to −2.48 N.N.: 4240 ± 130 B.P. (Bodenkunde Hamburg Nr. 765), its top: −1.08 to −1.03 m N.N.: 3210 ± 80 B.P. (Bodenkunde Hamburg, Nr. 776). The peat was covered with tidal sands, the degree of compaction is unknown (Linke, letter from 10.11.1976). The dating of the base is in accordance with the results obtained in Schleswig-Holstein.

If the curve (Fig. IV-24) had been based on the mentioned top layers of the sediments only (without regard to the vegetational development), the curve would have been nearly smooth. Then the main rise of the sea level would have stopped as early as 3500 B.P. The depressions in the curve are mainly based on changes in vegetation and on lithological changes. Haarnagel (1950) distinguished the following periods:

— Period of flooding from the birth of Christ up to A.D. 600.
— Interruption of the transgression from 300 B.C. to the birth of Christ (main base of the settlements on the flat marsh level: N.N. to + 1 m N.N.).
— Subatlantic transgression (starting 700–300 B.C.).
— Subboreal regression (starting 2500–2000 B.C.), base of peat mainly between −3.5 m N.N. and −1 m N.N.
— Advance of the sea at the end of the Atlantic transgression.
— Atlantic continental period (Jade–Weser area), phase of peat forming from 4000 to 3000 B.C., base of the peat mainly between −7 m N.N. and −5.5 m N.N.
— Atlantic transgression.

These results are roughly comparable to those obtained in Schleswig-Holstein. The transgression curve constructed by Müller (1962, Weser area) appears to be about congruent with Menke's curve (1976a) for Schleswig-Holstein. In the lowest part (before 7000 B.P.) Müller's curve seems to indicate too high sea levels (this part of the curve was drawn mainly according to Graul 1960). More distinct depressions in the curve are assumed by Menke (1976a) for the periods from 4400 to 4000 B.P. and from 3300 to 3000 B.P.

The most conspicuous difference between the curves of Müller (1962) and Menke (1976a) in the younger Holocene is about 2400 B.P.: Müller's curve shows a depression, whereas Menke's curve shows a culmination. Müller (1962:206) points to river banks or inversion ridges partly protecting the hinterland. Part of the Eider area was protected by the Lundener Nehrung (a barrier system) since 3900 B.P. Kolp (1974, 1976) concluded an interrupted sea level rise in the Early Holocene from terraces covered with organic sediments in the Baltic Sea. He thinks to have recognized part of these terraces in the Dogger Bank area. The deeper levels are beyond the area discussed in this paper. Holocene marine terraces from -24 m N.N. to -12 m N.N. are unknown so far from the North Sea coast of Schleswig-Holstein.

The reaction of man to the variations in sea level

The extension of areas of human habitation in former times is a useful criterion for determining sea level movements and sea level height. The limits of settlements are not imposed by the MHW level but by the storm flood level. The presence of prehistoric settlements in marsh clay areas indicates that in the past the rise of sea level was interrupted by periods of standstill or even of lowering of the water level. During such periods, previously flooded areas could be inhabited. For a long time man was able to react only passively to natural changes caused by the North Sea. Finds from early prehistoric times are scarce, because they are covered by sediments. In contrast to Holland, where several settlements of the Neolithic Vlaardingen Culture were found (Van Regteren-Altena et al. 1962), the oldest proved settlement in the marsh clay region along the German North Sea is not older than the Bronze Age. The oldest hitherto explored one west of Rodenkirchen in the Wesermarsh is a flachsiedlung (see above) of the Late Bronze Age, which originated on the level of a creek between -1 and -2 m N.N. and subsequently was covered with silt (Hayen 1972). This settlement and the following ones were built on top of the wide-spread sediments which correspond to the Dunkerque 0 deposits in the Netherlands.

Not until the Early Iron Age, i.e. after 700 B.C., do we find a greater number of settlements, three of them on the lower Ems: Jemgum (Haarnagel 1957), Boomborg/Hatzum (Haarnagel 1969) and Oldendorp (Brandt & Behre 1976). These settlements were situated between -1.0 m and -0.30 m N.N. on the levee of the river Ems. The environmental conditions have been established by botanical investigations. This perimarine area, between 700 and 300 B.C., must have had freshwater conditions, and there is no evidence of a tidal range (Behre 1970a). Another settlement of the same period is near Imsum (north of Bremerhaven) (Haarnagel 1973). Settlements from the time before Christ have not been discovered so far in the marsh area of Schleswig-Holstein.

After the above-mentioned Early Iron Age settlements had been abandoned, they were covered with a sediment layer between approximately 300 and 100 B.C. This can be distinctly observed in the Oldendorp settlement, where a pale clay layer of approximately 20 cm is found between the older settlement and a more recent habitation of the Roman period (Brandt & Behre 1976). This layer, found in many places along the German North Sea coast, corresponds to Dunkerque I of Brand et al. (1965). The Dunkerque I transgression period was followed by a retreat of the sea. Around the time of Christ, new settlements were established up to and even beyond the present coastline (Haarnagel 1961). Large areas must have been safe from floods for some time.

The resettlement on the lower Ems levee began already in the late 2nd or early 1st century B.C. (Jemgumkloster and Bentumersiel, see Brandt 1972, 1975). Several settlements on the flat marsh (flachsiedlungen) near Wilhelmshaven are from the time around the birth of Christ. In the Land Wursten north of Bremerhaven the resettlement of the marsh began in the middle of the 1st century B.C. (Haarnagel 1973), the marsh on the lower Elbe and in Schleswig-Holstein, however, was not resettled until after the birth of Christ; along the Elbe the settlements began during the 1st century A.D. (Haarnagel 1950), further north, in Dithmarschen and along the Eider (Tofting) still later, around A.D. 100 (Bantelmann 1955, 1966). Not until about A.D. 200 do we find settlements on the beach barriers of western Eiderstedt near Tating (Bantelmann 1970).

During the period of regression, for the first time settlements could be built on the flat marshes (so-called "flachsiedlungsphase"), which were then located beyond the storm flood level. This did not last very long. The renewed activity of the sea again caused flooding, and some settlements had to be abandoned (flachsiedlungen). Other villages remained occupied; the inhabitants protected themselves against the storm floods by raising their dwelling-places. This had to be repeated in the following years, thus leading to the so-called "wurten" (in Dutch "terpen") or dwelling mounds (Fig. IV-22), some of which reach a height of more than +5 m N.N. (some up to +7 m N.N.). These village wurten of the first wurt phase are widely distributed over the marshes. They contain several layers of habitation. In an excavation at the wurt Feddersen Wierde, seven complete villages were exposed in vertical succession (Haarnagel, in press).

The beginning of the building of wurten was generally during the first century A.D. Where the habitation of the marsh in Schleswig-Holstein first took place around A.D. 100 (Tofting, see Bantelmann 1955) or even later (Tating, see Bantelmann 1970), the lowest houses were already built on a clay elevation. Flachsiedlungen and wurten may serve as prehistoric tide gauges, from which the maximum flood level for each archaeologically dated period can be determined. A certain — varying — rate of compaction must be taken into consideration. A comparison of coastal settlements with those upstream shows different heights of the individual storm floods; the highest peaks are approximately located in the river mouth areas, and the level drops progressively upstream.

From the following prehistoric settlements the heights can be determined: Jemgumkloster/lower Ems, (Brandt 1972) 1st century B.C. +0.40 m N.N.; various flachsiedlungen of the 1st century B.C. in the Land Wursten, (Haarnagel 1973) ± 0 to +0.60 m N.N.; Tofting/Eiderstedt, (Bantelmann 1955) about A.D. 100 + 1.50 m N.N.; wurt (dwelling mound) Feddersen Wierde/Land Wursten, (Haarnagel 1973 and in press) 2nd century, +2.0 m N.N., 3rd century, +3.0 m N.N., 4th/5th century +3.50 to +3.80 m N.N.; wurt Tofting (see above), 5th century +4.60 m N.N.

The history of settlements can be supported and refined with the results of botanical investigations. The best known area is that around the mouth of the Ems (Behre 1970a, 1972, 1977). Fresh-water conditions prevailed in the area of the Ems settlements, Jemgumkloster and Bentumersiel, during Roman time. The first signs of brackish-water environments are found at Pogum; at Larrelt near Emden, the influence of brackish water is noticeable in the vegetation. The wurten Rysum and Loquard, which are located even closer to the sea, were partly surrounded by salt marsh vegetation, and partly by fresh-water vegetation. The wurt Pilsum, dating from the same time and ten km further north, shows the most extreme conditions; the vegetation indicates salt water environment. This confirms that the shape of the coastline in this region was approximately the same as today, although the Dollart estuary did not yet exist.

The investigations of Körber-Grohne (1967) of the surroundings of the Roman age wurt Feddersen Wierde likewise indicate predominant salt water conditions. The coastline there (i.e. the limit of the MHW) was further inland than today (Haarnagel 1973). The wurt Tofting on the lower Eider, dating from the same time, was also established in a salt meadow area, and the marsh became increasingly salty during the 3rd and 4th century (Behre 1976).

The general termination of settlements in the German North Sea marshes during the migration period (4th to 5th century) is probably not connected to changes in sea level; a similar termination is found for higher located inland settlements. A renewed general habitation of the clay area began in the early Middle Ages, from about the 8th century onwards. Most of the existing wurten (mounds) were re-inhabited, and new mounds were also built (Reinhardt 1969:256). In addition, villages were founded on the natural levees which were sufficiently high. Examples of flachsiedlungen from this time are found along the Ems (Brandt 1975), in the Land Wursten (Haarnagel 1973), and in Eiderstedt (Bantelmann 1975). The natural levee along the Eider had already reached a height of +1.80 to +2.20 m N.N., when resettlement near Elisenhof began in the 8th century.

The height of the storm floods at that time must have been somewhat lower than before. These conditions can only have been of short duration, as wurten were built soon again. Thus, the mounds of the second great "Wurt phase" in Northern Germany were built, indicating that the storm flood level was higher again. Botanical investigations of the wurt Elisenhof demonstrate that extremely halotrophic conditions existed around the area at the mouth of the river Eider during Viking time; the mounds were surrounded by salt marsh vegetation (Behre 1976).

The passive resistance of man against the encroaching sea, expressed by the construction of dwelling mounds, was replaced after A.D. 1000 by the active building of dikes. According to present knowledge, the building of dikes on the German North Sea coast began in the 11th century.

The diking caused a considerable increase in the storm flood levels; the MHW and the tidal range increased only slightly. In the beginning, the dikes were not high enough to protect the land against exceptionally high floods. With the construction of higher dikes, the overflow space for high storm floods was reduced as the dikes offered full protection against the sea. Particularly the northwest storms caused the water to pile up against the dikes and forced the water into the estuaries. In the river area upstream the reduction of overflow space by damming etc. was considerable, the highest flood water level rose quickly from century to century.

Dike breaks had a terrible effect (compare Fig. IV-25) and caused great losses of land. In Niedersachsen the Jade bay was first formed in the 12th century. About 1500 it reached far beyond its present limits. The Dollart existed since the 13th century and reached its maximum size around 1500. At that time the Ley bay was considerably larger than today, and the Harle bay reached deeply into the land (see the maps of Woebcken 1934 and Homeier 1962, 1969 and Fig. IV-25). Since about 1500 land was reclaimed in the above mentioned areas as a result of rediking. The mentioned bays are much smaller today than before; the Harle bay and several smaller bays have fully disappeared. On the other hand, by comparing Fig. IV-25 and the colour chart, it can be seen that some marsh areas, for instance the Krummhörn west of Emden, the Wilhelmshaven district, Butjadingen between Jade and Weser, and most of the marshes between Weser and Elbe, were never permanently flooded by the sea after the birth of Christ.

Schleswig-Holstein also suffered great losses of land during the late Middle Ages and in early modern time. During the storm floods of 1362 and 1634, Northern Friesland between Eiderstedt and Sylt was completely lost except for a few places. After 1362 the coast ran along the margin of the Pleistocene, from where reclamation started later (see Fischer 1955, Bantelmann 1966 and Prange 1967). In the more southern Dithmarschen and in parts of Eiderstedt, the marshes remained nearly unchanged since the beginning of our era (Fischer 1957, Lang 1975). In both areas several wurten of Roman age are known, whereas in Northern Friesland these are unknown.

An important anthropogenous factor in the land losses in the Middle Ages were the drainage measures, which were coupled with the diking. Particularly in the bog areas, drainage led to a compaction of the ground, leaving its surface often lower than the MHW on the other side of the dike. In some parts of Northern Friesland extensive peat digging was carried out. As a result of these measures, reclamation of such land, once it was flooded through a breach in the

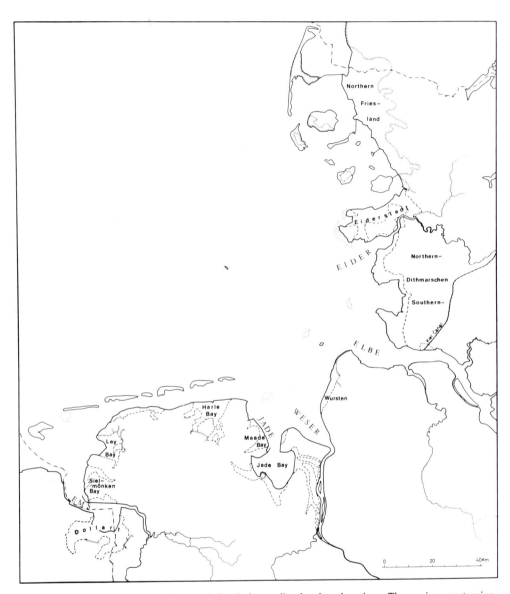

Fig. IV-25. Maximum extension of the North Sea during medieval and modern times. The maximum extension of the bays was not synchronous, but may be several centuries apart. The maximum extension of the North Sea, reached in many areas before the Middle Ages, is shown in the colour chart.
full line = present dikes
broken line = former bays
hatched line = margin of Pleistocene

dike, was often impossible. This is the main cause for the loss of Northern Friesland, where nowadays the remains of medieval settlements are frequently found in the tidal flat area (Fig. IV-23).

Before the building of dikes, raised bogs, in

Fig. IV-26. Erosional margin of the Sehestedter Moor at the Jade bay.
left hand = salt marsh
right hand = raised bog, covered with birches, and floating during storm floods

some places adjoining the coast, provided protection for the hinterland. During storm floods they floated on the water, and sank again as the flood retreated. In the high water period, a clay was deposited under the peat: this is called "inner sedimentation". A detailed description was given by Künnemann (1941), Haarnagel (1950) and Grohne (1957). The clay layers, a few millimetres to many centimetres thick, known as "Klappklei", are present in many coastal peat sections. Most of them are found in *Sphagnum* peat, some between *Sphagnum* peat and fen peat, and a few only in fen peat. The floating of such a bog during storm tide can still be observed in the Jade bay, in the "Sehestedter Moor", the remains of a raised bog on the seaward side of the dike (see Fig. IV-26). This bog is gradually destroyed by the sea.

Another factor was the digging of salt-bearing peat, described most recently by Bantelmann (1966), Prange (1967) and Marschalleck (1973). The sediment-covered peat on the seaward side of the dikes was cut and burned. The salt absorbed in the peat was won and sold as "Frisian salt". As a result, large areas formerly covered with salt marsh vegetation were lowered to below the MHW and became "Wadden" areas (tidal flats). The practice of cutting salt peat was widespread in Northern Friesland and also known in some areas on the East Frisian coast; it lasted until about 1800. Many of the areas from which peat was cut have been covered again with sediment. Such areas are known particularly in the North Frisian tidal flat area (see Fig. IV-27).

In recent time many coastal engineering projects have influenced the natural behaviour of the sea; they changed the tidal range, the height of storm floods, and the volumes of water. Deeper channels for shipping in the lower reaches of the large rivers provided also more room for tidal currents. Thus the Weser, a most

Fig. IV-27. Area in which salt-bearing peat was excavated. The long banks are remnants of the upper clay layer, thrown upside down into the pits left by excavating the salt-bearing peat. In the background Hallig Langeness — Nordmarsch (North Frisian Wadden Sea). Foto Bantelmann 1954

extreme example, had a tidal range of less than 0.5 m in Bremen about 1880 (before the correction of the Weser); the present tidal range is 3.60 m (Rohde 1970, 1975).

The measures of coastal protection have established a stable coast line. Loss of land is virtually unknown nowdays. Land reclamation for agricultural purposes is no longer practised. Land is reclaimed on a small scale, but only for industrial purposes or coastal protection.

The natural forces, which in the past imposed constant changes on the life of the coastal inhabitants, have been brought under control. The human activity has led to stability along the coastline of Northwest Germany.

REFERENCES

Averdieck, F.-R., 1967: Die Vegetationsentwicklung des Eem-Interglazials und der Frühwürm-Interstadiale von Odderade/Schleswig-Holstein. *Fundam.* 2, 2, 101—125, Köln-Graz.

Bantelmann, A., 1949: Ergebnisse der Marschen-archäologie in Schleswig-Holstein. *Offa 8*, 75—88, Neumünster.

Bantelmann, A., 1955: Tofting, eine vorgeschichtliche Warft an der Eidermündung. *Offa 12*, 1—90, Neumünster.

Bantelmann, A., 1966: Die Landschaftsentwicklung an der Schleswig-Holsteinischen Westküste, dargestellt am Beispiel Nordfriesland. *Die Küste 14*, 5—99, Heide in Holstein.

Bantelmann, A., 1970: Spuren vor- und frühgeschichtlicher Besiedlung auf einem

Strandwall bei Tating, Eiderstedt. *Probleme der Küstenforschung im südlichen Nordseegebiet 9*, 49–57, Hildesheim.

Bantelmann, A., 1975: Die frühgeschichtliche Marschensiedlung beim Elisenhof in Eiderstedt. Landschaftsgeschichte und Baubefunde. *Studien zur Küstenarchäol. Schleswig-Holsteins A, 1*, 190 pp., Bern and Frankfurt/Main.

Barckhausen, J., 1969: Entstehung und Entwicklung der Insel Langeoog – Beiträge zur Quartärgeologie und Paläogeographie eines ostfriesischen Küstenabschnittes. *Oldenburger Jahrb. 68*, 239–281.

Barckhausen, J., Preuss, H. & Streif, H., 1977: Ein lithologisches Ordnungsprinzip für das Küstenholozän und seine Darstellung in Form von Profiltypen. *Geol. Jahrb. A44*, 45–74, Hannover.

Bartenstein, H., 1938: Die Foraminiferenfauna des Jade-Gebietes. *Senckenbergiana 20*, 386–412.

Behre, K.-E., 1970a: Die Entwicklungsgeschichte der natürlichen Vegetation im Gebiet der unteren Ems und ihre Abhängigkeit von den Bewegungen des Meeresspiegels. *Probleme der Küstenforschung im südlichen Nordseegebiet 9*, 13–47, Hildesheim.

Behre, K.-E., 1970b: Die Flora des Helgoländer Süsswasser-"Töcks", eines Eem-Interglazials unter der Nordsee. *Flora 159*, 133–146, Jena.

Behre, K.-E., 1972: Kultur- und Wildpflanzenreste aus der Marschgrabung Jemgumkloster/Ems (um Christi Geburt). *Neue Ausgrabungen und Forsch. in Niedersachs. 7*, 164–184, Hildesheim.

Behre, K.-E., 1976: Die Pflanzenreste aus der frühgeschichtlichen Wurt Elisenhof. *Studien zur Küstenarchäol. Schleswig-Holsteins A, 2*, 183 pp., Bern and Frankfurt/Main.

Behre, K.-E., 1977: Acker, Grünland und natürliche Vegetation während der römischen Kaiserzeit im Gebiet der Marschensiedlung Bentumersiel/Unterems. *Probleme der Küstenforschung im südlichen Nordseegebiet 12*, 67–84, Hildesheim.

Behre, K.-E. & Menke, B., 1969: Pollenanalytische Untersuchungen an einem Bohrkern der südlichen Doggerbank. *Beiträge zur Meereskunde 24/25*, 123–129. Deutsche Akademie der Wissenschaften zu Berlin.

Benzler, J. H. & Geyh, M. A., 1969: Versuch einer zeitlichen Gliederung von Dwog-Horizonten mit Hinweisen auf die Problematik der ^{14}C-Datierung von Bodenproben. *Z. dtsch. geol. Ges. Jahrg. 1966, 118*, 361–367.

Boigk, H., 1961: Ergebnisse und Probleme stratigraphisch-paläogeographischer Untersuchungen im Buntsandstein Nordwestdeutschlands. *Geol. Jahrb. 78*, 123–134.

Brand, G., Hageman, B. P., Jelgersma, S. & Sindowski, K.-H., 1965: Die lithostratigraphische Unterteilung des marinen Holozäns an der Nordseeküste. *Geol. Jahrb. 82*, 365–384.

Brandt, K., 1972: Untersuchungen zur kaiserzeitlichen Besiedlung bei Jemgumkloster und Bentumersiel (Gem. Holtgaste, Kreis Leer) im Jahre 1970. *Neue Ausgrabungen und Forsch. in Niedersachs. 7*, 145–163, Hildesheim.

Brandt, K., 1975: Siedlungsarchäologische Untersuchungen im nördlichen Reiderland. *Nachr. des Marschenrates 23*, 17–18, Wilhelmshaven.

Brandt, K. & Behre, K.-E., 1976: Eine Siedlung der älteren vorrömischen Eisenzeit bei Oldendorp/Unterems mit Aussagen zu Umwelt, Ackerbau und Sedimentationsgeschehen. *Nachr. Niedersachs. Urgesch. 45*, 447–458, Hildesheim.

Brelie, G. von der, 1951: Die junginterglazialen Ablagerungen im Gebiet des Nordostseekanals. *Schr. naturwiss. Ver. Schleswig-Holstein 25*, 100–107, Kiel.

Brelie, G. von der, 1954: Transgression und Moorbildung im letzten Interglazial. *Mitt. Geol. Staatsinst. Hamburg 23*, 111–118.

Brelie, G. von der, 1959: Wattablagerungen des Eem-Meeres im Raum Norderney. *Z. dtsch. geol. Ges. 111*, 1–7.

Brockmann, C., 1932: Die Diatomeen aus dem Interglazial von Oldenbüttel. *Abh. Preuss. Geol. Landesanst., N.F., 140*.

Brockmann, C., 1950: Die Watt-Diatomeen der schleswig-holsteinischen Westküste. *Abh. Senckenb. Naturf. Ges. 478*, 1–26.

Cordes, H., 1967: Moorkundliche Untersuchungen zu Entstehung des Blocklandes bei Bremen. *Abh. naturw. Verein Bremen 37*, 147–196.

Dechend, W., 1950: Das Eem im Raum Norderney-Hilgenriede. *Z. Deutsch. Geol. Ges. 102/I*, 91–97.

Dechend, W., 1954: Eustatische und tektonische Einflüsse im Quartär der südlichen Nordseeküste. *Geol. Jahrb. 68*, 501–516.

Dechend, W., 1956: Der Ablauf der holozänen Nordsee-Transgression im oldenburgisch-ostfriesischen Raum, insbesondere im Gebiet von Jever i.O. *Geol. Jahrb. 72*, 295–314.

Dechend, W., 1958: Marines und brackisches Eem im Raum der Ems-Mündung. *Geol. Jahrb. 76*, 175–190.

Dittmer, E., 1938: Schichtenaufbau und Entwicklungsgeschichte des Dithmarscher Alluviums. *Westküste, I*, 105–150, Heide in Holstein.

Dittmer, E., 1941: Das nordfriesische Eem. Ein Beitrag zur Geschichte der junginterglazialen Nordsee. *Kieler Meeresforsch. 5*, 168–199.

Dittmer, E., 1951: Das Eem des Treene-Tales. *Schr. naturwiss. Ver. Schleswig-Holstein 25*, 91–99, Kiel.

Dittmer, E., 1952: Die nacheiszeitliche Entwicklung der schleswig-holsteinischen Westküste. *Meyniana, 1,* 138—167.
Dittmer, E., 1954: Interstadiale Torfe in würmzeitlichen Schmelzwassersanden Nordfrieslands. *Eiszeitalter und Gegenwart, 4/5,* 172—175.
Dittmer, E., 1960: Neue Beobachtungen und kritische Bemerkungen zur Frage der "Küstensenkung". *Die Küste 8,* 29—44, Heide in Holstein.
Dörjes, J., Gadow, S., Reineck, H. E. & Singh, I. B., 1969: Die Rinnen der Jade (südliche Nordsee). Sedimente und Makrobenthos. *Senckenbergiana maritima 1, 50,* 5—62.
Dümmler, H. & Menke, B., 1970: Der Einfluss der Holozänentwicklung auf Landschaft und Böden der Broklandsauniederung (Dithmarschen). *Meyniana 20,* 9—16.
Figge, K., 1974: Sediment distribution mapping in the German Bight (North Sea). *Mém. Inst. Géol. Bassin Aquitaine 7,* 253—257.
Fischer, O., 1955—1957: Das Wasserwesen an der schleswig-holsteinischen Nordseeküste. 3. Teil: Das Festland. Vol. 1—7, Berlin.
Forschungsstelle Norderney, 1961, 1962, 1963: Niedersächsische Küste, Historische Karte 1:50.000 Forschungsstelle Norderney der Niedersächsischen Wasserwirtschaftsverwaltung.
Geyh, M. A., 1969: Versuch einer chronologischen Gliederung des marinen Holozäns an der Nordseeküste mit Hilfe der statistischen Auswertung von ^{14}C-Daten. *Z. deutsch. geol. Ges., Jahrg. 1966, 118,* 351—360.
Geyh, M. A., 1971: Middle and Young Holocene sea-level changes as global contemporary events. *Geol. Fören. Stockh. Förh. 93,* 679—690.
Grahle, H. O., 1936: Die Ablagerungen der Holstein-See (Mar. Interglaz. I), ihre Verbreitung, Fossilführung und Schichtenfolge in Schleswig-Holstein. *Abh. Preuss. geol. Landesanst., N.F., 172,* 1—110.
Graul, H., 1960: Der Verlauf des glazialeustatischen Meeresspiegelanstieges, berechnet an Hand von C-14 Datierungen. *Verh. d. Dtsch. Geographentages 32,* 232—242, Wiesbaden.
Gripp, K., 1937: Die Entstehung der Nordsee. In *Werdendes Land am Meer.* Berlin.
Gripp, K., 1951: Die heutige Nordsee und ihre zwei eiszeitlichen Vorgänger, ein erdgeschichtlicher Vergleich. *Abh. naturwiss. Ver. Bremen 33,* 5—18.
Gripp, K., 1952: Auswertung der Bohrungen beiderseits des Nord-Ostsee-Kanals von Hochdonn bis Breidorf, Kilometer 19,2—49,5. *Meyniana, 1,* 58—106.
Gripp, K., 1964: Erdgeschichte von Schleswig-Holstein. 441 pp. Wachholtz, Neumünster.
Grohne, U., 1957: Zur Entwicklungsgeschichte des ostfriesischen Küstengebietes auf Grund botanischer Untersuchungen. *Probleme der Küstenforschung im südlichen Nordseegebiet 6,* 1—48, Hildesheim.
Grosse-Brauckmann, G., 1962: Moorstratigraphische Untersuchungen im Niederwesergebiet (über Moorbildungen am Geestrand und ihre Torfe). *Veröff. Geobot. Inst. Eidg. Techn. Hochsch., Stift. Rübel in Zürich 37,* 100—119, Bern.
Grosse-Brauckmann, G., 1963: Über die Artenzusammensetzung von Torfen aus dem nordwestdeutschen Marschen-Randgebiet (eine pflanzensoziologische Auswertung von Grossrestuntersuchungen). *Vegetatio 11,* 325—341, Den Haag.
Grosse-Brauckmann, G. & Dierssen, K., 1973: Zur historischen und aktuellen Vegetation im Poggenpohlsmoor bei Dötlingen (Oldenburg). *Mitt. florist. — soziolog. Arbeitsgem. N.F. 15/16,* 109—145, Todenmann, Göttingen.
Haake, F. W., 1962: Untersuchungen an der Foraminiferen-Fauna im Wattgebiet zwischen Langeoog und dem Festland. *Meyniana 12,* 25—64.
Haarnagel, W., 1950: Das Alluvium an der deutschen Nordseeküste. *Probleme der Küstenforschung im südlichen Nordseegebiet 4,* 1—146, Hildesheim.
Haarnagel, W., 1957: Die spätbronzefrüheisenzeitliche Gehöftsiedlung Jemgum b. Leer auf dem linken Ufer der Ems. *"Die Kunde" N.F. 8,* 2—44, Hannover.
Haarnagel, W., 1961: Die Marschen im deutschen Küstengebiet der Nordsee und ihre Besiedlung. *Ber. zur dtsch. Landeskunde 27,* 203—219.
Haarnagel, W., 1969: Die Ergebnisse der Grabung auf der ältereisenzeitlichen Siedlung Boomborg/Hatzum, Kreis Leer, in den Jahren von 1965 bis 1967. *Neue Ausgrabungen und Forsch. in Niedersachs. 4,* 58—97, Hildesheim.
Haarnagel, W., 1973: Vor- und Frühgeschichte des Landes Wursten. In v. Lehe, E., *Geschichte des Landes Wursten,* 19—128, Bremerhaven.
Haarnagel, W., (in press): Die Grabung Feddersen Wierde. *Feddersen Wierde 2,* Steiner-Verlag, Wiesbaden.
Hageman, B. P., 1969: Development of the western part of the Netherlands during the Holocene. *Geol. Mijnbouw 48,* 373—388.
Hartung, W., 1964: Helgoland — merkwürdigste Insel der Nordsee. 150 Jahre Naturforschende Gesellschaft Emden, 35—73, Emden.
Hayen, H., 1972: Siedlung der späten Bronzezeit und frühen Eisenzeit in der Marsch bei Rodenkirchen, Kr. Wesermarsch. *Nachr. Niedersachs. Urgesch. 41,* 261—262, Hildesheim.
Heck, H. L., 1932: Die Eem- und ihre begleitende

Junginterglazial-Ablagerungen bei Oldenbüttel in Holstein. *Abh. Preuss. geol. Landesanst. N. F. 140*, 1–80.
Hoffmann, D., 1969: The marine Holocene of Sylt — discussion of the age and facies. *Geol. Mijnbouw 48*, 343–347.
Hoffmann, D., 1974: Zum geologischen Aufbau der Hörnumer Halbinsel auf Sylt. *Meyniana 23*, 63–68.
Homeier, H., 1962: Historisches Kartenwerk 1: 50.000 der Niedersächsischen Küste (mit Beiheften), Norderney.
Homeier, H., 1969: Der Gestaltwandel der ostfriesischen Küste im Laufe der Jahrhunderte. Ein Jahrtausend ostfriesischer Deichgeschichte. *Ostfriesland im Schutze des Deiches 2*, 1–75, Pewsum.
Hummel, P. & Cordes, E., 1969: Holozäne Sedimentation und Faziesdifferenzierung beim Aufbau der Lundener Nehrung (Norderdithmarschen). *Meyniana 19*, 103–112.
Jelgersma, S., 1961: Holocene sea-level changes in the Netherlands. *Meded. Geol. Sticht. C, VI*, No. 7, 100 pp.
Jessen, K. & Milthers, V., 1928: Stratigraphical and Paleontological Studies of Interglacial Fresh-Water Deposits in Jutland and Northwest Germany. *Danm. Geol. Unders. II. 48*, 1–379.
Jong, J. D. de, 1972: The scenery of the Netherlands against the background of the Holocene Geology, a review of the recent Literature. *Rev. de Géogr. Phys. et de Géol. Dyn. 2 XIII*, 143–162.
Kersten, K. & La Baume, P., 1958: Vorgeschichte der nordfriesischen Inseln. 664 pp., Neumünster.
König, D., 1953: Diatomeen aus dem Eem des Treenetales. *Schr. naturwiss. Ver. Schleswig-Holstein 26*, 124–132, Kiel.
Körber-Grohne, U., 1967: Geobotanische Untersuchungen auf der Feddersen Wierde. *Feddersen Wierde 1*, 2 vols., Steiner-Verlag, Wiesbaden.
Kolp, O., 1974: Submarine Uferterrassen in der südlichen Ost- und Nordsee als Marken eines stufenweise erfolgten holozänen Meeresspiegelanstiegs. *Baltica 5*, 11–40, Vilnius.
Kolp, O., 1976: Submarine Uferterrassen der südlichen Ost- und Nordsee als Marken des holozänen Meeresanstiegs und der Überflutungsphasen der Ostsee. *Petermanns Geogr. Mitt. 120*, 1–23, Gotha.
Künnemann, Chr., 1941: Das Sehestedter Moor und die Ursachen seiner Zerstörung. *Probleme der Küstenforschung im südlichen Nordseegebiet 2*, 37–58, Hildesheim.
Kulczinski, S., 1949: Torfowiska Polesia. (Peat bogs of Polesie.) *Mein. Acad. Polon. Sci. lettr. Cl. Sci. math. nat. B. Sci. nat. 15*, 356 pp., Krakow.
Lafrenz, H. R., 1963: Foraminiferen aus dem marinen Riss-Würm-Interglazial (Eem) in Schleswig-Holstein. *Meyniana 13*, 10–45.
Lang, A.W., 1969 ff.: Historisches Seekartenwerk der Deutschen Bucht. Wachholtz, Neumünster.
Lang, A. W., 1975: Untersuchungen zur morphologischen Entwicklung des Dithmarscher Watts von der Mitte des 16. Jahrhunderts bis zur Gegenwart. *Hamburger Küstenforschung 31*, 154 pp.
Lange, W., 1962: Die Mikrofauna einiger Störmeer-Absätze (I-Interglazial) Schleswig-Holsteins. *Neues Jahrb. Geol. Pal., Abh., 115*, 222–242.
Lange, W. & Menke, B., 1967: Beiträge zur frühpostglazialen erd- und vegetationsgeschichtlichen Entwicklung im Eidergebiet, insbesondere zur Flussgeschichte und zur Genese des sogenannten Basistorfs. *Meyniana 17*, 29–44.
Linke, G., 1969: Die Entstehung der Insel Scharhörn und ihre Bedeutung für die Überlegungen zur Sandbewegung in der Deutschen Bucht. *Hamburger Küstenforschung 11*, 45–84.
Linke, G., 1970: Über die geologischen Verhältnisse im Gebiet Neuwerk/Scharhörn. *Hamburger Küstenforschung 17*, 17–58.
Little-Gadow, S. & Schäfer, A., 1974: Schwermetalle in den Sedimenten der Jade. Bestandsaufnahme und Vergleich mit der inneren Deutschen Bucht. *Senckenbergiana maritima 6, (2)*, 161–174.
Ludwig, G., Müller, H. & Streif, H., 1977: Neuere Daten zum holozänen Meeresspiegelanstieg im Bereich der Deutschen Bucht, (in press).
Madsen, V., Nordmann, V. & Hartz, N., 1908: Eem-Zonerne. *Danm. geol. Unders. V, 4*, 225 pp.
Marschalleck, K. H., 1973: Die Salzgewinnung an der friesischen Nordseeküste. *Probleme der Küstenforschung im südlichen Nordseegebiet 10*, 127–150. Hildesheim.
Menke, B., 1968a: Beiträge zur Biostratigraphie des Mittelpleistozäns in Norddeutschland (pollenanalytische Untersuchungen aus Westholstein). *Meyniana 18*, 35–42.
Menke, B., 1968b: Ein Beitrag zur pflanzensoziologischen Auswertung von Pollendiagrammen, zur Kenntnis früherer Pflanzengesellschaften in den Marschenrandgebieten der schleswig-holsteinischen Westküste und zur Anwendung auf die Frage der Küstenentwicklung. *Mitt. florist.-soziolog. Arbeitsgem. N.F. 13*, 195–224, Todenmann/Rinteln.
Menke, B., 1969a: Vegetationsgeschichtliche Untersuchungen und Radiocarbon-Datierungen zur holozänen Entwicklung der schleswig-holsteinischen Westküste. *Eiszeitalter und Gegenwart 20*, 35–45.
Menke, B., 1969b: Vegetationskundliche und vegetationsgeschichtliche Untersuchungen an Strandwällen. *Mitt. florist.-soziolog. Arbeitsg. N.F. 14*, 95–120, Toden-

mann/Rinteln.

Menke, B., 1970: Ergebnisse der Pollenanalyse zur Pleistozän-Stratigraphie und zur Pliozän-Pleistozän-Grenze in Schleswig-Holstein. *Eiszeitalter und Gegenwart 21*, 5—21.

Menke, B., 1976a: Befunde und Überlegungen zum nacheiszeitlichen Meeresspiegelanstieg (Dithmarschen und Eiderstedt, Schleswig-Holstein). *Probleme der Küstenforschung im südlichen Nordseegebiet 11*, 145—161, Hildesheim.

Menke, B., 1976b: Neue Ergebnisse zur Stratigraphie und Landschaftsentwicklung im Jungpleistozän Westholsteins. *Eiszeitalter und Gegenwart 27*, 1—16.

Müller, W., 1962: Der Ablauf der holozänen Meerestransgression an der südlichen Nordseeküste und Folgerungen in bezug auf eine geochronologische Holozängliederung. *Eiszeitalter und Gegenwart 13*, 197—226.

Nilsson, T., 1948: Versuch einer Anknüpfung der postglazialen Entwicklung des norddeutschen und niederländischen Flachlandes an die pollenfloristische Zonengliederung Südskandinaviens. *Medd. fran Lunds Geol.-Mineral. Inst. 112*, 79 pp.

Oele, E., 1969: The Quaternary Geology of the Dutch part of the North Sea, north of the Frisian isles. *Geol. Mijnbouw 48*, 467—479.

Oele, E., 1971: The Quaternary geology of the southern area of the Dutch part of the North Sea. *Geol. Mijnbouw 50*, 461—474.

Overbeck, F., 1975: Botanisch-geologische Moorkunde unter besonderer Berücksichtigung der Moore Nordwestdeutschlands als Quellen zur Vegetations-, Klima- und Siedlungsgeschichte. 719 pp., Neumünster.

Prange, W., 1963: Das Holozän und seine Datierung in den Marschen des Arlau-Gebietes, Nordfriesland. *Meyniana 13*, 47—76.

Prange, W., 1965: Die Höhe der Sturmflut vom 11. Oktober 1634 in Nordfriesland nach neuen Wasserstandsmarken. Zwischen Eider und Wiedau, 40—48, Husum.

Prange, W., 1967: Geologie des Holozäns in den Marschen des nordfriesischen Festlandes. *Meyniana 17*, 45—94.

Prange, W., 1968: Geologische Untersuchungen in den Marschen der alten Köge vor Bredstedt, Nordfriesland. *Neues Jahrb. Geol. Paläontol. Monatsh. 1968, 10*, 619—640.

Regteren-Altena, J. F. van, Bakker J. A., Clason, A. T., Glasbergen, W., Groenman-van Waateringe, W. & Pons, L. J., 1962—1963: The Vlaardingen Culture I-V. *Helinium 2—3*, Wetteren, Belgium.

Reinhard, H., 1974: Genese des Nordseeraumes im Quartär. *Fennia 129*, 95 pp.

Reinhardt, W., 1969: Die Orts- und Flurformen Ostfrieslands in ihrer siedlungsgeschichtlichen Entwicklung. *Ostfriesland im Schutze des Deiches 1*, 201—375, Pewsum.

Reineck, H. E. (editor), 1970: Das Watt. Ablagerungs- und Lebensraum. Senckenberg-Buch 50, 142 pp., Frankfurt a.M.

Reineck, H. E. & Singh, I. B., 1975: Depositional Sedimentary Environments. 439 pp., Springer, Berlin, Heidelberg, New York.

Richter, G., 1964—1965: Zur Ökologie der Foraminiferen I-III. *Natur und Museum 94*, 343—353, 421—430, *95*, 51—62.

Richter, K., 1937: Die Eiszeit in Norddeutschland. *Deutscher Boden 4*, 176 pp., Berlin.

Roeleveld, W., 1974: The Groningen coastal area. Thesis Vrije Universiteit Amsterdam, 252 pp., Amsterdam.

Rohde, H., 1970: Die Entwicklung der Wasserstrassen im Bereich der deutschen Nordseeküste. *Die Küste 20*, 1—44, Heide in Holstein.

Rohde, H., 1975: Wasserstandsbeobachtungen im Bereich der deutschen Nordseeküste vor der Mitte des 19. Jahrhunderts. *Die Küste 28*, 1—96, Heide in Holstein.

Rottgardt, D., 1952: Mikropaläontologisch wichtige Bestandteile recenter brackischer Sedimente an den Küsten Schleswig-Holsteins. *Meyniana 1*, 169—228.

Schäfer, W., 1962: Aktuo-Palaeontologie nach Studien in der Nordsee. 666 pp., W. Kramer, Frankfurt a.M.

Scheer, K., 1953: Die Bedeutung von *Phragmites communis* Trin. für die Fragen der Küstenbildung. *Probleme der Küstenforschung im Gebiet der südlichen Nordsee 5*, 15—25, Hildesheim.

Schmid, P., 1969: Die vor- und frühgeschichtlichen Grundlagen der Besiedlung Ostfrieslands nach der Zeitenwende. *Ostfriesland im Schutze des Deiches 1*, 107—200, Pewsum.

Schmidt, R., 1976: Das Küstenholozän der östlichen Meldorfer Bucht und angrenzender Marschen in Dithmarschen, Schleswig-Holstein. *Meyniana 28*, 69—85.

Schütte, H., 1933: Der geologische Aufbau des Jever- und Harlingerlandes und die erste Marschbesiedlung. *Oldenburger Jahrb. 37*, 1—39.

Schütte, H., 1939: Sinkendes Land an der Nordsee? *Schr. des Dtsch. Naturkundever., N.F. 9*, 144 pp., Öhringen/Württ.

Sindowski, K.-H., 1958: Das Eem im Wattgebiet zwischen Norderney und Spiekeroog, Ostfriesland. Geol. *Jahrb. 76*, 151—174.

Sindowski, K.-H., 1961: Die geologische Entwicklung des ostfriesischen Wattgebietes und der Inseln im Laufe des Quartärs. *Z. dtsch. geol. Ges. 112 (1960)*, 527—529.

Sindowski, K.-H., 1965: Das Eem im ostfriesischen Küstengebiet. *Z. dtsch. geol. Ges. 115*, 163—166.

Sindowski, K.-H., 1969: Erläuterungen zu Blatt Wangerooge Nr. 2213. Geologische Karte von Niedersachsen 1:25.000, 49 pp., Han-

nover.
Sindowski, K.-H., 1970: Das Quartär im Untergrund der Deutschen Bucht (Nordsee). *Eiszeitalter und Gegenwart 21*, 33—46.
Sindowski, K.-H., 1973: Das ostfriesische Küstengebiet — Inseln, Watten und Marschen. *Sammlung geologischer Führer 57*, 162 pp., Berlin-Stuttgart.
Schwabedissen, H., 1951: Zur Besiedlung des Nordseeraumes in der älteren und mittleren Steinzeit. Festschr. G. Schwantes, 58—77.
Streif, H., 1971: Stratigraphie und Faziesentwicklung im Küstengebiet von Woltzeten in Osttriesland. *Beih. geol. Jahrb. 119*, 59 pp.
Streif, H., 1975: Versuch einer Bilanzierung der Sedimentation im Küstenholozän Ostfrieslands. *Geol. Jahrb. A 28*, 3—14.
Succow, M., 1971: Die Talmoore des nordostdeutschen Flachlandes, ein Beitrag zur Charakterisierung des Moortyps "Niedermoor". *Arch. Naturschutz Landschaftsforsch. 11*, 133—168, Berlin.
Valentin, H., 1957: Glazialmorphologische Untersuchungen in Ostengland. *Abh. geogr. Inst. F.U. Berlin 4*, 1—86.
Voorthuysen, J. H. van, 1960: Die Foraminiferen des Dollart-Ems-Estuarium. III4 Symposium Ems-Estuarium (Nordsee). *Verh. K. Ned. Geol. Mijnbouwkd. Genoot., Geol. Serie, XIX*, 237—269.
Wagner, C. W., 1960: Ostracoden Biocoenosen und Thanatocoenosen im Ems-Estuarium (N. O. Niederlande). III13 Symposium Ems-Estuarium (Nordsee). *Verh. K. Ned. Geol. Mijnbouwkd. Genoot., Geol. Serie, XIX*, 221—236.
Wiermann, R., 1962: Botanisch-moorkundliche Untersuchungen in Nordfriesland. *Meyniana 12*, 97—146.
Wildvang, D., 1938: Die Geologie Ostfrieslands. *Abh. Preuss. Geol. Landesanst., N.F. 181*, 211 pp.
Woebcken, C., 1934: Die Entstehung des Jadebusens. *Schriftenr. Niedersächs. Ausschuss für Heimatschutz 7*, 62 pp., Aurich.
Woldstedt, P. & Duphorn, K., 1974: Norddeutschland und angrenzende Gebiete im Eiszeitalter. 2nd edition. 500 pp., Koehler-Verlag, Stuttgart.
Woszidlo, H., 1962: Foraminiferen und Ostrakoden aus dem marinen Elster-Saale-Interglazial in Schleswig-Holstein. *Meyniana 12*, 65—96.
Wurster, P., 1962: Geologisches Portrait Helgolands. *Die Natur 70*, Schwäbisch-Hall.
Ziegelmeier, E., 1957: Die Muscheln (Bivalvia) der deutschen Meeresgebiete. *Helgoländer Wissenschaftliche Meeresuntersuchungen 6*, 1—56.
Ziegelmeier, E., 1966: Die Schnecken (Gastropoda Prosobranchia) der deutschen Meeresgebiete und brackigen Küstengewässer. *Helgoländer wissenschaftliche Meeresuntersuchungen 13*, 1—61.

IV-e

Depositional History and coastal development in the Netherlands and the adjacent North Sea since the Eemian

SASKIA JELGERSMA, ERNO OELE & ALBERT J. WIGGERS

Jelgersma, S., Oele, E. & Wiggers, A. J., 1979: Depositional History and coastal development in the Netherlands and the adjacent North Sea since the Eemian. In E. Oele, R. T. E. Schüttenhelm & A. J. Wiggers (editors), The Quaternary History of the North Sea, 115–142. *Acta Univ. Ups. Symp. Univ. Ups. Annum Quingentesimum Celebrantis: 2*, Uppsala. ISBN 91-554-0495-2.

The depositional history of the Netherlands and the Dutch North Sea sector during the Eemian, the Weichselian glacial interval and the Holocene is described with an emphasis on coastal development. Datings, sea level stands, nature of sediments and inferred shorelines of the Eemian transgression are discussed. Although during the following glacial period, i.e. the Weichselian, the land ice did not reach the Netherlands or the Dutch sector of the North Sea, the resulting major fall of sea-level together with the climatic deterioration controlled erosion and sedimentation. Exclusively continental deposits were laid down. The subsequent climatic amelioration and sea level rise in the Holocene led to a renewed flooding of large parts of the area under consideration. The progress of the latter transgression with related sediments and inferred shorelines is depicted on the colour chart and on several palaeogeographical maps (Figs. IV-38, 39, 40).

Dr. S. Jelgersma and Dr. E. Oele, Rijks Geologische Dienst, Spaarne 17, Haarlem, the Netherlands; Dr. A. J. Wiggers, Instituut voor Aardwetenschappen, Vrije Universiteit Amsterdam, De Boelelaan 1085, Amsterdam, the Netherlands.

Introduction

The development of the Eemian sedimentation and the distribution of the various deposits is strongly related to the morphology caused by the earlier Saalian glaciation. During the Eemian

Interglacial the sea transgressed over the present coastal area of the Netherlands. Sea level rise and subsequent fall as well as climatic variations, are clearly reflected in the sediments, which settled in the marine and estuarine environments in Eemian time.

The predominantly marine sediments are comprised in the Eem Formation. Another significant area of deposition during the Eemian was present along the courses of Rhine and Meuse, where a series of fluviatile coarse sands with gravel was laid down. The fluviatile deposits are part of the Kreftenheye Formation, which ranges from Late Saalian to Preboreal in age. The latter formation will be discussed in the paragraph on the Weichselian. Locally some sedimentation took place in small lakes and along small streams resulting in the occurrence of peat and clay layers up to a few metres thick. The latter deposits are grouped together in the Asten Formation, which is time correlative with the Eem Formation.

The Weichselian glacial interval has been a period of considerable erosion and sedimentation, resulting in a further levelling of the Saalian and the subsequent Eemian relief. Especially the Saalian glacial depressions were filled up completely. The cold climatic conditions of Weichselian time are reflected in the deposits e.g. by the occurrence of frost wedges, which point to the presence of a permafrost. Temporary climatic ameliorations may be inferred from the occurrence of peat and gyttja. Results of pollen analysis and the study of sedimentary structures have provided detailed information on the environment of deposition during this glacial period.

The Weichselian sediments in the Netherlands may be subdivided in four types with regard to their mode of deposition: aeolian, lacustrine, meltwater and fluvial deposits. These types are included in the Twenthe Formation (cover sands, and fluvio-periglacial deposits together with slopewash-deposits and desert pavements), the Kreftenheye Formation (fluviatile deposits), and the Brown Bank Bed (lacustrine clay deposits).

Early in the Holocene, around 9000 B.P., the post-glacial transgressing sea entered the northern part of the Dutch North Sea sector. At around 7500 B.P. the sea locally reached the present coastline of the Netherlands, presenting evidence for a rapid relative sea-level rise. This sea-level rise is reflected in the nature and distribution of a large part of the Holocene sediments.

The marine, lagoonal and coastal deposits have been placed in the Westland Formation, comprising the older Elbow/Calais and the younger Dunkerque deposits. This formation includes also littoral deposits as coastal barriers and dunes, the so-called perimarine deposits and the Holland Peat Member. Fluviatile deposits are grouped together in the Betuwe Formation. Inland dunes, brook and moor deposits have been placed in separate formations of limited areal extent.

Eemian deposits in the onshore part of the Netherlands

Distribution and lithology

In the onshore part of the Netherlands two areas may be distinguished where marine and estuarine deposits are present. In the North and Northwest the Eemian sea transgressed over depressions and valleys of the glacial landscape. In the Southwest, outside the area previously covered by Saalian land ice, the marine and estuarine deposits are related to the river-mouths of Rhine, Meuse and Scheldt. The most complete sections of the Eem Formation, however, may be found in the glaciogenic depressions of the formerly glaciated area. Also the type locality is established in such a former glacier tongue basin, the "Gelderse Vallei" (Fig. IV-28).

In the deepest part of the glacial basins for example in former glacial valleys the base of the Eem Formation may occur at 80 metres below mean sea level. Outside the glacially influenced area the base is at approximately 40 metres below mean sea level. The top of the Eem Formation is between 8 and 25 metres below present sea level. The last mentioned figure represents either a Late Eemian marine level during sea level fall or an erosional feature

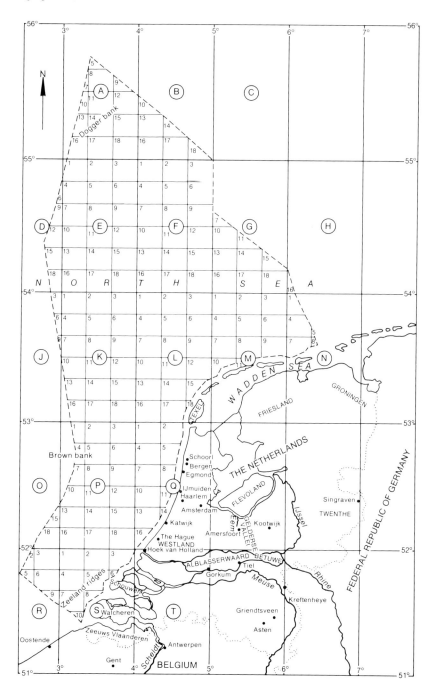

Fig. IV-28. Map of the Netherlands and the Dutch North Sea sector with localities mentioned in the text.

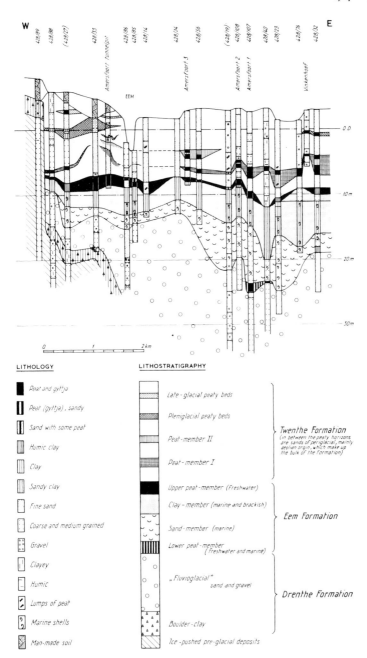

Fig. IV-29. Section through the Late Pleistocene infill of the "Gelderse Vallei" — a glacial tongue basin — with the type locality of the Eem Formation (after Zagwijn 1961).

caused by Early Weichselian fluviatile activities.

Whereas sediments in the basins consist of medium sands and clays, the marine-estuarine deposits of the SW area are predominantly coarse sands. Organic deposits like gyttja and peat occur rather infrequently in the former coastal area of the Eemian sea.

Stratigraphy

The micro- and macrofauna have been studied by various investigators. Reference should be made to descriptions of the faunal content, published by Harting (1874), Lorié (1906), Burck (1949), Spaink (1958) and van Voorthuysen (1958). Pollen-analytical investigations have been of particular value for a more detailed subdivision of the Eemian and a better understanding of its geological history.

A complete vegetational history of the Eemian Interglacial has been derived from the examination of samples collected in the area of the type locality and other glacial depressions (Zagwijn 1961). In Fig. IV-30 a diagram showing the vegetational history and pollen zones is presented.

A cross-section running through the "Gelderse Vallei" with the type locality shows the sedimentary sequence of the interglacial (Fig. IV-29). On the western side the depression is flanked by an ice-pushed ridge, composed of (pre-Saalian) fluvial deposits. The deposits in the lower part of the depression are fluvioglacial sands and gravel. The overlying Eem Formation consists of marine medium-grained sands. These deposits rich in shells are part of a transgressive series. They are overlain by a brackish-marine clay covered by a peat (regressive series).

The transgression as observed here, took place at the transition from pollen zone E3b to E4a. The dating is inferred from the pollen-analysis of a basal brackish gyttja (Zagwijn 1961). The above mentioned sands may be placed in the E4 zone, and the overlying brackish clay in pollen zone E5. The withdrawal of the sea created favourable conditions for peat growth during pollen zone E6. The highest marine level i.e. the top of the E5 clay layer is at present approximately 8 metres below mean sea level.

In the glacial basins of Amsterdam and Flevoland (Fig. IV-28) the Saalian meltwater clay is overlain by marine Eemian sediments which may reach a thickness of more than 60 metres. Locally, fresh-water gyttja and lake marl are sandwiched between the lacustroglacial and the marine Eemian clays. In this basin the transgression started in pollen zone E3b (Zagwijn, unpubl.) The base of the marine deposits may be as deep as 80 metres below the present sea-level. Based on pollen-analytical results these marine clays were deposited just before the climatic optimum of the Eemian Interglacial. Similar results have been obtained from investigations on Eemian deposits in the Ijmuiden glacial basin. The sea level changes during this E3 zone will be discussed in a following paragraph.

In the Southwest i.e. in the area where fluvial activities during the Eemian are apparent, it is difficult to reconstruct the depositional history due to the coarseness of the material. Between 26 and 32 metres below mean sea level a shell bed is intercalated. The poorly developed *Cardium* fauna may indicate cooler depositional conditions, presumably during the Late Eemian. It would imply, that the level with a marine fauna corresponds with a stage during sea-level fall at the end of the Eemian. Previously, these deposits were included in the Schouwen Formation (Van Rummelen 1965); at present they are considered to be a part of the Eem Formation.

Close to the Dutch-Belgian border, in the western part of Zeeuws-Vlaanderen a freshwater gyttja has been apparently formed in pollen zones E5b and E6. It occurs on top of coarse, estuarine sands, which may be assumed to correlate with the "sables marine à *Corbicula fluminalis* de l'assise d'Ostend" in adjacent Belgium.

Eemian deposits in the Dutch part of the North Sea

Distribution and lithology

The initial results of examination of widely-spaced shallow drilled holes led to the assumption that Eemian sediments were absent in the northern part of the Dutch North Sea sector. However, shortly after publication of the first tentative geological map of the area (Oele 1969) a geophysical survey and additional borings amongst others for site investigations for drilling rigs, revealed the presence of Eemian deposits in glacial depressions (Oele 1971a). Based also on more recent information concerning the southern region it may be concluded that Eemian deposits occur in the entire Dutch North Sea sector. They are apparently lacking only in the extreme NW part near the Dogger Bank and in the extreme South (Oele 1971b). The distribution pattern resembles that on land in the following respects: in the area previously covered by Saalian inland ice the Eemian deposits have filled glacial depressions. South of that area a continous layer of marine sands of Eemian age is present.

The top of the Eemian is generally at about 40 metres below mean sea level. Near the West Frisian Isle of Texel and just north of it the upper boundary is at relatively shallow depths (10—20 metres below mean sea level). Here, the occurrence of Late Eemian littoral to tidal sediments is in accordance with the observations in the German bight (Sindowski 1970). Part of the coarse sands in this area has been reworked during the Holocene.

The nature of Eemian sediments varies strongly i.e. from coarse sands to clay. In the southern part coarse sands are predominant due to the supply of clastic material by the rivers Rhine and Meuse, in the North the sands are mostly fine to medium grained and frequently clay layers occur. Peat has not been observed.

Stratigraphy

The age of the marine series may be inferred from its stratigraphic position in relation to morphologic features, which are undoubtedly of Saalian age. For instance, the deposits may be present on ice-pushed ridges or in glacial valleys and overlain by a series of sediments indicative of a cold climate (for example cover sands). The Eemian sediments have also been dated by means of the macrofauna (cf. Spaink 1970) and the microfauna (cf. van Voorthuysen 1958, Toering 1970). However, pollen analysis could provide a more detailed subdivision of the Eemian Interglacial (cf. Zagwijn 1970).

To date the sediments sampled, range in age from Middle Eemian (E3) to Late Eemian (E6). Zagwijn (1971) has recognized fine sands on the flank of a glacial valley in the L-blocks of the Dutch sector that were formed during pollen zone E4. In the adjacent boring L2-5, situated in the glacial valley itself, the pollen zones E3 and E5a were distinguished, both developed as marine clay, on top of which E5b marine sands are present. Zagwijn draws the attention to the similarity with regard to thickness and lithology of these Eemian sediments and those of the Eemian deposits in the Amsterdam basin. The present structurally lower position in the L-blocks of the entire Eemian sequence (approx. 20—25 metres) suggests more pronounced downwarping in the North (Zagwijn 1977).

In the Northern part in boring G13-2 typical molluscs indicating climatic conditions similar to at present were not found in the basal part (approx. 60 m below sea level) of the Eemian deposits (Spaink 1970). The Eemian age, however, cannot be ascertained. In the strata from 50 to 60 metres below mean sea level a lusitanic fauna points to slightly warmer environmental conditions, which may correspond to those during pollen zone E4-5. As mentioned by Oele (1971b) definite Eemian deposits have not been noticed in the southernmost part. Over a vast area in the central part of the Southern bight of the North Sea the Eemian sequence is overlain by the up to 5 metre thick Brown Bank clay bed, which marks the transition from the marine to a lacustrine environment.

Course of events

It is evident that the transgression of the sea in the present southern North Sea area was not noticeable earlier in pollen zone E3. The same applies to the mainland of the Netherlands. However, earlier, the sea level must have been already considerably above the level, which at present marks the base of the Eemian in the glacial basins (80 metres below mean sea level). This may be explained as follows:

The glacial basins may be considered to be closed features i.e. depressions without connections to the open sea. The transgressive sea, therefore, had to overflow plateau-like areas before entering the deeper basins. Consequently, the depth of the base of the marine Eemian, does not represent the relative sea level. It only gives the date of overflow of the plateau-like areas. During this zone the sea level had risen so much that the sea covered the shallower parts as well as the basins. Erosion by the transgressing sea has been apparently of minor significance. It is believed that the sea level during pollen zone E3 was about 40 to 45 metres below the present one. This phenomenon will be discussed in more detail by Zagwijn (1977).

The course of events in the North Sea is similar to that known from the present land area: the sea invaded the glacial depressions, resulting in deposition of a marine series. In the southern part of the North Sea and southwestern Holland a relatively thin marine sequence of coarse sands was laid down. It is considered to have taken place at a late stage of the interglacial. The regression started after pollen zone E5. The maximum sea level stand in the present land area was 8 metres below present mean sea level. The presence of Early Weichselian fresh water clay at about 40 metres below mean sea level in the North Sea indicates a strong drop in sea level at the end of the Eemian.

Weichselian deposits in the onshore part of the Netherlands

Distribution and lithology

Twenthe Formation. — Deposits of this formation crop out or occur close to the surface at various places. A Holocene cover of maximum 20 m be present. The Twenthe Formation is generally up to 10 metres thick; in some places it may reach a thickness of 25 metres; outside previously formed glacial depressions the thickness is generally restricted to a few metres.

As stated earlier the Twenthe Formation comprises various types of sediments. The aeolian deposits consist of fine sandy loam (loess) and fine sands (cover sands). Whereas the loess is generally restricted to the southeastern part of the country, cover sands are predominant in other parts of the Netherlands. The mean grain size of the aeolian sands is around 150 microns.

The meltwater deposits are mainly fine sands, although locally coarse material with some gravel may be present. Loam and peat occur locally. Slopewash-deposits vary strongly in composition, depending on the source material rather than on the transporting agents. Slopewash-deposits occur on the flanks of the Saalian ice-pushed ridges. Thin gravel strings observed as intercalations in the cover sands, are considered to represent desert pavements.

Kreftenheye Formation. — The distribution of the formation is confined to a zone along the present Rhine-Meuse course and a zone along the former Rhine-IJssel course across the northern Netherlands towards Bergen. The Kreftenheye Formation is sloping towards the West. In the eastern part of the Netherlands it occurs at or near the surface. In the coastal area the top of the formation is at about 25 metres below the present sea level. The thickness varies from about 10 to 15 m.

In the upstream area the Kreftenheye Formation consists of coarse sands with gravel. The upper part of the series is finer grained. In many places the series is capped by a clay or loam bed (Bennema & Pons 1952, Wiggers 1955). Sand ridges are present in various parts of the country overlying the fluviatile sands. They are considered to be river dunes (Vink 1954, Ente 1971). Verbraeck (1974) described and dated such dunes in the Alblasserwaard area. He has assigned a Late Dryas age to these dunes.

Stratigraphy

In Fig. IV-30 the chronostratigraphy of the

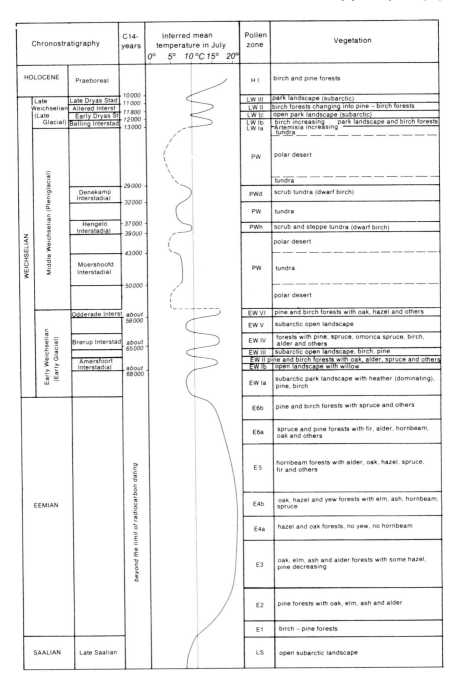

Fig. IV-30. Chronostratigraphy, pollen zonation and climatic curve of the Late Pleistocene (after Zagwijn 1975).

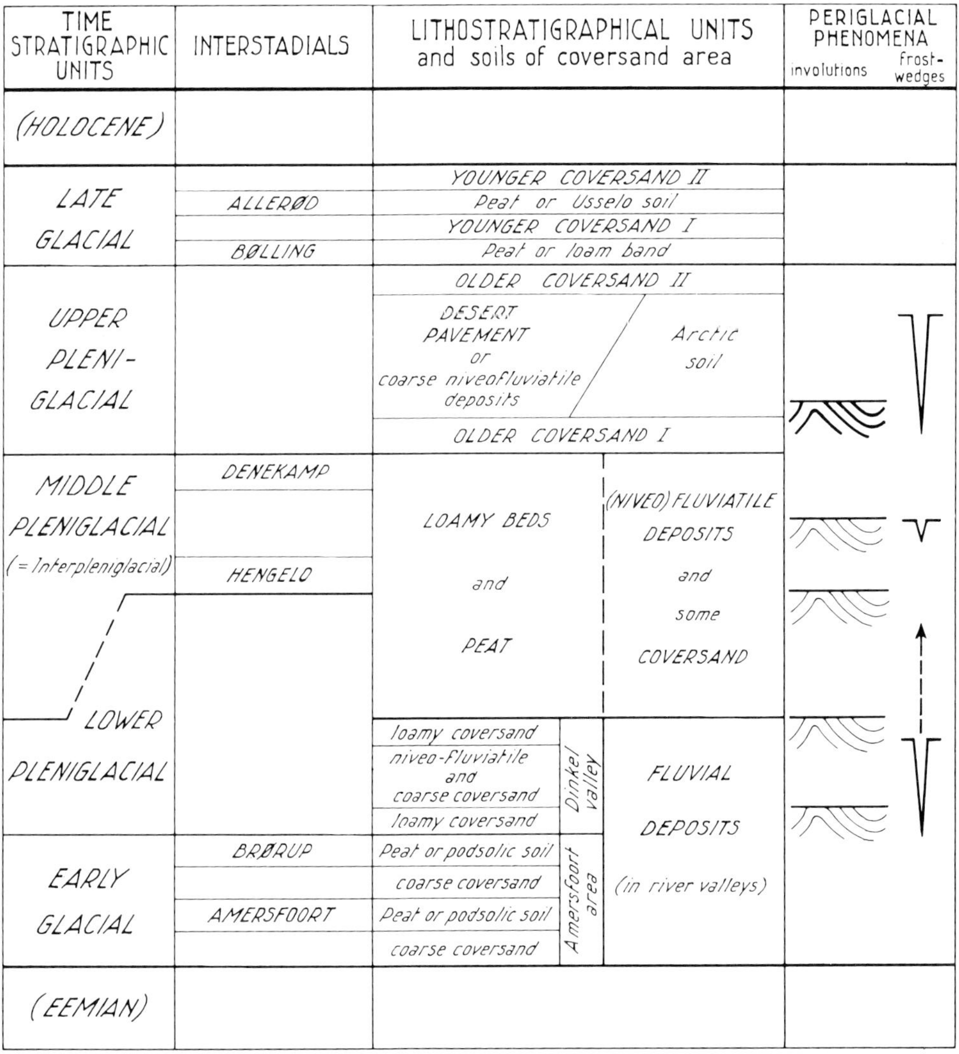

TIME STRATIGRAPHIC UNITS	INTERSTADIALS	LITHOSTRATIGRAPHICAL UNITS and soils of coversand area		PERIGLACIAL PHENOMENA involutions / frost-wedges
(HOLOCENE)				
LATE GLACIAL	ALLERØD	YOUNGER COVERSAND II		
		Peat or Usselo soil		
	BØLLING	YOUNGER COVERSAND I		
		Peat or loam band		
UPPER PLENI- GLACIAL		OLDER COVERSAND II		
		DESERT PAVEMENT or coarse niveofluviatile deposits	Arctic soil	
		OLDER COVERSAND I		
MIDDLE PLENIGLACIAL (=Interpleniglacial)	DENEKAMP	LOAMY BEDS and PEAT	(NIVEO)FLUVIATILE DEPOSITS and some COVERSAND	
	HENGELO			
LOWER PLENIGLACIAL		loamy coversand niveo-fluviatile and coarse coversand loamy coversand	Dinkel valley / FLUVIAL DEPOSITS	
EARLY GLACIAL	BRØRUP	Peat or podsolic soil	Amersfoort area / (in river valleys)	
		coarse coversand		
	AMERSFOORT	Peat or podsolic soil		
		coarse coversand		
(EEMIAN)				

Fig. IV-31. Chronostratigraphy, lithology and informal lithostratigraphy of the Weichselian glacial interval in the Netherlands (after van der Hammen et al. 1967).

Weichselian is presented, also shown are the pollen zones, the type of vegetation and a curve of the climate (Zagwijn 1961, 1974, Van der Hammen et al. 1971). The coldest part of the last glacial period is the Middle Weichselian (Pleniglacial); during very cold intervals (stadials) a polar desert was present. During the improved climates of Middle Weichselian interstadials tundra or shrub tundra vegetations had developed. In contrast the Early Weichselian interstadials are characterized by pine, birch and spruce forests and during the stadials subarctic park landscapes were present. In the Late Weichselian again subarctic park landscapes and pine-birch forests coincide with stadials and interstadials respectively. It should be noted that size and abundance of frost wedges and cryoturbate structures vary, depending on the stratigraphic position of the levels in which they occur (Fig. IV-31).

The Kreftenheye Formation, ranging in age from Late Saalian to Late Weichselian, has been defined by Zonneveld (1947). The sands are characterized by high amounts of augite. The clay, which is found on top of the formation has been dated as Late Weichselian. Incidentally, pumice, a local constituent, may provide a dating of the sediment, since according to Frechen (1959) this pumice is related to Late Weichselian volcanic activities in the Eifel area in Western Germany.

Weichselian deposits in the Dutch part of the North Sea

Distribution and Lithology

Weichselian sediments are present in extensive areas of the Dutch sector of the North Sea. They are absent in the E-blocks, the northern K-blocks, part of the L-blocks and in the extreme south where Holocene marine sands rest directly upon Tertiary or Eemian strata. In the North, the overlying Holocene cover is thin, at some places even less than 1 metre. South of 53° N. Lat. the cover gradually increases in thickness southward and may reach a maximum of 10 metres. It is difficult to mention general figures for top and base of the Weichselian sediments, in relation to mean sea level as they vary considerably.

Brown Bank Bed. — Over a large part of the Southern Bight of the North Sea the Eemian transgression-regression sequence is covered by the Brown Bank Bed which may have a maximum thickness of 5 m. In the Dutch sector, the Brown Bank occurs in various K and P blocks. It consists of a very tough, brownish-grey clay with minor fine sand laminae, representing deposition in a lacustrine environment. The Brown Bank clay is of Early Weichselian age, as interpreted by Zagwijn (1970). The top of the Brown Bank Bed varies from 37 to 42 metres below mean sea level.

Twenthe Formation. — The aeolian cover sands, which are extensively distributed on land, occur also wide-spread in the North Sea area. Their distribution is shown by Oele (1969, 1971). In the past, the sands of the Dogger Bank were assumed to be aeolian cover sands (Oele 1969). Recently collected and studied samples necessitate a change of this assumption, since much of the material has appeared to be of fluvio(peri)glacial origin and may even be of Saalian age. Although on land a distinction is possible between older and younger cover sands (Fig. IV-31), such a distinction is not feasible in the off-shore area due to insufficient data.

The cover sands are relatively well-sorted with a mean grain size varying from about 105 to 150 microns. Locally, loamy and peaty seams are intercalated. A peat horizon occurring close to the Zeeland coast and another in block G1 were dated by means of pollen analysis as Middle Weichselian (Pleniglacial) (Zagwijn & de Jong 1969, 1970). A humic clay from block L16 also of Pleniglacial age according to pollen analysis, found at a depth of 29 m below mean sea level, revealed an age of $47100 \, ^{+5300}_{-3200}$ years B.P., as based on ^{14}C dating (GrN-6766). Humic horizons and peat of Late Glacial age were observed in blocks F17 and F18.

The fluvial deposits from small (ephemeral) streams in the periglacial zone are also fine-grained, however, less well-sorted, due to the occurrence of some coarser grains. These deposits may occur as intercalations and are more frequently present in a rather restricted area in the North (54°20' N Lat./5° E Lat.) and in the L and F blocks. Contrary to their occurrence on land thick aeolian deposits of Weichselian age have not been found in the present North Sea area, which may be attributed to the levelling effect on the pre-existing Saalian topography of the transgressing Eemian sea that hardly extended beyond the present North Sea area.

Kreftenheye Formation. — The Kreftenheye Formation (Late Saalian to Late Weichselian in age) consists of fluviatile deposits of the Rhine and Meuse. The presence of Kreftenheye deposits has been recorded from the southern part of the area under consideration. In the offshore area the formation is overlain by a Young Sea-sand cover, with in some places older Holocene tidal and organic deposits between the Kreftenheye sands and the latter sand cover. The top of the formation is at 20 metres below N.A.P. close to the coast and about 40 m —N.A.P. about 50 km offshore. The thickness of the formation is limited to about 10 m.

The sediments consist predominantly of medium and coarse sands. The top of the Kreftenheye Formation shows a decrease in grain size; locally a loam layer is present. The clay and peat overlying the above mentioned series on land, appears to extend into the present marine area. Its occurrence seems to be restricted to the area off Hoek van Holland up to several tens of km offshore (see de Jong 1966).

Stratigraphy

Many of the datings available rely on pollen analysis (Zagwijn 1961, 1974b). The Brown Bank Bed dates from the Early Weichselian, in particular from pollen zone EWIa. Deposits from the Twenthe Formation have ages ranging from Early to Late Weichselian, though they predominantly date from the Middle Weichselian. The deposits of the Kreftenheye Formation (brownish coarse sand of Rhine provenance) may be easily distinguished from those of the Twenthe Formation. The minimum age of the Kreftenheye Formation could be established by a pollen-analytical dating of the clay layer in the upper part of the formation. The clay may date from the Late Weichselian or from the earliest part of the Holocene (Preboreal). Although on land younger ages have been recorded for this clay layer, this is not yet the case in the offshore area.

As mentioned earlier the presence of pumice also enables a dating. In various borings the pumice occurs in fluvial sands, which sands therefore may be placed in the Late Weichselian. Where the Kreftenheye Formation overlies Eemian strata a distinction between marine Eemian deposits and the Kreftenheye Formation may be difficult, because the marine material frequently has been reworked in various places during sedimentation of the Kreftenheye deposits. The worn appearance of shell fragments may be in some places indicative of reworking.

Course of events during the Weichselian

The fall of the sea level during the Weichselian glacial interval resulted in complete withdrawal of the sea from the Dutch sector of the North Sea. Subaerial conditions during that time are demonstrated by the occurrence of desiccation cracks in the Early Weichselian Brown Bank clay in the North Sea, filled with marine deposits of Holocene age.

Neither the land area nor the Dutch sector of the North Sea were covered by inland-ice. The mentioned areas were situated in the periglacial zone. As may be deduced from the pollen-analytical data (Fig. IV-30) trees were scarce or absent during the stadials in the Early and Late Weichselian as well as during the Pleniglacial. The paucity of vegetation gave rise to the formation of wind-blown sand deposits over vast areas. These sediments, the so-called cover sands, are considered to be niveo-aeolian in origin. Locally, fluvial sediments (fluvio-periglacial sediments) were deposited by small streams. The combined Rhine — Meuse river system continued to supply material for the Kreftenheye Formation, the lower part of which is of Late Saalian age. The river system changed its course during the Weichselian, however, within certain limits. Relatively small amounts of water were available in the winters of the Pleniglacial when severe climatic conditions prevailed. In summer, large amounts of meltwater caused the river to swell considerably. Consequently the river will have been intermittent. The considerable distance to the erosion base certainly has limited the erosive capacity.

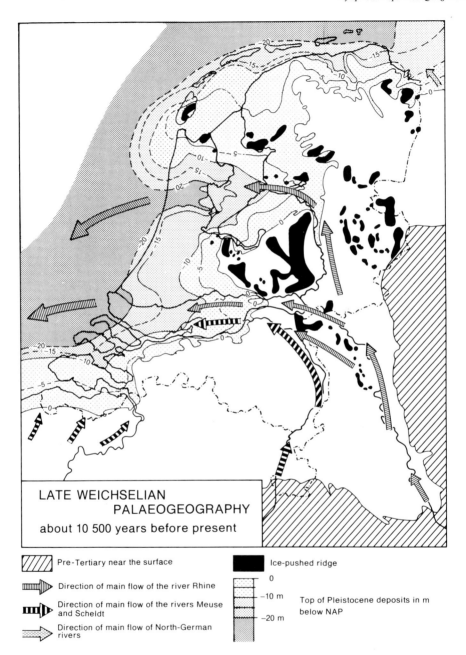

Fig. IV-32. Paleogeographical map of the Netherlands at about 10500 B.P. with outcropping Saalian ice-pushed ridges and the top Pleistocene in relation to the Dutch Ordance Datum Level (modified after Zagwijn 1974).

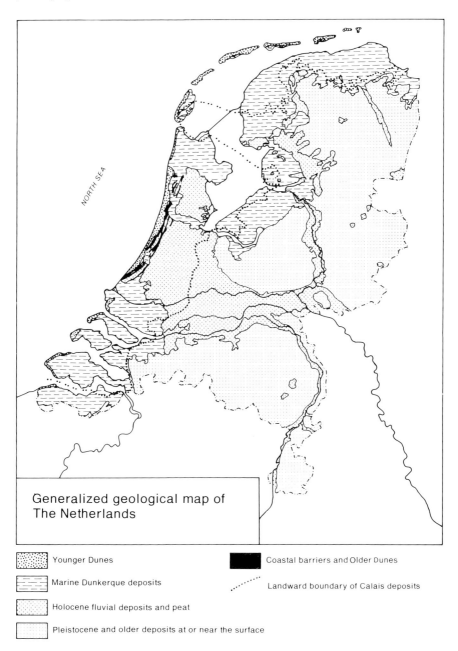

Fig. IV-33. Generalized geological map of the Netherlands showing the maximum extension of the Calais and Dunkerque transgressions.

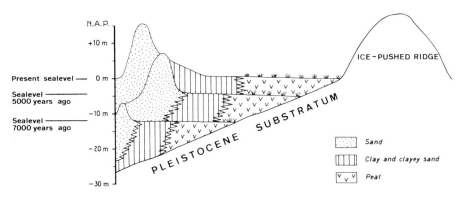

Fig. IV-34. Scheme of lithologic succession related to the Holocene rise in sea level (after Zagwijn 1974a).

It is difficult to trace the course of the river system into the North Sea area. However, there are no indications that the river was flowing northward. The distribution of the Kreftenheye deposits and of marine sands with heavy minerals of Rhine provenance (Baak 1936) suggests that the river discharged into the Atlantic Ocean through the British Channel (Fig. IV-32).

Holocene deposits in the onshore part of the Netherlands

Distribution and lithology

At present the western part of the Netherlands is a low plain, mainly at or below mean sea level and protected against the sea by coastal dunes and dikes. The rise of sea level during the last 10000 years, resulted in the formation of coastal barriers. During the Holocene, the deposits in the coastal basin behind the barriers were in part supplied by rivers and in part derived from the North Sea. Except for deposition of clastic material, peat was formed in an extensive area, mainly along the inland margin of the depositional basin.

The simplified geological map (Fig. IV-33) shows that a large part of the Netherlands is covered by Holocene deposits. These deposits increase in thickness to the West and NW. They may reach a thickness of more than 20 m in the western part of the coastal plain (Fig. IV-34).

In the inland areas the Holocene deposits include fluvial sands and clays, brook deposits, peat i.e. frequently highmoor peat, and inland dunes. In the coastal plain large tracts with complex alternations of peat and clastic layers are indicative of a lateral shifting of the boundary between the marine environment where clastic sediments were deposited and the fresh-water environment where peat was formed.

A schematic cross-section (Fig. IV-34) through the Holocene deposits in the western part of the Netherlands shows near the outcropping Pleistocene a zone with peat, in the central part a zone with lagoonal and tidal flat deposits consisting of clay and clayey sand and along the present coastline a zone with barriers and sand dunes. The coastal barriers have played an important role in the Holocene development. Barriers protected the coastal area against inroads of the sea and thereby facilitated the formation of peat and the deposition of lagoonal sediments.

Coastal barriers and coastal dunes. — The development of the coastal zone with barriers and dunes has been described in recent times by amongst others van Straaten (1965), Zagwijn (1965) and Jelgersma et al. (1970).

The coastal barriers consist of fine to

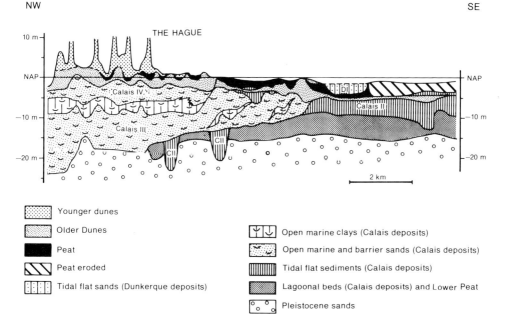

Fig. IV-35. Section across the coastal barrier complex near The Hague (after Jelgersma et al. 1970).

medium-grained sand with shells; intercalated clay deposits may be present. They are overlain by dune sand: the so-called Older Dunes deposits. In the western part of the Netherlands, between Schoorl and Hoek van Holland, the landscape of the Older Dunes is made up of a series of low ridges more or less parallel to the present coast and separated from each other by depressions filled with peat and clay. The low ridges are covered by small irregular dune forms. As mentioned earlier coastal barrier deposits are present below the Older Dunes deposits; the topography of the coastal barriers is concealed by the overlying Older Dunes. In the western part the Older Dunes deposits are covered by the Younger Dunes (Fig. IV-35).

Along the remaining part of the coast, coastal barriers and Older Dunes are not clearly developed (Fig. IV-33). In most places the so-called Younger Dunes only are present overlying Dunkerque III (see below) tidal deposits.

Lagoonal deposits. — As a result of the rapid rise in sea level of approximately 50 cm in a century during the Atlantic (8000—5000 B.P.), a relatively thick sequence of lagoonal and tidal flat sediments was laid down behind the coastal barrier system. The sediments include low and high tidal flat deposits, marine and brackish marsh clays intercalated in peat layers. The lagoon behind the barriers became locally for the greater part filled with fresh-water and clastic material from the rivers Rhine, Meuse and IJssel which discharged into the lagoon from the East. The outlets for the river water through the coastal barriers were also inlets for the sea, especially during periods of increased transgressive activity. During transgressive phases the influence of the sea extended farther eastward resulting in sedimentation of clayey sheets with sandy channel-fills.

Perimarine deposits. — Perimarine deposits are formed in an environment where fluvial sedimentation resulted directly from a relative sea level rise and where marine or brackish clastic sediments are absent (Hageman 1963). The sediments mainly consist of sand, clay and

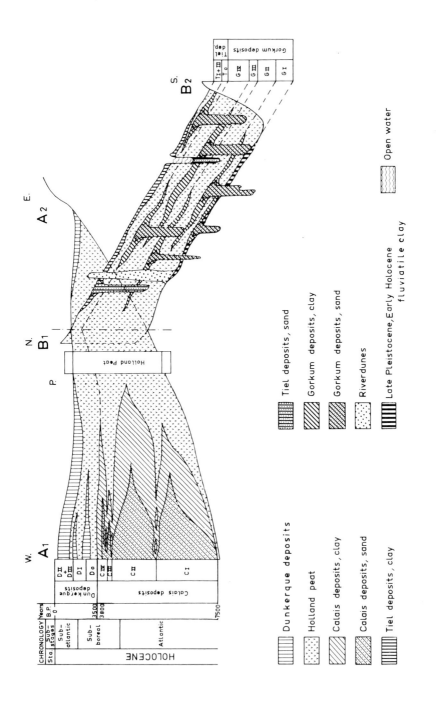

Fig. IV-36. The relation between perimarine deposits and marine deposits of the Calais and Dunkerque transgressions (modified after Hageman 1969).

peat. Fig. IV-36 indicates the occurrence of numerous channels filled with sand or sandy clay and of clay sheets and thick peat layers.

Stratigraphy

In the Netherlands the marine, coastal, lagoonal and perimarine deposits of Holocene age have been grouped together in the Westland Formation (Doppert et al. 1975). This formation includes the Calais and Dunkerque Members consisting of marine, tidal flat and lagoonal deposits laid down during the Calais and Dunkerque transgressions. Also the Older and Younger Dunes, the Basal Peat member and the Holland Peat Member, and the Gorkum and Tiel Members with perimarine deposits are parts of the Westland Formation. Fluvial sediments which were not deposited in a perimarine environment are included in the Betuwe Formation. The Singraven Formation is of restricted occurrence and consists of brook deposits and deposits in small rivers. Inland peat, mainly sphagnum peat, is included in the Griendtsveen Formation. Inland dunes, mainly reworked Weichselian cover sands, are placed in the Kootwijk Formation.

The stratigraphic sequence of Holocene deposits in the coastal and fluvial area of the Netherlands and adjacent countries shows an alternation of clastic layers and peat. In the coastal area the varying intensity of marine influence as reflected by this alternation is commonly considered to represent transgressions and regressions.

In most areas in the coastal plain, four main stratigraphic units have been described; a lower peat bed (Basal Peat), a lower clastic layer (Calais deposits), an intermediate peat layer (Holland peat) and an upper clastic layer (Dunkerque deposits). The Geological Survey of the Netherlands distinguishes four main transgressive phases represented by so-called Calais deposits (Brand et al. 1965, Zagwijn 1975):

Calais IV 2700–1800 B.C. (C IV)
Calais III 3300–2700 B.C. (C III)
Calais II 4300–3300 B.C. (C II)
Calais I 6000–4300 B.C. (C I)

The main Dunkerque transgressive phases and related deposits are:

Dunkerque III after 800 A.D. (D III)
Dunkerque II 250–600 A.D. (D II)
Dunkerque I 500–200 B.C. (D I)
Dunkerque 0 1500–1000 B.C. (D 0)

The perimarine Gorkum deposits are in time equivalent to the Calais deposits, whereas the Tiel deposits correspond to the Dunkerque deposits.

Based on differences in lithology and stratigraphic position this scheme has been further subdivided for mapping purposes. This had led to the distinction of such transgressive sub-phases as Calais IIa, IIb, IIIa, IIIb, IVa1, IVa2, IVb, Dunkerque Ia, Ib, IIIa, IIIb (amongst others Hageman 1963, Ente 1969, Ente et al. 1975, Roeleveld 1976). The various intercalated peat layers represent regressive phases.

The available data point to cyclic variations with relatively short periods (500 to 550 years) during the geological development of the coastal and perimarine fluvial area (Bakker 1953, Bennema 1954, Pons & Wiggers 1959/1960, Jelgersma et al. 1970). Roeleveld (1976) analyzed the problems arising by the application of the Calais – Holland Peat – Dunkerque terminology in areas where this succession is indistinct or incomplete, due to a poorly developed coastal barrier system (northern Netherlands, northwest Germany). Based on the concept of cyclicity, he arrived at 11 transgressive and 10 regressive Holocene intervals in this area. For more details reference is made to Roeleveld (1976).

Below the clastic Calais deposits a peat layer is generally present. Early in the Atlantic (about 7500 B.P.) the shoreline of the North Sea approached the present coastline of the Netherlands. Generated by a favourable climate and a rising groundwater table as a result of the rising sea level, a backswamp area where peat growth developed, was formed in front of the invading sea. Due to continued sea-level rise marine sediments gradually covered the peat layer and the backswamp moved landward to the East over

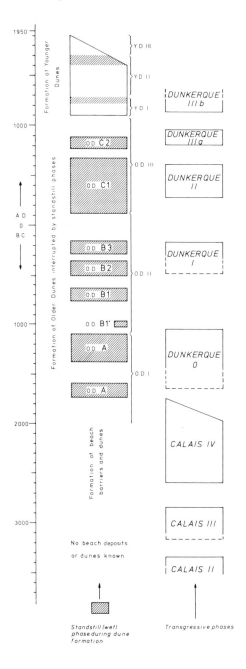

the gentle slope of the Pleistocene surface. The growth of this so-called Lower Peat or Basal Peat started in the Late Boreal and Early Atlantic in the lower and depressed western and northern parts of the coastal area, whereas on the higher parts peat development started at a later stage namely in the second part of the Atlantic.

The data on age and depth of the basis of the Basal Peat have been used by Jelgersma (1961, 1966) to construct part of the relative sea level curve. Only in those areas where it could be ascertained that the growth of the Lower Peat is triggered by the rise of sea level, the age data are considered reliable. For further considerations reference is made to Jelgersma (1961).

The history of the formation of coastal barriers and dunes has been described by van Straaten (1965) and Jelgersma et al. (1970). According to this last study, most of the first phase of the formation of the barrier system has been correlated with the Calais III phase (5300–4700 B.P.), the second with the Calais IV phase (4700–3800 B.P.). After completion of the beach barrier sequences they were overlain by Older Dunes deposits. The oldest Older Dune sands known to date, were already laid down before 4100 B.P., as follows from finds of Late Neolithic inhabitation. In the East aeolian deposition continued for a short time only, however, in the more western part deposition of dune sand continued at least until Roman times. The barriers and the Older Dunes were partly covered by Younger Dunes deposits, which did not extend as far inland as the Older Dunes. The formation of these Younger Dunes did not start earlier than Medieval time (12th century A.D.).

Detailed research has pointed out that especially in the Older Dunes deposits soil horizons occur, indicating wet phases, alternating with dry phases of active dune-building. Climatic changes are thought to have been responsible for this cyclic development, which apparently may be related to the succession of transgressions and regressions in the coastal area (Fig. IV-37).

Fig. IV-37. The standstill phases during the formation of the Older and Younger Dunes deposits and their relation to the Calais and Dunkerque transgressions (after Jelgersma et al. 1970).

Holocene deposits in the Dutch sector of the North Sea

Distribution and lithology

Since the sea-floor is affected by wave action, currents, burrowing organisms and fishing activities (trawls), at least a layer of recent material is present over the entire area. In the North this recent or subrecent material may be directly underlain by Weichselian cover sands or other Pleistocene deposits. Generally the thickness of the Holocene sequence is limited here, even when older Holocene deposits are present. In the South, i.e. S of the 53° N Lat. the Holocene sequence is thicker and may comprise at least 5 metres of sediment.

Like on land the Holocene series in the North Sea may be subdivided into two parts: a lower part, with a lithology comparable to the Calais deposits and an upper part consisting of sands, which have settled in an open marine environment.

At the base of the Holocene a peat layer may be present. Jelgersma (1961), Oele (1969, 1971), Behre & Menke (1969) and Kolp (1974) have reported on this Lower Peat, present in the North Sea in a zone along the coast of the western Netherlands and in the central part of the northern area. In the northern part the peat generally occurs at shallow depths below the sea-floor, in the South it is covered by 5–15 metres of younger material. In some places in the southern part the peat is exposed and has been locally dredged. Exposed peat beds are rapidly eroded by boring clams (*Pholas*) and fishery activities. In this respect it is interesting to note that Reid (1913) mentioned the former wide-spread occurrence of moorlogs (peat boulders) on the flanks and part of the top of the Dogger Bank.

The above mentioned reports present some depths of occurrence of the Lower Peat: in the North a depth of about 45 metres is recorded, in the coastal zone of the southern part the peat surface gradually rises in a landward direction. Its occurrence ranges here from about 34 to 14 metres below mean sea level.

Additional information has become available from site investigation borings. The peat appears to have a more extensive distribution and occurs at more variable depths than may be concluded from the above mentioned reports. Borings, which have yielded peat samples are amongst other located in blocks A18, B13, E11, K7, K12 and K14. In E13 the peat has been found at 51 metres below mean sea level.

The lower clastic part of the Holocene on top of the Lower Peat (Elbow/Calais deposits) consists of clays and fine generally clayey sands. Intercalation of peat layers as recorded in the Calais deposits on land, has not been reported at sea.

Both the sands and the clay are gray in colour. The generally clayey sands have a median value of the sand grains of about 150 microns. The thickness of the deposits ranges from 0 to 20 metres in the North; in large areas the average thickness varies from 2 to 8 m (channel-fills excluded). In the South the thickness of the older Holocene deposits is generally limited to a few metres.

A survey with the "Sonia" subbottom profiler in an offshore area south of Hoek van Holland revealed well-developed cross-bedding and channelling features. The sedimentary structures support the paleontologic indications for a previously present tidal flat environment. One of the channels is filled with 20 m of fine sandy material. The base of the channel is at 40 metres below mean sea level, implying that the gully has deeply cut into Pleistocene fluviatile deposits.

The upper clastic part of the Holocene (Dunkerque deposits) consists predominantly of sand. The sands have a yellowish colour, which points to oxydizing conditions. They are named Young Sea-sands (Oele 1969, 1971a, 1971b). The sandy material varies strongly in grain size. In general, the grain size recorded at the sea-floor, is representative of the entire Young Sea-sand layer. In the North the median value is about 150 microns, which is roughly similar to that for the Weichselian cover sands. In the southern part both the thickness and the grain size increase in a southerly direction. South of the 52° N Lat. the median value is about 300 microns and the thickness is generally 5 to 10 metres. The area off the coast of the SW

Netherlands, f.i. off Walcheren where smaller grain sizes (approx. Md 100 microns) occur, forms an exception.

Ripples and banks composed of Young Sea-sands will be described by Eisma et al. in chapter V of this volume.

Stratigraphy

The three units mentioned above (Lower Peat, lower part and upper part of the clastic Holocene sequence) will be discussed in this paragraph. Again the age determinations are mainly based on variations in pollen assemblages; faunal remains have permitted some conclusions on the depositional environment.

The Lower Peat. — Peat has been collected at various sites in the northern and southern part. In blocks F14 and F15 the peat has been sampled from 47 and 46 metres below mean sea level respectively. Radiocarbon dating revealed ages of 9445 ± 80 B.P. (GrN 5759) and 9935 ± 55 B.P. (GrN 5758). This is in agreement with the pollen-analytical dating of the material (Preboreal) bij Zagwijn & De Jong (1969) and with a similar dating of Kolp's material by Behre & Menke (1969) from a somewhat shallower depth i.e. 45 metres below mean sea level.

In the southern area, where peat occurs at shallower depths the age has appeared to be younger e.g. a peat found off Schouwen at 27 metres below mean sea level has been dated by means of pollen analysis as Boreal (De Jong 1970).

The Elbow/Calais deposits. — In the North the peat is overlain by a thin layer of fresh water clay. A stratigraphic hiatus between the clay and the peat has not been observed. A break, however, is present between the fresh water clay and the overlying brackish-marine clay, dated as Boreal. This is in accordance with the results of Behre & Menke (1969). The brackish-marine environment of deposition is indicated by characteristic ostracods and molluscs. The upper part of the clay as well as the overlying sandy sediments contain assemblages, which are indicative of deposition in a tidal flat area (amongst others *Hydrobia ulvae*, *Cardium edule* and its breed are present). Kolp (1974) mentions the presence of a clearly marked *Hydrobia* horizon, which he could distinguish after a close investigation of an area near the Dogger Bank.

Comparable to occurrences of Lower Peat, marine tidal flat deposits are also younger in age, where they are present at shallower depth. In the coastal zone they are Atlantic in age coinciding with a depth of occurrence of 20 metres. Previously the correlative onshore lithostratigraphic unit has been named "Calais deposits", however, with the restriction, that it should be Atlantic or Early Subboreal in age (Hageman 1963). For the older deposits in the offshore area a new name has to be introduced. Oele (1969) has proposed the term Elbow deposits for the non-terrestrical sediments ranging in age from Preboreal to Boreal, which occur in the Netherlands sector of the North Sea and adjacent regions.

The Dunkerque deposits. — The clay seams found in the above mentioned Young Sea-sands from the northern part could be dated by means of pollen analysis as Late Medieval. The sand rests directly upon Elbow/Calais deposits or older sediments, which implies the presence of a hiatus.

In the southern part the yellowish Young Sea-sands are too coarse to yield pollen. Near the coast of the SW Netherlands the Young Sea-sands overlie a series of gray, fine sands with some clay laminae. Fortunately, this fine material could also be dated. Generally, ages ranging from 2000 B.P. to 500 B.P. have been determined (pollen zone Vb, amongst others de Jong 1972).

Consequently, it may be concluded, that the Dunkerque deposits comprise the Young Sea-sand cover as well as the gray sands, which occur near the Zeeland islands. It is difficult to distinguish the gray Dunkerque sands from the older tidal flat sands of the Elbow/Calais deposits, other than by pollen analysis.

Course of events during the Holocene

The Holocene events are governed by the climatic improvement, which induced a sea-level rise and subsequent transgressions and which also led to a considerable change in the vegetational cover, resulting in the formation of dense birch and pine forests which prevented to some extent further erosion. In the inland areas aeolian deposits were locally formed throughout the Holocene, particularly since human interference. Besides the mentioned aeolian reworking of the Pleistocene sandy material, fluvial deposits of limited areal distribution were laid down by small streams, whereas in some depressed areas conditions were favourable for peat growth. However, the Holocene history is expressed mainly in the deposits of the North Sea area and the coastal area, and to a minor extent of the perimarine zone.

In the greater part of the region a basal peaty layer (peat or gyttja) was formed at the onset of the Holocene. The peat is considered to have been formed by rising groundwater related to a significant rise in sea level. The layer is a lithostratigraphic unit, which is younger the higher its base occurs. Found at a depth of 46 metres below mean sea level the basal peat from the North Sea block F15 has revealed the oldest age recorded in the Netherlands i.e. 9935 ± 55 B.P. (GrN 5758). The peat is generally composed of *Phragmites* or fen wood peat, the latter type occurs in the perimarine area.

Shortly before 9000 B.P. the sea entered the Dutch sector of the North Sea, however, too little information is available to draw even a tentative shoreline of that time.

The process of peat inundation and subsequent clastic sedimentation started in the lowest areas. The oldest fresh-water and brackish/marine clays overlie the peat in the more depressed areas of the northern North Sea. The sea invaded the northern part of the Dutch sector of the North Sea both from the NE, penetrating to the SW and S and from the SW, transgressing to the NE. In the northern part the total thickness of the Holocene clastic sediments is generally restricted to a few metres only, which may be the result of very rapid sea level rise and limited supply of sediments. It is noted here, that strong erosional features like tidal gullies have not been observed in the central northern part of the Dutch sector. Possibly it may be concluded that neither strong tidal movements nor narrow inlets were present in that area. The area may have been a kind of lagoon, which became filled in the Early Boreal as based on pollen-analytical evidence (Oele 1971b). Channels are present in the area east of the Outer Silver pit.

In a short lapse of time (see Jelgersma, this volume) the sea level had risen to a height of about −36 metres N.A.P. in the later part of the Boreal (8700 B.P.). The resulting shoreline is depicted on the colour chart. A few hundred years later around 8000−8500 B.P. almost the entire Southern Bight was below sea level; a connection between the Straits of Dover and the northern waters had been established. The Dogger Bank was an island. Tidal conditions reigned over extensive areas.

In the following period up to 7500 B.P. the entire Southern Bight submerged and the shoreline reached the present shoreline. Most likely in this or a preceding period the banks, present in the North Sea were formed. It is still uncertain whether or not the Brown Bank and Zeeland ridges with crests at 18 and less than 10 metres below mean sea level represent drowned coastal barriers as suggested by Jelgersma (1961). The situation at 7500 B.P. is reflected on the colour chart.

By means of some additional paleogeographical maps (Figs. IV-38 to IV-40) the change of the configuration of the shoreline may be followed rather closely. The sedimentation pattern at about 7500 and 7000 B.P. (Fig. IV-38) shows that the marine environment was divided into a marine depositional area of the North Sea with inlets and estuaries and an intertidal brackish depositional area. Along the latter area a zone where peat growth developed has been indicated. The assumed beach barrier, tentatively indicated is still farther west of the present coastline (Zagwijn 1974).

In the coastal and perimarine zone the peat growth was interrupted at times as a result of marine transgressions, which showed a cyclic pattern. Such interruptions in the peat formation were followed by the deposition of fresh-water clays on top of the peat and, in the coastal zone,

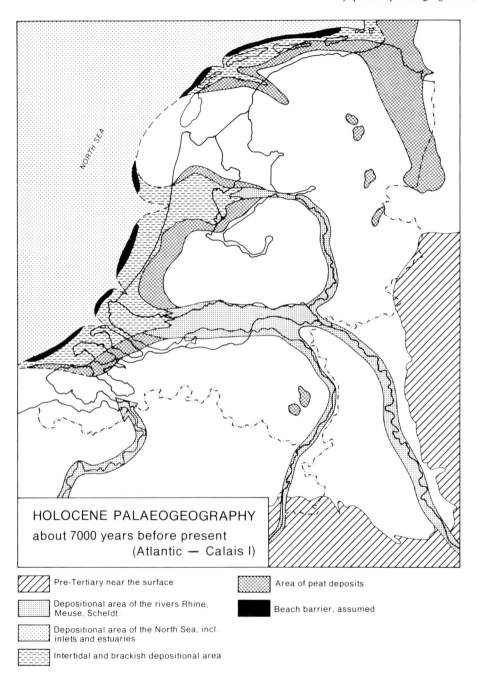

Fig. IV-38. Palaeogeographical map of the Netherlands at about 7000 B.P. (Atlantic–Calais I) (modified after Zagwijn 1974a).

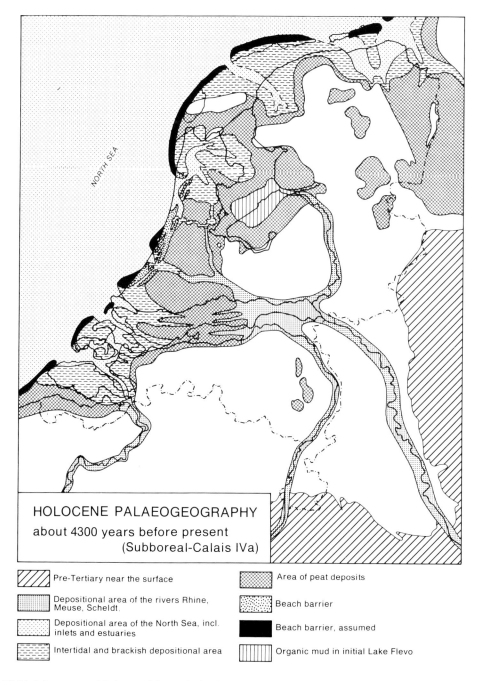

Fig. IV-39. Palaeogeographical map of the Netherlands at about 4300 B.P. (Subboreal–Calais IVa) (modified after Zagwijn 1974a).

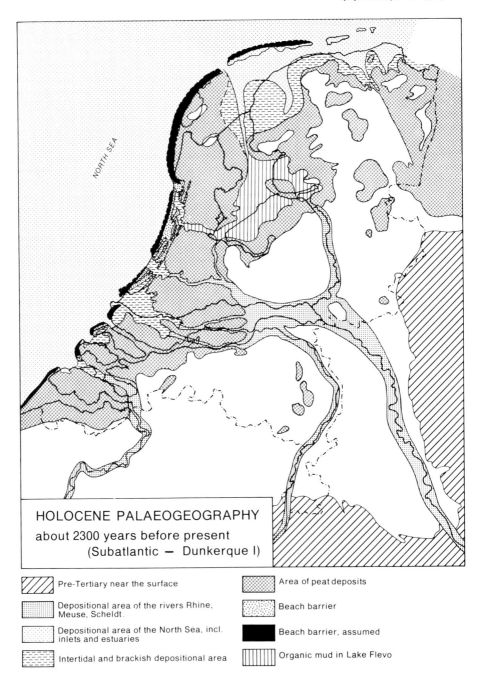

Fig. IV-40. Palaeogeographical map of the Netherlands at about 2300 B.P. (Subatlantic— Dunkerque I)

deposition of marine and/or brackish sediments. During transgressive phases extensive land erosion occurred. However, it should be noted that the erosional effects during the various phases differ strongly in intensity.

Essentially the changes in shoreline configuration until the shoreline at 5000 B.P. was formed (see the colour chart) are of minor importance. Shortly after 5000 B.P. the beach barriers that are still present in western Holland, started to form. Around 4500 B.P. a first series may have been completed, which were soon afterwards covered by aeolian sands, the Older Dunes deposits (Jelgersma et al. 1970). Shortly after completion of this coastal barrier system new ridges were built at the western side of the former ones. This process ended in most places before 3500 years ago. Fig. IV-39 shows the configuration during the Calais IV phase (about 4300 B.P.). A simplified pattern of beach barriers and assumed beach barriers has been indicated. These barriers were formed during the Calais III phase and close to the beginning of the Calais IV phase; the marine sediments belong to the Calais IVa deposits, dated between 4500 and 4200 B.P. Compared with Fig. IV-38 the area covered by peat has increased in size. In the central part of the Netherlands a lake is shown with deposits of organic mud. This lake may be considered to be the initial Lake Flevo, well-known from Roman times.

As mentioned earlier, in a second phase between 4200 and 3500 B.P. the coastal barrier system had become more extensive. This barrier system was also covered by Older Dunes rather soon after formation. Similar to the age of the barriers that of the Older Dunes deposits becomes progressively younger proceeding to the West. Formation of the Older Dunes ceased around the beginning of Roman times, 2000 B.P. but for the most western part, where incidentally some wind-blown deposits were laid down until Medieval times (Jelgersma et al. 1970).

In the area behind the coastal barrier system marine sedimentation discontinued at about 3700 B.P. (end of the Calais transgressive phases). The lagoonal and tidal flat area became covered by a layer of peat of considerable thickness. This sequence of lagoonal and tidal flat deposits covered by the main peat layer of the Holland peat is the classical picture of the Calais deposits overlain by the Holland peat, which is present behind the coastal barrier system in the western part of the Netherlands.

There is a short interval only between the last Calais transgressive subphase (about 4000 to 3700 B.P.) and the earliest Dunkerque transgressive phase (3500 to 3000 B.P.). During the Dunkerque transgressions the sea broke through weak places in the coastal barriers. During the Dunkerque 0 phase sediments were deposited at rather narrow inlets in the western part, whereas in the northern part of the Netherlands and particularly in the province of Groningen marine deposits were laid down over an extensive area.

In some parts of the Netherlands a clear distinction between deposits from a late subphase of the Dunkerque 0 and Dunkerque I is possible. After the late subphase of the Dunkerque 0 phase (around 2600 B.P.) a vast salt marsh developed in the Groningen coastal district as a result of a progressive decrease in the rate of sea level rise; this marsh area was not subsequently overgrown by peat as in earlier phases. These large areas now became inhabitable. It appears that the inhabitants of the sandy region south of the salt marsh moved towards the North as soon as soil conditions in the newly developed salt marshed became favourable for inhabitation (Roeleveld 1976, Van Es 1965/1966).

Fig. IV-40 shows the paleogeography at about 2300 B.P. during the Dunkerque I transgressive phase, The southwestern part of the coastal district was a vast peat area behind a barrier system. The course of the river Scheldt is clearly visible. In this southwestern area only narrow inlets were present. Near Hoek of Holland the map reveals the presence of marine and brackish sediments at the mouth of the rivers Meuse and Rhine. The inlet of an older branch of the river Rhine at Katwijk is narrow, whereas more to the North a small inlet is visible with a connection between the North Sea and the large inland lake, Lake Flevo. The outlet of this lake to the North is in fact unknown due to intensive reworking of related sediments in the Wadden Sea area. In the provinces of Friesland and Groningen vast salt marsh areas are present.

In the last 2000 years of our history younger

Dunkerque transgressive phases led to sedimentation of marine and brackish deposits on top of the above mentioned peat bed. Like the Calais transgressive phases the Dunkerque phases had affected various places differently. Especially the Dunkerque III transgressive phase of Late Medieval times, was strongly erosive. An important coastal retreat occurred during this period. The transgression was accompanied by storm surges, which penetrated far land inward. Due to the coastal erosion especially in the 15th and 16th century considerable quantities of sand became available for aeolian transport and were blown from 'the beach over the Older Dunes deposits. Thus an important natural barrier against the sea was created by these Younger Dunes, the building of which started already in the 12th century. The Younger Dunes closed the mouths of the Old Rhine at Katwijk and of the Utrecht-Vecht river between Egmond and Bergen (Jelgersma et al. 1970). Accordingly during the last 1000 years important changes in the shape of the coastline occurred.

Acknowledgements. – The authors are grateful to Drs. J. W. Chr. Doppert for reviewing the manuscripts and to Dr. W. H. Zagwijn for his critical comments. Dr. R. T. E. Schüttenhelm composed the final text. The drawings were prepared by Mr. A. Walkeuter.

REFERENCES

Baak, J. A., 1936: Regional petrology of the southern North Sea. Veenman & Zonen, Wageningen.
Bakker, J. A., 1953: Zijn de bijzonder hoge vloeden in ons land in vroghistorische en historische tijd aan bepaalde perioden gebonden?. *Folia Civitatis,* March 1953.
Behre, K.-E. & Menke, B., 1969: Pollenanalytische Untersuchungen an einem Bohrkern der südlichen Dogger bank. *Beiträge zur Meereskunde 24/25,* 123–129, Deutsche Adakemie der Wissenschaften zu Berlin.
Bennema, J., 1954: Bodem – en Zeespiegelbewegingen in het Nederlands kustgebied. Thesis Wageningen University, *Boor en Spade 7,* 1–96.
Bennema, J. & Pons, L. J., 1952: Donken, fluviatiel laagterras en Eemzee afzettingen in het westelijk gebied van grote rivieren. *Boor en Spade 5,* 126–137.

Brand, G., Hageman, B.P., Jelgersma, S. & Sindowski, K.-H., 1965: Die lithostratigraphische Unterteilung des marinen Holozäns an der Nordseeküste. *Geol. Jahrb. 82,* 365–384.
Burck, H. D. M., 1949: Het geologische profiel Breukelen – Deventer. *Meded. Geol. Sticht., Nieuwe Ser. 3,* 5–17.
Doppert, J. W. Chr., Ruegg, G. H. J., van Staalduinen, C. J., Zagwijn, W. H. & Zandstra, J. G., 1975: Formaties van het Kwartair en Boven-Tertiair in Nederland. In W. H. Zagwijn & C. J. van Staalduinen (editors), toelichting bij geologische overzichtskaarten van Nederland, 11–56, Rijks Geologische Dienst, Haarlem.
Eisma, D., Jansen, J. H. F. & van Weering, Tj. C. E., 1977: Sea-floor morphology and recent sediment movement in the North Sea. this volume.
Ente, P. J., 1969: De bodemgesteldheid en de bodemgeschiktheid van het Balgzand en de Breehorn. Flevobericht 66, Rijksdienst voor de IJsselmeerpolders, Lelystad.
Ente, P. J., 1971: Sedimentary geology of the Holocene in Lake IJssel region. *Geol. Mijnbouw 50,* 373–382.
Ente, P. J., Zagwijn, W. H. & Mook, W. G., 1975: The Calais deposits in the vicinity of Wieringen and the geogenesis of northern North Holland. *Geol. Mijnbouw 54,* 1–14.
Es, W. A. van, 1965–66: Friesland in Roman times. *Ber. Rijksdienst Oudheidk. Bodemonderz. 15–16,* 37–68.
Frechen, J., 1959: Die Tuffe des Laacher Vulkangebietes als Quartärgeologische Leitgesteine und Zeitmarken. *Fortschr. Geol. Rheinland Westfalen 4,* 363–370.
Hageman, B.P., 1963: De profieltype-legenda van de nieuwe geologische kaart voor het zeeklei- en rivierengebied. *Tijdschr. K. Ned. Aardrijkskd. Genoot. 80,* 217–229.
Hageman, B.P., 1969: Development of the western part of the Netherlands during the Holocene. *Geol. Mijnbouw 48,* 373–388.
Hammen, T. van der, Maarleveld, G. C., Vogel, J. C. & Zagwijn, W. H., 1976: Stratigraphy, climatic succession and radiocarbon dating of the last glacial in the Netherlands. *Geol. Mijnbouw 46,* 79–95.
Hammen, T. van der & Wijmstra, T. A. (editors), 1971: The Upper Quaternary of the Dinkel Valley (Twente, Eastern Overijssel, The Netherlands). *Meded. Rijks Geol. Dienst, Nieuwe Ser. 22,* 55–213.
Harting, P., 1874: De bodem van het Eemdal. *Versl. K. Akad. Wet., Afd. N, II, 8,* 282–290.
Jelgersma, S., 1961: Holocene sea level changes in the Netherlands. *Meded. Geol. Sticht., C, VI, 7,* 100 pp.
Jelgersma, S., 1966: Sea level changes during the last 10,000 years. *R. Meteorol. Soc. Proc. Int.*

Symp. World Clim. from 8000 to 0 B.C., 54—71.
Jelgersma, S., 1977: Sea-level changes in the North Sea basin. this volume.
Jelgersma, S., de Jong, J., Zagwijn, W. H. & van Regteren Altena, J. F., 1970: The coastal dunes of the western Netherlands; geology, vegetational history and archaeology. *Meded. Rijks Geol. Dienst, Nieuwe Ser. 21*, 93—167.
Jong, J. de, 1966: Pollenanalytisch onderzoek van boringen afkomstig van de Maasvlakte (Noordzee—Maasmond). *Geol. Sticht., Palaeobot. Lab., Intern. Rep. 452*, 3 pp.
Jong, J. de, 1972: Pollenanalytisch onderzoek van een aantal kernen, afkomstig van het gebied voor de kust van Goeree. *Rijks Geol. Dienst, Afd. Palaeobot., Intern. Rep. 660*, 1 p.
Kolp, O., 1974: Submarine Uferterrassen in der südlichen Ost- und Nordsee als Marken eines stufenweise erfolgten Holozänen Meeresanstiegs. *Baltica 5*, 11—40, Vilnius.
Lorié, J., 1906: De Geologische Bouw der Gelderse Vallei, benevens beschrijving van eenige nieuwe Grondboringen — VII. *Verh. K. Akad. Wet., Tweede Sectie, XIII*, 1, 1—100.
Oele, E., 1969: The Quaternary geology of the Dutch part of the North Sea, north of the Frisian Isles. *Geol. Mijnbouw 48*, 467—480.
Oele, E., 1971a: Late Quaternary geology of the North Sea southeast of the Dogger Bank. *Inst. Geol. Sci., Rep. No. 70/15*, 29—34.
Oele, E., 1971b: The Quaternary geology of the southern area of the Dutch part of the North Sea. *Geol. Mijnbouw 50*, 461—474.
Pons, L. J. & Wiggers, A. J., 1959/1960: De holocene wordingsgeschiedenis van Noordholland en het Zuiderzeegebied. Part I, *Tijdschr. K. Ned. Aardrijkskd. Genoot. 76* (1959), 104—152; Part II, *Tijdschr. K. Ned. Aardrijkskd. Genoot. 77*, (1960), 3—57.
Reid, C., 1913: Submerged forests. 129 pp., Cambridge University Press, Cambridge.
Roeleveld, W., 1976: The Holocene Evolution of the Groningen Marine — Clay District. *Ber. Rijksdienst Oudheidk. Bodemonderz. 24*, (1974), 133 pp.
Rummelen, F. F. F. E. van, 1965: Toelichting bij de Geologische kaart van Nederland 1:50.000, Bladen Zeeuwsch-Vlaanderen West en Oost. 79 pp., Geol. Sticht., Haarlem.
Sindowski, K.-H., 1970: Das Quartär im Untergrund der Deutschen Bucht (Nordsee). *Eiszeitalter und Gegenwart 21*, 33—46.
Spaink, G., 1958: De Nederlandse Eemlagen I. *Wet. Meded. K. Natuurhist. Ver. 29*, 1—44.
Spaink, G., 1970: Unnamed report (in Dutch) on the mollusc fauna of boring G13—2 (G13—4). *Rijks Geol. Dienst, Afd. Macropalaeontolog., Intern. Rep. 481*, 5 pp.
Straaten, L. M. J. U. van, 1965: Coastal barrier deposits in South- and North-Holland, in particular in the areas around Scheveningen and IJmuiden. *Meded. Geol. Sticht., Nieuwe Ser. 17*, 41—75.
Toering, K., 1970: Onderzoek Fugro-boring G-13 (Noordzee). *Rijks Geologische Dienst, Afd. Micropalaeontologie, Intern. Rep. 904*, 2 pp.
Verbraeck, A., 1974: The genesis and age of the riverdunes (donken) in the Alblasserwaard. *Meded. Rijks Geologische Dienst, Nieuwe Ser. 25*, 1—8.
Vink, T., 1954: De rivierstreek. 807 pp., Bosch & Keuning, Baarn.
Voorthuysen, J. H. van, 1958: Foraminiferen aus dem Eemian (Riss-Würm-Interglacial) in der Bohrung Amersfoort I (Locus typicus). *Meded. Geol. Sticht., Nieuwe Ser. 11*, (1957), 27—40.
Wiggers, A. J., 1955: De wording van het Noordoostpoldergebied. Thesis Amsterdam University, 216 pp., Tjeenk Willink, Zwolle.
Zagwijn, W. H., 1961: Vegetation, climate and radiocarbon datings in the Late Pleistocene of the Netherlands. Part I: Eemian and Early Weichselian. *Meded. Geol. Sticht., Nieuwe Ser. 14*, 15—45.
Zagwijn, W. H., 1965: Pollen-analytic correlations in the coastal barrier deposits near The Hague (The Netherlands). *Meded. Geol. Sticht., Nieuwe Ser. 17*, 83—88.
Zagwijn, W. H., 1970: Pollenanalytisch onderzoek van Fugro Boring P5B (Noordzee). *Rijks Geol. Dienst, Afd. Palaeobot., Intern. Rep. 560*, 3 pp.
Zagwijn, W. H., 1971: Stratigrafische interpretatie van boringen tot circa 90 m onder zeeniveau in het Noordzeegebied (blokken L2, F17, F18, F14, F11, F3.) *Rijks Geol. Dienst, Afd. Palaeobot., Intern. Rep. 615*, 6 pp.
Zagwijn, W. H., 1974a: The Palaeogeographic Evolution of the Netherlands during the Quaternary. *Geol. Mijnbouw 53*, 369—385.
Zagwijn, W. H., 1974b: Vegetation, climate and radiocarbon datings in the Late Pleistocene of The Netherlands, Part II: Middle Weichselian. *Meded. Rijks Geol. Dienst, Nieuwe Ser. 25*, 101—111.
Zagwijn, W. H., 1975: Indeling van het Kwartair op grond van veranderingen in vegetatie en klimaat. In W. H. Zagwijn & C. J. van Staalduinen (editors), Toelichting bij geologische overzichtskaarten van Nederland, 109—114, Rijks Geologische Dienst, Haarlem.
Zagwijn, W. H., 1977: Sea Level changes during the Eemian in the Netherlands. Abstracts 10th INQUA Congress, Birmingham.
Zagwijn, W. H. & Jong, J. de, 1969: Pollenanalytisch onderzoek van een aantal kernen en grijpmonsters van de Noordzee. *Rijks Geol. Dienst, Afd. Palaeobot., Intern. Rep. 516*, 9 pp.
Zagwijn, W. H. & Jong, J. de, 1970: Pollenanalytisch onderzoek van een aantal gestoken kernen

afkomstig van de Noordzeebodem. *Rijks Geol. Dienst, Afd. Palaeobot., Intern. Rep.* **567**, 2 pp.

Zonneveld, J. I. S., 1974: Het Kwartair van het Peelgebied en de naaste omgeving (een sediment-petrologische studie). *Meded. Geol. Sticht.*, Ser. C, **VI**, 3, 223 pp.

IV-f

The Belgian coastal plain during the Quaternary

ROLAND PAEPE & CECILE BAETEMAN

Paepe, R. & Baeteman, C., 1979: The Belgian coastal plain during the Quaternary. In E. Oele, R. T. E. Schüttenhelm & A. J. Wiggers (editors), The Quaternary History of the North Sea, 143–146. *Acta Univ. Ups. Symp. Univ. Ups. Annum Quingentesimum Celebrantis: 2*, Uppsala. ISBN 91-554-0495-2.

In Belgium the Middle Pleistocene marine deposits of Cromerian and Holsteinian age are restricted to the western part of the coastal area. In Eemian time the marine influence moved eastward and had a landward extension along the Flemish valley. During the Holocene the coastal plain was covered with sediments of the Vlaanderen Formation, with a maximum thickness of 30 m, and with distinct regional environmental differences. The underlying Calais Member is relatively thick compared to the overlying Dunkerque Member. Both members were deposited during transgressive phases. Peat belonging to the Holland Peat Member interfingers with he Calais Member and is also present between the Calais and Dunkerque Members.

Dr. R. Paepe & Dr. C. Baeteman, Belgische Geologische Dienst, Jennerstraat 13, B-1040 Brussel, Belgium

Introduction

The Belgian coastal plain, about 15 km wide, is part of the narrow strip of low-lying land between the Sangatte Cliff in France and the large polder area of the Netherlands. The southwestern part of the plain, however, has an extension into the mainland along the river IJzer; it has been called the IJzer Golf (IJzer embayment) (Rutot 1897). The landward boundary of the coastal plain is defined by the line of maximum extension of Holocene marine and brackish deposits. This line is nearly conformable with the contourline of +3 m N.A.P., while the most general level of the Holocene ranges from +0.5 m to +2 m. The plain is separated from the present coast by an almost continuous dune barrier. The Belgian Ordance Datum level at Oostende (O.P.) is 2.33 m lower than the Dutch Ordance Datum Level (N.A.P.).

Pleistocene marine transgressions

At the end of the Tertiary, the southern boundary of the North Sea had moved from the vicinity of the French-Belgian border towards the area of the Belgian-Dutch border, north of Antwerp. Laga (1973), on the basis of the foraminiferal content, assigned the so-called Merksem Sands to the Pliocene. Overlying these are the Sands of Mol of continental origin; the lower part beneath the Arendonk lignite layer is still of Pliocene age, the upper part with cryoturbation structures (Tavernier 1943) is of Pleistocene Pretiglian age. The whole formation may reach a thickness of more than 70 m.

The Sands of Mol are followed by a series of continental and estuarine deposits called the Campine Formation. The upper part is subdivided into a lower Rijkevorsel (estuarine) clay Member and an upper Turnhout (estuarine) clay Member of Tiglian and Waalian age respectively (Paepe & Vanhoorne 1970). Between these two members is a continental Beerse Member, of Eburonian age (Dricot 1961, Paepe & Vanhoorne 1971), consisting of sand and peat. The age of these deposits was based on palaeobotanical and palaeomagnetic investigations (Van Montfrans 1971, Hus et al. 1976). The alternating continental and estuarine deposits in this northern part of Belgium, from the surface at about +30 m O.P. (Belgian Ordance Datum level) to a maximum depth of 40 m, are proof of transgressions and regressions in this area during the Early Pleistocene.

Marine influence is virtually absent in this area after the beginning of the Middle Pleistocene. It may be assumed that the formation of the Campine High Terrace started during Menapian times and that deposition continued during the "Cromerian". In the meantime, the North Sea apparently withdrew definitely from its eastern parts. Tilting of the North Sea basinal margin to the West seems a plausible explanation for the renewed marine deposition since the "Cromerian" in the French-Belgian border area. These marine deposits are situated around +15 m O.P.

The Herzeele Formation (Sommé 1973) described elsewhere in this volume by Sommé, contains two marine deposits separated by a peat layer formed during a warm period, and a marine crag unconformably overlying the deposits mentioned above. The two marine deposits are assumed to belong to the second and third "Cromerian" stages (sensu Zagwijn), whereas the upper crag is believed to be of Holsteinian age according to its fauna. Baeteman (unpublished) found these deposits extending eastward, into Belgian territory. None of them, however, were found east of the line Oostende-Gistel (Paepe 1970).

On the contrary, marine deposits of Eemian age are present all over the eastern Belgian coastal plain. The landward boundary of these deposits is further inland than the margin of the coastal plain. The Eemian deposits are generally present at a depth of 5 m below the surface. Towards the present coastline, Eemian deposits are rapidly sloping to 12—15 m in the middle of the plain and to about 20—25 m near the shore. A gradual change in lithology can be observed: the landward clayey lagoonal or tidal flat facies becomes increasingly sandy towards the sea.

Estuarine Eemian deposits are found at 12—20 m and far inland along the so-called Flemish Valley east of Ghent (Tavernier 1973). In these deposits and in many of the tributary valley deposits such as the Scheldt, Lys and Zenne, *"Corbicula fluminalis"* was found (Rutot 1910, Halet 1921). Recently, double shelled specimens of *Tapes senscens* var. *eemiensis* have been collected from Eemian deposits, most of them in the Eemian of the coastal plain.

The distribution of Eemian and Middle Pleistocene marine deposits indicates a considerable change in the position of the coastline.

West of Oostende, up to the Cliff of Sangatte, a large bay seems to have existed throughout Cromerian and Holsteinian times; later, it disappeared. The shoreline west of Oostende was then generally located off the present coast. West of Gravelines, however, the coastline turned landward to the Sangatte Cliff. During Eemian times, the coastline extended east of Oostende, with a landward extension along the Flemish valley and further to the North, towards the Delta area of the southern Netherlands.

Holocene marine transgressions

Stratigraphical units

The Holocene sediments in the coastal plain attain a maximum thickness of 30 m near the present coastline and of only 2 m near the mainland. This must be attributed to the steeply dipping top of the Tertiary substratum. The contourlines of the top Tertiary are generally parallel to the present coastline. Along the IJzer embayment, however, the top Tertiary is −20 m.

The coastal plain is covered with sediments of the Vlaanderen Formation, which consists in this area of the Calais, Dunkerque, and Holland Peat Members (Paepe, Sommé, Cunat & Baeteman 1976). In the western part of the plain, these deposits rest directly upon the Tertiary substratum. In the eastern part, the Holocene (Flandrian) deposits are resting upon marine Eemian deposits and in some areas upon sediments of Weichselian age.

In the IJzer Golf (IJzer embayment) Pleistocene deposits of the Herzeele Formation are absent over a large part of the area. However, the Tertiary substratum is covered by continental deposits consisting of reworked Pleistocene, mainly derived from the Herzeele Formation, and peat remnants of Atlantic age. This Lampernisse Member belongs to the Vlaanderen Formation. The Calais Member occurs in two small channel-fills. The marine Holocene is only represented by the Dunkerque Member resting upon the main peat layer of the Holland Peat Member.

The Calais Member

The Calais Member is found all over the coastal plain except in the IJzer embayment. It is generally covered by the main peat layer belonging to the Holland Peat Member and by deposits of the Dunkerque Member. In the area of De Moeren (near the French border) Calais deposits are found directly at the surface. The thickness ranges from 0.5 m on the landward side of the coastal plain to nearly 25 m at the present coast.

The lithology of the Calais Member in the inland area and in the area near the coast is distinctly different. Near the landward margin of the coastal plain, the Calais Member is a bluish-grey clay with numerous molluscs as *Cardium edule, Hydrobia ulvae, Macoma balthica, Mactra* and others. The clay contains serveral peat or peaty layers at −11.5 m, −9.5 m, −7.5 m, −6 m, −5 m and −2 m N.A.P. The peat layers are local and discontinuous, except the one at −2 m. A brown loamy sand is the main sediment at a depth of −8 m N.A.P.

In the area east of Oostende and in other places in the middle of the coastal plain, the bluish-grey clay passes into grey sand with sporadic peaty intercalations. In an area about 5 km wide along the present coast, the Calais Member has a different lithology. In certain areas (especially in the westernmost), the Calais Member consists of grey sand, up to 25 m thick. Clayey or peaty intercalations are missing; numerous molluscs (especially *Cardium edule* and their fragments, concentrated in layers, are found in many places. The sandy sediments have been considered in the Belgian literature as the typical Calais facies, erroneously named Atlantic tidal flat sediments. In doing so, the beach character of part of the deposits was neglected.

The Holland Peat Member

The Calais Member is separated from the Dunkerque Member by the main peat layer (so-called surface peat) that belongs to the Holland Peat Member. The peat is the most distinct layer of the Holland Peat Member, as regard its distribution and its thickness of 2 m. It is found almost everywhere in the coastal plain at a depth of about −2 m; it may be absent as a result of erosion in former tidal channel areas belonging to the Dunkerque transgressions. The main peat layer is dated 4150 B.P. (mean ^{14}C dating) and the top is dated 2900 B.P. It does not contain clayey intercalations, contrary to the clayey Dunkerque 0 layer, which is usually found in equivalent Dutch deposits. A second but minor peat layer is situated at −5 m and dated 5830 B.P. (mean ^{14}C dating). Finally, the peat layer at −11.5 m has been found only locally; it corresponds most probably to the so-called Basal Peat or Lower Peat. See the contributions by Jelgersma, Oele & Wiggers and Behre, Menke & Streif elsewhere in this volume.

The presence of several peat or peaty layers points to several transgressive and regressive phases during deposition of the Calais Member. The available data from radiocarbon dating and pollen analysis are insufficient to establish a chronological timescale for the area.

The Dunkerque Member

The Dunkerque Member, overlying the so-called surface peat, has the largest extension of all members of the Vlaanderen Formation. As stated above, it is absent in the De Moeren area. The predominantly clayey facies reaches a thickness of 2 m, whereas a sandy facies generally found in tidal channels may be more than 4 m thick. This member contains only sporadic root levels, and not the peat or peaty layers, as found interfingering in the Calais Member.

On the basis of archaeological data, lithological characteristics and sedimentological studies, several transgressive phases can be distinguished, i.e. Dunkerque I, II, IIIA, and IIIB. The greatest lateral extension was reached during the Dunkerque II transgressional phase, the sedi-ments of which cover nearly the entire coastal plain (J.B. Amerijckx, 1960).

REFERENCES

Amerijckx, J.B., 1960: De jongste geologische geschiedenis van de Belgische zeepolders. *Technische-Wetenschappelijk Tijdschrift* 29, 1−10.

Baetman, C., Lambrechts, G. & Paepe, R., 1974: Autosnelweg Brugge − Calais − Kb. Nieuwpoort, Lampernisse, Veurne, Hoogstade. Prof. Paper 1, Belg. Geol. Survey: 55, Brussels.

Dricot, E., 1961: Micro-stratigraphie des argiles de Campine. *Bull. Soc. belg. Géol.* 20, 113−141.

Halet, F., 1933: Sur la présence de couches à *Corbicula fluminalis* MULLER aux environs de Saint-Denis-Westrem − *Bull. Soc. belg. Géol.* 43, 111−116.

Halet, F., 1939: Sur la précense de *Corbicula fluminalis* dans le Pleistocène des environs d'Escanaffles. *Bull. Soc. belg. Géol.* 69, 233−234.

Hus, J., Paepe, R., Geeraerts, R., Sommé, J. and Vanhoorne R., 1976: Preliminary magnetostratigraphical results of pleistocene sequences in Belgium and Northwest France. *Quaternary glaciations in the Northern Hemisphere, I.G.C.P. 24 report 3*, 99−128, (Bellingham, 1975), Prague.

Laga, P., 1973: The neogene deposits of Belgium − Guide book for the Field Meeting of the Geologists' Association London.

Paepe, R., 1971: Autosnelweg Brugge − Calais − Kb Houtave, Bredene, Gistel − Prof. Paper 9, Belg. Geol. Survey: 59, Brussels.

Paepe, R. & Vanhoorne, R., 1970: Stratigraphical position of periglacial phenomena in the Campine Clay of Belgium, based on palaeobotanical analysis and palaeomagnetic dating. *Bull. Soc. belg. Géol.* 79, 201−211.

Paepe, R., & Vanhoorne, R., 1972: Marine Eemian Deposits at Meetkerke (Belgian Coastal Plain). Prof. Paper 7, Belg. Geol. Survey: 12, Brussels.

Paepe, R., Sommé, J., Cuñat, N. & Baeteman, C., 1976: Flandrian, a formation or just a name? *Stratigr. Newsl.* 5, 18−30.

Rutot, A., 1897: Les origines du Quaternaire de la Belgique. *Bull. Soc. belg. Géol.* 11, 140.

Rutot, A., 1910: Sur la découverte de *Corbicula fluminalis* à Hofstade. *Bull. Acad. R. Belg. Sci.* 7, 163−169, Brussels.

Sommé, J., 1974: Les formations quaternaires de la région du Nord − *Ann. Sci. Univ. Besançon, Géol.*, 3e série, 21, 97−102.

Tavernier, R., 1942: L'age des argiles de la Campine. *Bull. Soc. belg. Géol.* 51, 193−209.

Van Montfrans, H. M., 1971: Palaeomagnetic dating in the North Sea Basin. Acad. Thesis, University of Amsterdam, 113 pp., Princo N.V. Wageningen.

Zagwijn, W. H., Montfrans, H. van & Zandstra, J M., 1971: Subdivision of the "Cromerian" in the Netherlands; pollen analysis, palaeo magnetism and sedimentary petrology. *Geol Mijnbouw* 50, 41−58.

IV-g

Quaternary coastlines in northern France

JEAN SOMMÉ

Sommé, J., 1979: Quaternary coastlines in northern France. In E. Oele, R. T. E. Schüttenhelm & A. J. Wiggers (editors), The Quaternary History of the North Sea, 147–158. *Acta Univ. Ups. Symp. Univ. Ups. Annum Quingentesimum Celebrantis: 2*, Uppsala. ISBN 91-554-0495-2.

Marine ingressions in the coastal plain of northern France are known from the Eocene, the Late Miocene and the Middle and Late Quaternary.
 The Pleistocene marine formations of the coastal region (Herzeele Formation, Loon Formation) prove that the origin of the southern North Sea may be placed as early as the beginning of the Middle Pleistocene. The Sangatte cliff and its deposits were formed probably prior to the Eemian.
 The Holocene coastal plain was invaded by the sea as from the Early Atlantic. The lower and sandy Calais deposits are related to the shingle spit system (Pierrettes). Marine sedimentation was interrupted by the formation of the "tourbe de surface" (surface peat), essentially of Subboreal age. The uppermost peat layer just prior to the Dunkerque I transgression separates the Calais deposits from Dunkerque deposits in their type areas. The sandy Dunkerque deposits cover almost the whole coastal plain except in the southwestern marshy area where peat (tourbe de surface) is exposed.

Dr. J. Sommé, Institut de Géographie, Univesité des Sciences et Techniques de Lille, B.P. 36, F-59650 Villeneuve d'Ascq, France.

Introduction

The "plaine maritime" of northern France (Calaisis and Flandre maritime) forms the southwestern extremity of the Holocene coastal area extending along the southern North Sea.

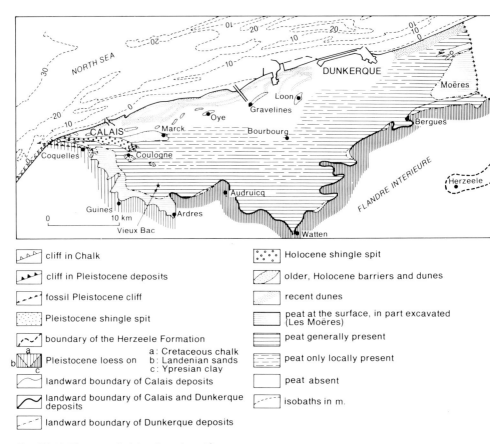

Fig. IV-41. The coastal plain of northern France.

This area deserves special attention, as it is the type region for the stratigraphical subdivision of the Flandrian coastal sequence in Calais and Dunkerque deposits (Dubois 1924) (Fig. IV-41).

In parts of this area and in the adjacent region, also Pleistocene marine deposits are exposed or present in the subsoil. These deposits provide information on ancient North Sea shorelines and on the origin of the Strait of Dover, discussed by Sommé (1975), who compared these areas to the Manche (Channel) coast of Boulonnais and Picardy.

Pleistocene development

Late Tertiary and Early Pleistocene

During the Eocene the greater part of the region was covered by the sea. Variations in the position of the shoreline were mainly caused by tectonic activities in the Paleozoic Brabant and Artois anticlines.

One of the first of these activities, which resulted in the uplift of the Artois anticline, may be assigned to the Middle Lutetian (Middle Eocene) on the basis of faunal differences between the northern basin and the Paris basin. A regression may be assumed for the Late Eocene, as the Bartonian beds of the Flemish Hills are interpreted as nearshore deposits (Pomerol 1965, Blondeau et al. 1965).

During the Oligocene and the greater part of the Miocene, continental conditions prevailed (Riveline-Bauer 1970). Transgressive deposits are known from the Diest Formation (Late Miocene) that overlies the Eocene beds of the Flemish Hills (Mont Cassel, Mont des Cats). The lower part of the Diest Formation, which may be correlated to the Deurne beds of the Antwerp area, is definitely marine (Tavernier 1954, Tavernier & De Heinzelin 1962a).

In the French area there is no evidence of the Late Miocene coastline having extended further south than the line connecting the Flemish Hills (Sommé 1975). The question whether the Strait of Dover existed already at that time is still unsolved. Generally speaking, the Miocene sea must have covered approximately the same area as that covered by the sea at the end of the Eocene.

During the Miocene and the Pliocene, the sea retreated to the North. At the transition from Pliocene to Pleistocene, marine influence is known only from the northern part of Belgium (Paepe & Sommé 1975, Zagwijn 1974).

Middle and Late Pleistocene

Pleistocene marine sands have been known in southwestern Belgian Flanders for a long time. They were mentioned by Rutot in 1897 and studied more recently by Tavernier & De Heinzelin (1962b) (Cardium sands of West-Flanders) and Vanhoorne (1962) (Lo Peat); a more complex marine to continental sequence was recognized at Herzeele (France) in the same area (Sommé 1975, Paepe & Sommé 1975).

The Herzeele Formation. — Some km west of the Belgian border, the section in the Herzeele quarry, situated south of the Yser valley at +13

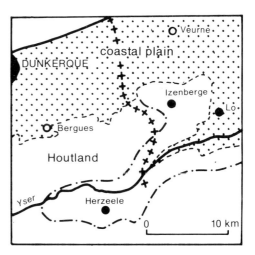

Fig. IV-42. Situation map of the Herzeele area.

m N.G.F. (Fig. IV-42), contains the following stratigraphical units (Fig. IV-43):

— cover of aeolian silt and silty sand (1 and 2) with paleosols (pseudogley to grey-brown podzolic type), separated by stone pavements with frost wedges. These deposits are covered by a thin yellow-brown layer of (Weichselian) loess;

— an upper marine silty to sandy complex (4) with numerous shells (*Cardium edule, Macoma balthica,* etc) indicating nearshore lagoonal conditions. This is the so-called Izenberge Crag. In the eastern part of the section, the crag becomes thicker and directly overlies the Ypresian clay. The marine complex is overlain by clayey and sandy, laminated sediments (3);

— a silty and clayey bed (5) with two hydromorphic horizons;

— a middle silty to clayey marine lagoonal series

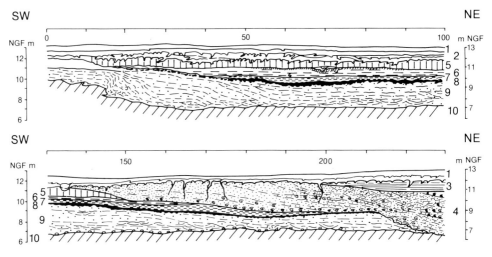

Fig. IV-43. Section in the Herzeele quarry. For explanation see text.

with some shells (7), the topmost part of which is a humic clay (6);

- a disturbed peat layer (8), indicating warmer climatic conditions; abundant *Taxus* and *Alnus*, and fewer *Corylus, Pinus, Quercus, Ulmus, Tilia, Betula, Fraxinus, Picea, Salix*. *Hedera* is common. Typical Tertiary elements are missing. *Abies* appears in the overlying humic clay (Vanhoorne, unpubl.; Hus et al. 1976);

- a lower marine sandy tidal flat series (9) with characteristic point-bar structures overlying the Ypres Clay (10) of Eocene age.

The entire sequence below the aeolian sediments, from layer 3 and the crag down to the Tertiary substratum, is called Herzeele Formation. The Izenberge crag was considered of Holsteinian age (Tavernier & De Heinzelin 1962b), particularly for its fauna (*Cardium edule, Macoma balthica, Scrobicularia plana, Mytilus edulis, Theodoxus fluviatilis, Hydrobia stagnalis, Bela plicifera, Lymnaea trunculata, Retusa alba, Nucella lapillus*). Typical Eemian elements are absent; the presence of *Bela plicifera* points to a Hoxnian (=Holsteinian) interglacial age, the presence of *Macoma balthica* indicates a deposit younger than Baventian (Norton 1970). The position of the crag below the loess with fossil soils of an interglacial type supports the assumed Middle Pleistocene age.

The peat might be of "Cromerian" age. The lowermost marine deposits may correspond to an interglacial sea level rise, after which the peat developed. The humic clay (6) indicates a second rise of sea level. The interpretation of a "Cromerian" age does not seem to be in disagreement with the palaeobotanical contents of the peat. Therefore, it was proposed to place the middle and the lower marine beds in the "Cromerian" III and II respectively (Paepe & Sommé 1975). According to the first results of palaeomagnetical studies, the beds may be older, as they show negative inclination, except for the topmost part of the aeolian cover and the lower sand complex, which have positive inclinations (Hus et al. 1976). In the Netherlands the youngest sediments with negative inclinations have been found between interglacials I and II of the "Cromerian complex". Additional studies may clarify this contradiction.

The section at Herzeele definitely indicates marine conditions in several phases of the Middle Pleistocene in the southern part of the North Sea from East Anglia to Flanders. The sea level must have been close to the present level. In view of the large distance between these sediments and the coast, the southern North Sea must have been much larger than at present, and the Strait of Dover undoubtedly existed already at that time.

An early Middle Pleistocene (or Early Pleistocene) marine transgression from the southwest is proved by the recent find of marine shelly sands with "Icenian" fauna (Loon Formation) at a depth of −20 m N.G.F., as far southwest as Gravelines (Vanhoorne, Leplat & Sommé, unpubl.).

The sudden change in direction of the transgressions from NE to SW seems to be related to pre-existing (Tertiary) tectonic structures.

The Sangatte Cliff. — Other relicts of a Pleistocene shoreline are known near Calais: the Sangatte cliff and raised beach, covered with slope deposits, and related units in the Holocene coastal plain (Briquet 1905, 1906, 1923, 1930, Dubois 1924, Sommé 1969a, 1969b, 1975, 1976).

In the area west of Calais the present shoreline obliquely intersects successively: Cretaceous chalk (Cap Blanc Nez), the Pleistocene cliff of Sangatte with its slope deposits, and the Holocene deposits with the shingle spit of "Banc des Pierrettes" (Fig. IV-41). In the Sangatte cliff, more than one km long, can be distinguished from SW to NE:

— the ancient cliff in the Lower Turonian chalk; the fossil shore platform is covered with flint shingle, and laterally passes into a sandy beach with shells (*Cardium edule, Tellina balthica,* etc.). These marine deposits represent an interglacial high sea level (relative level +5 m, located about +10 m N.G.F.);

— a thin peat layer, containing *Pinus* pollen (Vanhoorne in Sommé 1975), filling a small depression in the fossil beach (^{14}C dating; older than 50300 B.P., GrN 5868); chalky landslide deposits at the base of the cliff.

These continental deposits may have been formed at the end of the interglacial;

— periglacial deposits covering the older forms; a lower complex of calcareous, regularly bedded deposits, the lower part mainly consisting of chalky material overlain by layers with flint and these covered by loess layers; an upper complex of irregular beds of flint pebbles, sand and silt, non-calcareous at the base and calcareous at the top; finally, a sandy to silty cover.

Many authors dated the Pleistocene shoreline as Eemian, following Dubois and Zeuner (1959). Briquet and other authors assigned the Sangatte Cliff to an earlier interglacial, the Mindel-Riss or Holsteinian (Bourdier 1969b, Baudet 1971). This opinion is more in agreement with the nature of the cliff deposits, which are too complex to be considered as Weichselian only. In the cliff deposits, at least two major periglacial sedimentary cycles can be recognized below the sandy to silty cover (Sommé 1975).

Signs of Acheulean industry are known from the periglacial deposits (Lefebvre 1968, 1969, 1976), but there is no evidence of Mousterian artefacts (Tuffreau, 1971, 1974). The Levalloisian artefacts, discovered in the top part of the peat and in the fossil beach sand, considered as Early Wurm (Agache 1968, Lefebvre 1968, 1969) might be older (Bourdier 1969a, 1969b).

Therefore, a correlation may be possible between the Sangatte cliff and the upper part of the Herzeele Formation. The buried geomorphological structure of the coastal plain adds evidence for a pre-Eemian age of the shoreline at Sangatte. The Weichselian deposits underneath the marine Holocene sediments have filled deep valleys, which separated the remnants of ancient spits (Petite-Rouge-Cambre, Coulogne, Attaques) related to the Sangatte cliff (Fig. IV-41). This indicates heavy erosion before the Late Pleistocene, also known from the valleys in the interior region.

Near Gravelines, a peaty layer (depth −17 m N.G.F.), intercalated in continental deposits, has been dated as Middle Weichselian (33110 ± 740 B.P., GrN 5869) (Sommé 1975). It contains a

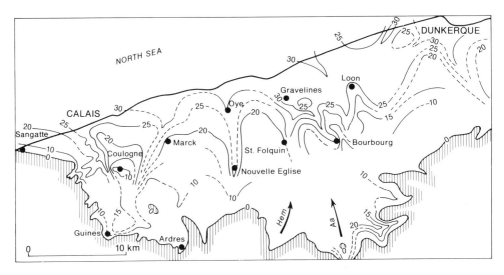

Fig. IV-44. Isobaths (in metres below N.G.F.) of the pre-Quaternary surface.

woody steppe flora (Vanhoorne, unpubl.). In many borings, similar deposits have been found about at the same depth. Dubois (1924, 1926) described a boring near Coquelles, and assigned the lower layers of peat and sand with *Elephas primigenius* to the "Assise d'Ostende", which were considered as Early Flandrian. Near the present shoreline north of Gravelines, the pre-Quaternary surface has a steep slope. The Loon Formation was found in this area. A great number of borings indicate the presence of marine Pleistocene deposits lower than -17 m N.G.F. The overlying continental peaty silts may be of various ages.

There is no clear evidence of subsurface marine Ostende deposits of Eemian age. In the area near Sangatte, sandy shingle deposits have been related to a pre-Holocene marine phase (Briquet 1923, Dubois 1921, Sommé 1975).

Holocene development

General description of the Holocene coastal plain

The Holocene coastal plain of northern France roughly occupies a triangular area, extending approx. 60 km, from Sangatte a few km east of Cap Blanc Nez to the Belgian frontier in the East.

The largest width of about 20 km in the central part near Gravelines is related to the presence of former estuaries (Hem, Aa).

The coastal plain which consists of marine and brackish Holocene deposits, coincides approximately with the +2.5 m N.G.F. contour line. The Holocene sediments cover a larger area than the Pleistocene marine sediments, except in two marginal areas.

To the west the plain wedges out against the chalk hills of Artois, from which it is separated by continental Pleistocene deposits covering the ancient Pleistocene cliff of Sangatte. To the east the plain narrows to about 10 km at the Belgian frontier. To the South the plain is bordered by the hills of Artois consisting of Cretaceous chalk and Landenian sands, and by the hills and the plain of Inner Flanders (Houtland) consisting of Ypresian clay.

The subsurface boundary between Cretaceous chalk and Eocene clay coincides with a line from Calais to Guines. Beyond the coastal plain, the marine Herzeele Formation is present under an aeolian cover along the Yser valley (Fig. IV-42).

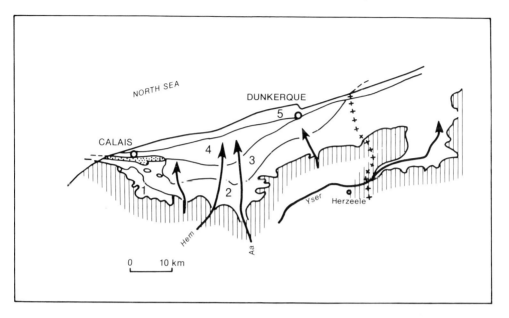

Fig. IV-45. Zones of the coastal plain with main fluvial axes. For explanation see text.

Stratigraphical units

The Quaternary sediments are 30 m thick near the present shoreline (Fig. IV-44). Generally, they are of Holocene age, but older sediments may be incorporated.

Stratigraphically, the Holocene sediments are combined into the Flandrian (Paepe et al. 1976). Dubois (1924) who defined the subdivisions, considered the Flandrian as a complete sequence deposited during the last major sedimentary cycle in Flanders. He named the lower part of the Flandrian the "Assise d'Ostende"; it is found between −15 and −30 m N.G.F. and contains abundant *Corbicula fluminalis*. He proposed the term "Assise de Calais" for the thick marine, grey-blue sandy deposits underlying the upper peat, between 0 and −15 m and containing a faunal assemblage similar to the present one *(Zirphaea crispata, Ostrea edulis)*. He related the Assise de Calais to the shingle spit system of Pierrettes (type locality: southern part of the town of Calais).

Dubois also proposed the term "Assise de Dunkerque" (upper part of the Flandrian) for the "historical" beds, about at sea level, near Dunkerque containing a recent fauna with *Mya arenaria*. As had been established before by Gosselet (1893, 1903), Debray (1872) and Blanchard (1906), the Dunkerque deposits at their base contain a Gallo-Roman habitation level above the upper peat. The Dunkerquian transgression must have started in the fourth century, after the Roman habitation. Briquet (1930), of the same opinion, recognized a major break in the postglacial sea level rise at the main peat layer; this break separates Flandrian and Dunkerquian transgressions and sediments.

The present definition of the Flandrian subdivisions on the geological maps (Sommé 1975) for the French sector differs from the original one of Dubois in two respects:

— the "Assise d'Ostende" is of Eemian age, as established by Tavernier in Belgium;

— the "Assise de Dunkerque" is still defined as marine sediments overlying the main peat layer, but it includes pre-Roman deposits (Dunkerque I). The deposits of the Dunkerque 0 transgression according to the Dutch nomenclature are in the French area still considered to belong to the "Assise de Calais".

The Flandre Formation. — The Flandre Formation in the type region near Calais contains five stratigraphical units:

— a shingle spit system of flint and chalk. The main element is the "Banc des Pierrettes", exposed near Sangatte and ending in a number of curved spits S and SE of the town of Calais. The most conspicuous of these spits near the surface are those near Virval. Several stages of spit formation are known: one at −20 m N.G.F. under the city of Calais, a second one at −15 m and further south, and a third above −10 m N.G.F.

— the tidal deposits of the "Assise de Calais" below the uppermost bed of the "tourbe de surface". The sediments are predominantly sandy ("sables pissarts"), but silty and clayey Calais deposits may be intercalated in the upper peat layers. This is particularly the case in the southern part of the coastal plain where the Holocene sediments are less than 10 m thick. A layer of peat is only locally present at the base of the Calais deposits. Some organic beds or thin peat layers are interbedded with the marine deposits at −13 m, −10 m, −7,5 m and −5 m. At −3 m, the complex of the "tourbe de surface" appears. Most of the Calais deposits and the shingle spit system correspond to the Atlantic and Subboreal transgression phases.

— the mainly Subboreal "tourbe de surface" complex; the base of the main bed of the upper peat has been dated 3500—3400 B.P. (Fig. IV-46). The complex is generally present in the southern marginal belt of the plain (Fig. IV-45), zones 1 and 2), but only in the "Marais" (zone 1) near Ardres and Guines

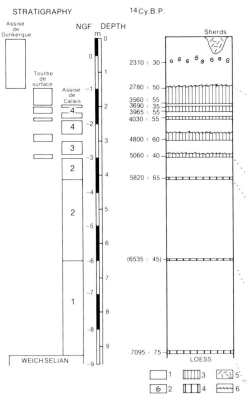

Fig. IV-46. The Vieux-Bac boring: stratigraphy, depth, radiocarbon datings and lithology. Legend: 1. Sand and clay, 2. Shells, 3. Peat, 4. Humic horizon, 5. Reworked (gallo-roman) soil, 6. Erosional boundary (from W. Roeleveld et J. Sommé, 1975).

this complex is present at the surface. In this area the peat attains its maximal thickness of more than one metre; locally it contains a layer of lacustrine marl, called "Calcaire à Limnées". In many places the "tourbe de surface" contains intercalated marine beds (Sommé 1975, Roeleveld & Sommé 1975) (Fig. IV-46). The number of individual peat beds tends to decrease towards the sea, and the stratigraphically higher beds are apparently extending further seaward than the lower ones (zone 3); the complete absence of peat in the northern part of the coastal plain (zone 4) may be attributed to non-deposition.

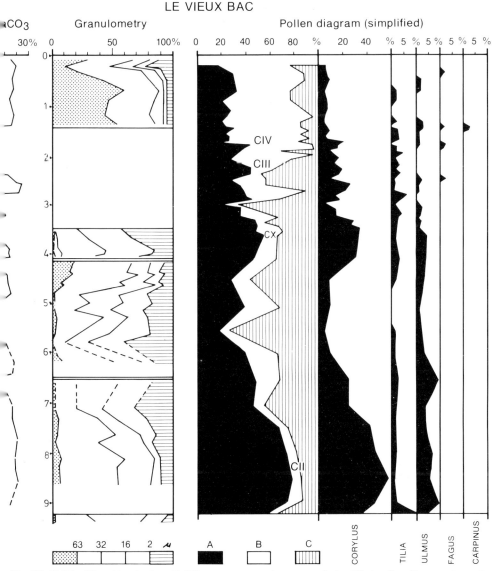

Fig. IV-47. The Vieux-Bac boring: $CaCO_3$-content, granulometry and simplified pollen diagram. Legend: A. Arboreal pollen, B. and C. non-arboreal pollen, C. Halophilous plants (from H. Claus, R. Diriken, 1974; compiled by A. V. Munaut et J. Sommé).

— the deposits of the "Assise de Dunkerque" above the main upper peat layer. The facies is generally sandy, even in the southern part of the plain, in contrast to the clayey facies of the uppermost beds of the Calais deposits (Diriken 1974; Fig. IV-47). East of Ardres, particularly in the area south of Dunkerque, the Dunkerque II sands may overlie Weichselian loess (Paepe 1956, 1960, Coque-Delhuille 1972). In the coastal plain the

"Assise de Dunkerque" is found practically everywhere at the surface, except in the marshy area (zone 1) with peat.

— the dune ridges. In the coastal zone (zone 4), a series of older barriers is covered by low dunes, stabilized by vegetation. The littoral zone (zone 5) is a series of dune ridges dating from the Middle Ages and modern time.

Evolution of the coastal plain

As the deepest part of the Strait of Dover lies at 60—70 m, it may be assumed that the ingression of the sea from the South took place in the very Early Holocene. According to Jelgersma (1961) the sea level was about 65 m below the present one at 10300 B.P.

In the northern part of the Channel a peat at —55 m N.G.F., dated 9700 B.P., indicates a mean sea level of about —58 m (Ters 1973). A connection with the southern part of the North Sea was formed between 8600 and 9000 B.P. (Prentice 1972). About 8700 B.P., the southern part of the North Sea and the Channel were connected by a rather narrow channel, and sea level was about —36 m. According to Morzadec-Kerfourn & Delibrias (1972) and Ters (1973), the sea did not reach the —32 m line until 8250 B.P.

From the depth of the pre-Holocene it may be concluded that the sea reached its present limit towards the end of the Boreal. A sandy to pebbly deposit at a depth of —25 to —30 m may be related to a coastal barrier system at the beginning of the Atlantic, corresponding to the "Le Havre sea level" of Ters (1973). At a depth of about —15 m, a similar deposit further south may be related to the Pierrettes shingle system near Calais (Fig. IV-41). The landward migration of the coastal barrier system may be compared with that in the Dutch coastal plain (Hageman 1969) and in the Picardy coastal area (Le Fournier 1974).

From a depth of —10 m N.G.F. the Pierrettes shingle system grew upward, which points to a stable position of the spit system during the Atlantic, and this in its turn points to a slow rate of sea level rise. During the Atlantic the orientation of the whole depositional system changed as a consequence of the erosional retreat of the Chalk and the Pleistocene cliff. At the same time, the clayey tidal sediments reached the southern part of the plain. Indications have been observed of transgressive and regressive phases. The presently available radiocarbon datings permit a tentative transgression and regression chronology (Roeleveld & Sommé 1975), which is in agreement with pollen-analytical data (Claus 1974, Munaut, pers. comm.). In the profile of Vieux-Bac (NW of Ardres, Fig. IV-41) the diminishing number of tree pollen from bottom to top may be attributed to the southward migration of the shoreline. On the other hand, the whole sequence, up to the main peat layer (base = 3500 B.P.), represents a brackish environment, as concluded from the percentages of halophilous plants (Figs. IV-46, 47).

The alternating transgressive and regressive phases and the radiocarbon datings are remarkably congruous to the Dutch sequence, although the Middle and Late Subboreal oscillations are less well comparable. The latter period is characterized by the growth of the Pierrettes shingle spit, the continuation of which is the sandy barrier of Marck. The pebble basis is located at —2.5 m. Next in the Holocene development is the formation of older dune ridges, following the seaward advance of the coastal barrier (Baraques near Calais, Oye, Loon, Ghyvelde). Henceforth, the WSW-ENE direction of the shoreline was established.

In the central part of the coastal plain, the main peat layer of the upper peat (dated between 3500 and 2700 B.P.) attained its greates extension. It contains a fresh-water flora, which indicates an important regression. The maximal extension of this peat layer, deposited just before the Dunkerque I transgression, separates the Calais and Dunkerque deposits in the French coastal plain.

The large extension and the sandy facies of the "Assise de Dunkerque", compared to the older silty and clayey deposits, indicate a rather significant change in the coastal environmental conditions. The Roman regression permitted widespread human habitation in the southern part of the plain (Sommé & Cabal 1972) on the peat and on the Dunkerque I deposits.

The most extensive marine ingression corresponds to the Dunkerque II transgression, which reached the base of the hills in many areas. It slightly exceeded the area affected by the Dunkerque I transgression. The clayey Dunkerque III deposits show a more limited areal distribution; they are mainly restricted to the former Hem-Aa estuaries.

Information on a more recent evolution of the coastline, as presented by Briquet (1930), indicates a seaward progression of the younger dune ridges during medieval and modern times, at least between Dunkerque and Calais. In the Sangatte sector, however, and east of Dunkerque, the shoreline has retreated, and older and younger dunes form one single system.

REFERENCES

Agache, R., 1968: Informations archéologiques, circonscription Nord-Picardie. *Gallia Préhistoire 11*, 267–309.
Baudet, J. L., 1971: La Préhistoire ancienne de l'Europe septentrionale. Paris, 257 pp.
Blanchard, R., 1906: La Flandre, étude géographique de la plaine flamande en France, Belgique et Hollande. Lille, 530 pp., reprint 1970.
Blondeau, A., Cavelier, C., Feugueur, L. and Pomerol, Ch., 1965: Stratigraphie du Paléogène du bassin de Paris en relation avec les bassins avoisinants. *Bull. Soc. Géol. France 7, 7*, 200–221.
Bourdier, F., 1969a: Etude comparée des dépôts quaternaires des bassins de la Seine et de la Somme. *Bull. Inf. Géol. Bassin Paris 21*, 169–231.
Bourdier, F., 1969b: Sur la position chronologique du Paléolithique de Sangatte, Wissant et Wimereux (Pas-de-Calais). *Bull. Soc. Préhist. France 66*, 230–231.
Briquet, A., 1905: Extension de la plage soulevée de Sangatte. *Ann. Soc. Géol. Nord 34*, 109–111.
Briquet, A., 1906: Notes sur quelques formations quaternaires du Pas-de-Calais. *Ann. Soc. Géol. Nord 35*, 211–236.
Briquet, A., 1923: Les formations pleistocènes des environs de Sangatte. *Ann. Soc. Géol. Nord 48*, 176–184.
Briquet, A., 1930: Le littoral du Nord de la France et son évolution morphologique. Orléans, 439 pp.
Claus, H., 1974: Pollenanalytisch onderzoek van de Holocene mariene afzettingen in de Noordfranse kustvlakte. Kathol. Univ. Leuven, Belgium, 75 pp.
Coque-Delhuille, B., 1972: Recherches sur les formations quaternaires et le modelé de la Flandre maritime dunkerquoise. *Cah. Géogr. Phys. Lille 1*, 45–63.

Debray, H., 1872: Etude géologique et archéologique de quelques tourbières du littoral flamand et du département de la Somme. *Mém. Soc. Sc. Agr. Arts Lille 3, 11*, 433–486.
Diriken, P., 1974: Sedimentologische studie van de Holocene afzettingen in de Noordfranse kustvlakte. Kathol. Univ. Leuven, Belgium 100 pp.
Dubois, G., 1921: Résultats d'une campagne de sondages à travers les terrains quaternaires et récents du Calaisis (note préliminaire). *Ann. Soc. Géol. Nord 46*, 67–78.
Dubois, G., 1924: Recherches sur les terrains quaternaires du Nord de la France. Thesis Lille, 355 pp.
Dubois, G., 1926: Dunes et cordons littoraux dans l'agglomération de Calais. *Ann. Soc. Géol. Nord 51*, 129–136.
Gosselet, J., 1893: Géographie physique du Nord de la France et de la Belgique: la Plaine maritime. *Ann. Soc. Géol. Nord 21*, 119–137.
Gosselet, J., 1903: Esquisse géologique du Nord de la France et des contrées voisines, 4: Terrains quaternaires. Lille, 78 pp.
Hageman, B. P., 1969: Development of the western part of the Netherlands during the Holocene. *Geol. Mijnbouw 48*, 373–388.
Hus, J., Paepe, R., Geeraerts, R., Sommé, J. & Vanhoorne, R., 1976: Preliminary magnetostratigraphical results of pleistocene sequences in Belgium and northwest France. *Quaternary Glaciations in the Northern Hemisphere, I.G.C.P. report 3*, 99–128, (Bellingham, 1975), Prague.
Jelgersma, S., 1961: Holocene sealevel changes in the Netherlands. *Med. Geol. Sticht. C, VI, 7*, 100 pp.
Lefebvre, A., 1968: Repères chronologiques pour l'étude de la falaise de Sangatte. *Bull. Hist. artist, Calaisis*, 161–165, Calais.
Lefebvre, A., 1969: Aperçu sur quelques gisements préhistoriques de la région côtière du Nord de la France. *Septentrion 1*, 57–67, Calais.
Lefebvre, A., 1976: Les industries lithiques du Calaisis. In Guide-book excursion A 10: Northwest France, IXth congress Union Intern . Sc. Préh. Protohist., Nice, France, 186–190.
Le Fournier, J., 1974: La sédimentation holocène en bordure du littoral picard et sa signification dynamique. *Bull. Centre Rech. Pau-SNPA 8*, 327–349.
Morzadec-Kerfourn, M. T. & Delibrias, G., 1972: Analyses polliniques et datations radiocarbone des sédiments quaternaires prélevés en Manche centrale et orientale. In Colloque sur la Géologie de la Manche. *Mém. Bur. Rech. Géol. Min. 79*, 160–165.
Norton, P. E. P., 1970: The Crag Mollusca. A conspectus. *Bull. Soc. belg. Géol. 79*, 157–166.
Paepe, R., 1956: De kustvlakte tussen Duinkerken en de Belgische grens. Een fysisch-geographische studie. Mém. Lic. Gent, Belgium, 98 stenciled pp.
Paepe, R., 1960: La plaine maritime entre Dunkerque et la frontière belge. *Bull. Soc. belg. Etud.*

Géogr. 29, 47—66.
Paepe, R. & Sommé, J., 1975: Marine pleistocene transgressions along the flemish coast (Belgium and France). *Quaternary Glaciations in the Northern Hemisphere, I.G.C.P. project 2*, 108—116, (Salzburg, 1974), Prague.
Paepe, R., Sommé, J., Cunat, N. & Baeteman, C., 1976: Flandrian, a formation of just a name? *Stratigr. Newsl.* 5, 18—30.
Pomerol, Ch., 1965: Les sables de l'Eocène supérieur des bassins de Paris et de Bruxelles. Mém. Expl. Carte Géol. dét. France, Paris, 214 pp.
Prentice, J. E., 1972: Sedimentology of the North Sea and the English Channel. In Colloque sur la Géologie de la Manche, *Mém. Bur. Rech. Géol. Min.* 79, 229—232.
Riveline-Bauer, J., 1970: Contribution à l'étude sédimentologique et paléogéographique des sables de l'Oligocène des bassins de Paris et de Belgique. Thesis, 3d cycle, Geology, Paris, 2 vol., 164 pp.
Roeleveld, W. & Sommé, J., 1975: The North Sea coastal plain of Northern France. In Guidebook, Meeting INQUA Holocene commission.
Rutot, A., 1897: Les origines du Quaternaire de la Belgique. *Bull. Soc. belg. Géol.* 11, 140.
Sommé, J., 1969a: La plaine maritime. *Ann. Soc. Géol. Nord* 89, 117—126.
Sommé, J., 1969b: La falaise de Sangatte et la plaine maritime. *Septentrion 1*, 43—56, Calais.
Sommé, J., 1975: Les plaines du Nord de la France et leur bordure, étude géomorphologique. Thesis Paris, 790 stenciled pp.
Sommé, J., 1976: La falaise de Sangatte et les formations Quaternaires du Calais. In Guidebook excursion A 10: Northwest France, IXth congress Union Intern. Sc. Préh. Protohist., Nice, France, 182—186.
Sommé, J. & Cabal, M., 1972: La plaine maritime dans la région d'Ardres (Pas-de-Calais) et le site archéologique des Noires-Terres. *Cah. Géogr. Phys. Lille 1*, 29—43.
Tavernier, R., 1954: Le Néogène, le Quaternaire. In Prodrome d'une description géologique de la Belgique, 533—589, Liège.
Tavernier, R. & de Heinzelin, J., 1962a: Introduction au Néogène de la Belgique. In symposium sur la stratigraphie du Néogène nordique. *Mém. Soc. belg. Géol.* 6, 7—28.
Tavernier, R. & de Heinzelin, J., 1962b: De Cardium-lagen van West-Vlaanderen. *Natuurw. Tijdschr.* 44, 49—58, Gent.
Ters, M., 1973: Les variations du niveau marin depuis 10.000 ans le long du littoral atlantique francais. Le Quaternaire, Géodynamique, Stratigraphie et Environnement, 114—135, Paris.
Tuffreau, A., 1971: Quelques aspects du Paléolithique ancien et moyen dans le Nord de la France. *Bull. Soc. Préh. Nord, No sp.* 8, 98 pp., Amiens.
Tuffreau, A., 1974: Contribution à l'étude du Paléolithique ancien et moyen dans le Nord de la France et le bassin oriental de la Somme. Thesis 3e cycle, Paris, 2 vol., 324 stenciled pp.
Vanhoorne, R., 1962: Het interglaciale veen te Lo (België). *Natuurwet. Tijdschr.* 44, 58—64, Gent.
Zagwijn, W. H., 1974: The palaeogeographic evolution of the Netherlands during the Quaternary. *Geol. Mijnbouw* 53, 369—385.
Zeuner, F., 1959: The Pleistocene period. London, 447 pp.

IV-*h*

The western (United Kingdom) shore of the North Sea in Late Pleistocene and Holocene times

W. GRAHAM JARDINE

Jardine, W. G., 1979: The western (United Kingdom) shore of the North Sea in Late Pleistocene and Holocene times. In E. Oele, R. T. E. Schüttenhelm & A. J. Wiggers (editors), The Quaternary History of the North Sea, 159—174. *Acta Univ. Ups. Symp. Univ. Ups. Annum Quingentesimum Celebrantis: 2*, Uppsala. ISBN 91-554-0495-2.

When the sea stood at its highest levels in Eemian (Ipswichian) and Holocene times, the position of much of the western shoreline of the North Sea was approximately the same as at present. Exceptionally, the coast was located several kilometres landwards of its present position near the Humber estuary and The Wash (in Holocene and Eemian times), and adjacent to the Firths of Tay and Forth, and the Moray Firth (in Holocene times). During the Weichselian (Devensian) glaciation the "western" shoreline on the floor of the North Sea fluctuated between locations very remote from and locations rather closer to its present position. Late in Weichselian and early in Holocene times the coast lay inland from its present position in the valley of the River Forth and in the Firth of Tay area. Contemporaneously in the southern North Sea area the shoreline was located far to the east and northeast of its present position. The present configuration of the western shoreline of the North Sea had been attained by approximately 5500 B.P. About 4400 B.P. a short-lived marine transgression occurred in the Fenland area to the south of The Wash.

Dr. W. G. Jardine, Department of Geology, The University of Glasgow, Glasgow G12 8QQ, United Kingdom.

Introduction

The western (British) coast of the North Sea extends discontinuously for more than 1250 km from the northern extremity of the Shetland Islands at 60°50′ N latitude to near Dover at

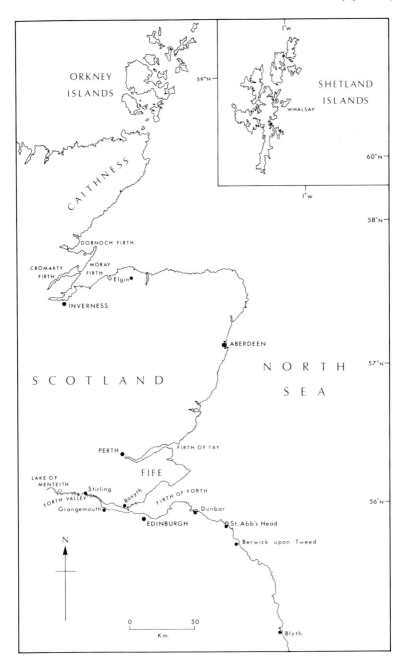

Fig. IV-48. The parts of Scotland adjacent to the North Sea, showing the main locations mentioned in the text.

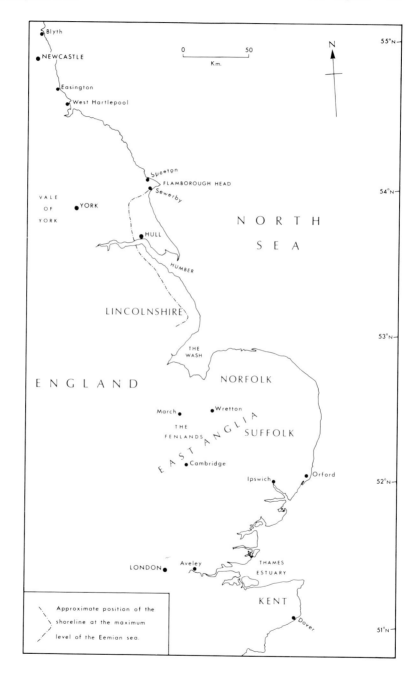

Fig. IV-49. The parts of England adjacent to the North Sea, showing the main locations mentioned in the text. The approximate position of the highest Eemian shoreline is shown in the area between Flamborough Head and The Wash.

51°07′ N latitude (Figs. IV-48 and IV-49). Along its length, the present shore varies in character from high, steep rocky cliffs to low, gently-inclined intertidal mudflats, and the rocks exposed cover almost the whole range of geological divisions from Precambrian to Holocene.

The detail in which the Quaternary history of the seaboard, especially coastal evolution since the Eemian (Ipswichian) interglacial interval, has been studied varies greatly from place to place. The parts of the coast that have been closely investigated are widely scattered and, although the Late Weichselian and Holocene history of such areas as southeast Scotland and the Thames estuary are comparatively well known, the Late Quaternary history of many parts of the coast is still obscure. Perhaps because of this, a synthesis of what is known of the coastal history of the western margin of the North Sea has not hitherto been presented. In this conspectus, in addition to conclusions based on information derived from investigation of the visible landmass, data recently obtained from the western part of the North Sea floor are incorporated because, together with other evidence of Late Quaternary movements of sea level, they are relevant in a study concerned with the changing position of the coastline since the Eemian interglacial.

All altitudes quoted are referred to British Ordnance Datum (Newlyn), which is approximately equivalent to present mean sea level within the North Sea area (Jardine 1975a, 1976).

General principles

The period of coastal evolution considered here was concurrent with a complete cycle of glaciation and deglaciation of the northern part of Britain, a cycle that was approximately contemporaneous with the last major glaciation and deglaciation of Fennoscandia and North America. Coastal development in the British sector of the North Sea area was influenced by two major effects of the cycle of glaciation and deglaciation:

1. On a global scale, lowering and raising of sea level within the range of approximately -150 m. O.D. to present mean sea level occurred. Global sea level at the times of the climatic optima of the Eemian and Holocene warm intervals was approximately the same as at present; at the maximum of the Weichselian (Devensian) glaciation sea level was low, perhaps around -130 m O.D. (Curray 1965:724).

2. On the land surface of northern Britain and adjacent parts of the continental shelf (including the North Sea area), ice loading and unloading resulted in regional terrestrial depression and rebound that, at its maximum, amounted certainly to several decametres and possibly to more than 100 metres.

As a result, in the glaciated area the *relative* level of the sea at the glacial maximum was very much higher than on the unglaciated or less-heavily glaciated areas to the south and north, and to the east on the floor of the North Sea. The two major effects, operating sometimes in unison, at other times in opposition, and complicated by a third effect — hydroisostatic loading and unloading — produced a series of shorelines some remnants of which now occur below O.D., whilst other remnants now are preserved above O.D.

Ideally the "shoreline" of the sea at any given time is the line of intersection of mean sea level at that time and the contemporaneous terrestrial surface. For practical purposes, however, especially along marine margins such as the British North Sea coast where anomalous tidal conditions may result (paradoxically) in the "ideal" shoreline being at a level lower than low water mark of ordinary spring tides at some locations (cf. Jardine 1975b: 190—191), the "shoreline" at any given time is the line joining points located at high water mark of ordinary spring tides on the contemporaneous coast. It is this latter shoreline whose position is painstakingly measured in many studies of shoreline changes, and it is this shoreline whose development is considered here. Movements of sea level affect the position of this shoreline just as they affect the position of the "ideal" shoreline. For this reason movements of sea level in the North Sea area since Eemian times are discussed below as an integral part of the consideration of the development of ancient shorelines.

On theoretical grounds it is to be expected that any given Weichselian (Devensian) shoreline although originally a line joining points within an altitudinal range of perhaps as much as 4.5 m, now may be markedly warped. Thus, its highest remnants may be preserved in or offshore from mid-Scotland (e.g. in the Firth of Tay or Firth of Forth areas) at altitudes several decametres above others of its remnants in or offshore from Yorkshire, East Anglia and the Thames Estuary, or in or offshore from Caithness, Orkney and Shetland. The remnants of some Weichselian shorelines now are wholly submerged, whilst those of others are partly submerged and partly above O. D. Commonly the oldest shorelines now have the steepest gradients (Sissons 1976:120).

The Eemian and Holocene shorelines also may have been warped since the time of their formation, because terrestrial rebound following the Saalian (Wolstonian) and Weichselian (Devensian) glaciations persisted respectively into the Eemian and Holocene time intervals. In general, warping of Holocene shorelines may be expected to be, and is, less than that of Weichselian shorelines, but the degree of warping of Eemian shorelines in relation to that of Weichselian and Holocene shorelines is rather uncertain. One reason for this is that since Eemian times the eastern part of East Anglia has been downwarped (tectonically rather than glacio-isostatically) as much as 9 m in relation to the western part of East Anglia (West 1972:95). Another is that the loading and unloading effects of the Wolstonian ice sheet in comparison with those of the Devensian ice sheet in (mainly) northern Britain are not accurately known.

It may be worth noting, however, that because the extent of the Wolstonian ice sheet was greater (? and the weight therefore also was greater) than that of the Devensian ice sheet, terrestrial glacio-isostatic rebound into the Ipswichian (Eemian) interglacial may have been greater than the corresponding rebound extending into the Holocene epoch (Jardine 1977:104). The maximum altitude of the main Holocene shoreline in Britain is approximately +15 m O.D. (in the upper Forth Valley in east central Scotland), and contemporaneously with formation of that shoreline relative sea level may have stood at −6 m O.D. or lower in the Humber area (Gaunt & Tooley 1974:fig. 6) and at −9 m O.D. or lower in the Shetland Islands (Sissons 1976:131). It is possible, therefore, that Eemian shorelines now occurring close to O.D. in the Humber area (see below) are the correlatives of coastal features (as yet unrecognised as Eemian) in east central Scotland, at altitudes higher than +15 m O.D. Examples may be the (presumed-interglacial) till-covered remnants of rock platforms at +18 to +25 O.D. that occur inland from the coast in south-east Scotland near Dunbar (Sissons 1976:119).

In the period extending from the Eemian interglacial through the Weichselian glaciation to the Holocene epoch, changes in the lateral position of the shoreline in relation to the position of the present coast were more marked in the southern part of the North Sea area than in the more northerly part adjacent to the ice-loaded land mass (cf. the coloured chart). This was because in the southern area, where ice loading and unloading and therefore isostatic depression and rebound were almost negligible, movements of sea level were imposed upon the comparatively-static surface of low relief that constituted the floor of the North Sea. In contrast (the rather vaguely-known time-lag effects attributed to isostasy being discounted), in the more northerly area, where ice loading and unloading and therefore isostatic depression and rebound were considerable, movements of sea level were accompanied by movements of the terrestrial mass in the same (downward or upward) sense as the surface of the sea was moving. As a result, the position of the shoreline in that area remained comparatively stable when sea-level movements occurred.

Eemian shorelines and movements of sea level

Traces of Eemian (Ipswichian) shoreline

Fig. IV-50. Eemian coastal forms, and sediments possibly deposited in the vicinity of the Eemian shoreline. Locations are shown in Fig. IV-49.

Location, and reference in literature	Nature of evidence	Approximate altitude	Ipswichian pollen zone (where applicable)	Inferred mean sea level
Easington Smith and Francis 1967: 208 and 242–243	calcreted gravel with marine shells on bevelled Magnesian Limestone	+27.5 m O.D.		
Speeton (lower shell bed) Catt and Penny 1966: 400–401 Gaunt et al. 1974: 21–22	shell bed	+2 m O.D.		
Sewerby Catt and Penny 1966: 385 & 390 Gaunt et al. 1974: 21–22	beach gravel buried cliff and shore platform	+2.5 m O.D.	III	+0.9 to +1.5 m O.D.
Vale of York Gaunt et al. 1974: 21–22 (a) Westfield Farm (b) Langham	silty clay gravels, sands and clays	+2.5 m O.D. −11.7 to −7.4 m O.D.	III/IV II (b)	rising from −12 m O.D. to O.D.
Lincolnshire Straw 1961: fig. 1	buried cliff			
March Baden-Powell 1934: 195–197 West 1972: 94–95	sands and gravels with marine organisms	+10 to +12 m O.D.		+7 m O.D.
Wretton Sparks and West 1970: 26–28	brackish-water deposits	−2 to +0.5 m O.D.	IIb	
Stutton (near Ipswich) Sparks and West 1963: 430–431 Sparks and West 1972: 281	brackish-water deposits	+1 m O.D.	III	−1 m O.D.
Aveley West 1969: 273 and fig. 1 Sparks and West 1972: 282	aggradation deposits	+7 to +15 m O.D.	IIb, III	

features and coastal deposits have been identified at a number of sites between the Thames estuary in the south of England and Easington in the north. The data are summarised in Fig. IV-50 and the approximate position of part of the shoreline at the maximum level of the Eemian sea is shown in Fig. IV-49. The following points should be noted:

1. In places the shoreline of the Eemian sea lay up to several kilometres inland from the present shoreline, especially between Flamborough Head and The Wash. South of The Wash, the position of the shoreline is known only in the broadest of terms and, as Sparks & West (1972:283) suggested, it may be misleading rather than helpful to express on a map inferences concerning that area that are based on rather scanty and uncertain evidence.

2. The shoreline is most readily traceable in the vicinity of the Humber estuary. To the south of the estuary the shoreline is located near the eastern extremity of exposed Chalk, whereas to the north of the estuary a buried (coastal) cliff, occasionally with associated (shore) platform, exists within the Chalk outcrop. The cliff and platform may have been formed initially long before Eemian times (in places, as at Sewerby, pre-Eemian till rests on the platform), but there is little doubt that these features formed the shoreline to the north of the Humber estuary when the Eemian sea stood at or near its maximum level.

3. The Eemian age of the relevant deposits at Speeton and Easington has been questioned. It is possible, therefore, that the Eemian shoreline did not intersect the present land surface north of Flamborough Head.

Suggestions concerning changes in the level of the surface of the North Sea during the Eemian interglacial have been made for East Anglia and the Humber area. In summary, it may be said that Eemian (Ipswichian) coastal deposits occur in southeast England in two main areas, the first being thought to have remained comparatively stable since Eemian times, the second to have experienced a net subsidence of several metres since Eemian times. The first area includes Selsey Bill on the English Channel coast and the March gravel deposits of the western part of East Anglia, together with terrace gravels of the River Cam near Cambridge and the deposits at Wretton (Fig. IV-50 and Fig. IV-49). The second area comprises the eastern part of East Anglia, including the site at Stutton, near Ipswich (Fig. IV-50).

Broadly, the data from western sites suggest that the level of the Eemian sea rose between zone Ip IIb and zone Ip III to a maximum during which, in zone Ip III, mean sea level stood about 7.5 m above O.D. (West 1972:94—95). In contrast, the data from eastern sites suggest that by zone Ip III mean sea level probably stood at about +1.2 m O.D. In addition to this evidence of downwarping of the eastern area since zone Ip III, there is some indication, in truncation of aggradation deposits assigned to zone Ip IIb, of sea-level oscillation in the middle of the Eemian (Sparks & West 1970:28, 1972:283) similar to that found in the Netherlands by van der Heide (1957).

In the vicinity of the Humber estuary there is evidence that about the time of the transition from the Saalian (Wolstonian) to the Eemian, i.e. about the beginning of zone Ip I, mean sea level stood at least 13 m lower than its present level (Gaunt et al. 1974:20). Thereafter, about the beginning of zone Ip IIb mean sea level began to rise from about −13 m O.D. in the course of zone Ip IIb to reach a maximum of about +1.5 m O.D. during zone Ip III. The evidence of sea-level position provided by sites in the vicinity of the Humber is similar to that existing in the eastern part of East Anglia. Apparently not previously indicated in the literature, it is suggested here that the Humber area, like eastern East Anglia, was subject to post-Eemian downwarping. If this was not the case, it is difficult to fit together the evidence from the Humber area, East Anglia and the English Channel coast.

It should be noted that the altitude of the (low-level) shell beds at Speeton provides no major problems of explanation, but that the altitude of the coastal deposits and marine bench at Easington (+27 m O.D.), if indicative of an Eemian sea level, is not readily explained despite the remarks made above concerning the possible post-formational warping of Eemian shorelines.

Weichselian shoreline development

Early and Middle Weichselian development

At the time of writing (mid-1976) there is little published information on the position of the western shoreline of the North Sea in Early and Middle Weichselian times, although intensive investigation of the continental shelf surrounding the British Isles has been in progress for a decade, and some of the relevant results were reported orally at a meeting in London in October 1975. Global sea level around 40000 B.P. is thought to have been low, perhaps standing at approximately −145 m O.D. (Curray 1965: 724). If glaciation of the British Isles in Early and Middle Weichselian times was minimal or non-existent (cf. Coope 1975:fig. 1), contemporaneously glacio-isostatic depression of the landmass must have been almost nil, and the east coast of both Scotland and England around 40000 B.P. must have been located many kilometres or tens of kilometres to the east of its present position at levels markedly lower than O.D.

The known sequence of Quaternary deposits on offshore parts of the British sector of the North Sea floor probably relates mainly to Late Weichselian marine events but, because there are unconformities in the sequence, it is possible that some of the lowermost sediments were deposited in Early or Middle Weichselian times. The Aberdeen Ground Beds are more than 200 m thick and their upper half consists of carbonaceous sandy clay with well-preserved marine bivalves (Fannin 1975a, 1975b). The sediments occur now at approximately −200 to −450 m O.D., depths that are consistent with their deposition several kilometres or tens of kilometres offshore should their accumulation have occurred when sea level stood about 145 m lower than at present.

Late Weichselian development

The maximum of the Late Weichselian (Late Devensian) glaciation in Britain is generally agreed to have occurred around 20000 to 18000 B.P. Contemporaneously global sea level may have stood at about −130 m O.D. (Curray 1965:724; Milliman & Emery 1968:1121). Because of the effects of ice loading, however, the *relative* level of the marine waters within parts of the British area of the North Sea may have been much higher than the "absolute" level (see above, and Jardine (1977) in relation to southwest Scotland and the Scottish Hebrides). Clearly, the relative level of the surface of the western part of the North Sea at the time of the Late Weichselian glacial maximum varied from place to place, but the figure of −60 to −70 m O.D. suggested by West (1975) may be a good estimate of the mean relative level (cf. −47 m O.D., the lowest Late Devensian base-level reached in the outer Thames area; Greensmith & Tucker 1971:315−316). If around −70 m O.D. indeed was the mean level of the Late Weichselian sea, contemporaneously the western coastline of the North Sea must have stood many kilometres east of the present shore although, for reasons discussed above, in the part affected by ice loading its distance from the present coast would be less than in Early and Middle Weichselian times when global sea level was low but ice-load depression of the British land surface and the adjacent sea floor was negligible.

Athough controlled mainly by movements of sea level, the position of the Late Weichselian shoreline on the floor of the North Sea also was dependent to some extent upon the position of the westwards-receding ice front. The exact position of the ice front on the sea floor at the maximum of the last glaciation is not known, but it is generally agreed that at the glacial maximum ice extended continuously between Britain and Norway (cf. Veenstra 1970:176). The ice cover probably was thin near the mid-line of the North Sea, however, and by 16000 to 17000 B.P. the British and Scandinavian ice masses may have become disunited (B. G. Andersen 1977, personal communication). The southern part of the North Sea, although not ice covered, may have been dry land for much or all of Late Weichselian time because the lowest parts of its (present) floor lie at elevations higher than −60 to −70 m O.D. Some evidence in support of this suggestion occurs in the vicinity of Sandettie Bank, a short distance to the north of the Straits of Dover, where a Weichselian sand deposit, overlain by a Late Weichselian clay, contains a freshwater molluscan fauna (Oele & Kirby 1976:15).

Farther north, in the approaches to the Firth of Forth, there is sedimentary evidence of the proximity of Weichselian ice on the North Sea floor. The relevant sediments are the Wee Bankie Beds, composed of till preserved in hollows in rockhead, and their correlatives to the east, the Marr Bank Beds comprising shelly, gravelly sands and clays — reworked marine beds accumulated in front of the ice sheet (Fannin 1975a, 1975b:6−7). The Wee Bankie Beds are overlain by the St. Abb's Beds, clays with an arctic-type marine fauna, the sedimentary sequence thus indicating penetration of cold marine waters into an area recently vacated by the receding ice front.

Changes in the position of the coast that followed the maximum of the Late Weichselian glaciation have been studied in greater detail and are much better known than those that preceded or were contemporaneous with the glacial maximum. The St. Abb's Beds of submarine areas are thought to be equivalent to the well-known "Errol Beds" that are exposed up to altitudes of +30 m O.D. at a number of sites along the east coast of Scotland in addition to occurring at numerous submarine inshore sites. The "Errol Beds" are characterised by ice-rafted stones and a fauna of low diversity (comparable with that of present-day Spitzbergen and West Greenland; Peacock 1975:45). The age of these sediments is not certain, but they are succeeded in the Forth and Tay estuaries and in the inner Cromarty Firth by representatives of the "Clyde Beds", deposition of which is known to have commenced prior to 12600 B.P. Like the "Errol Beds", the clays, silts and sands of the "Clyde Beds" (so-named because they were first described at sites on the Firth of Clyde in western Scotland) contain ice-rafted debris, but they also characteristically contain a boreal fauna, the present-day temperature equivalent occurring broadly in southern Iceland, The Faroes and Norway south of the Lofoten Islands (Peacock 1975:46). It has been suggested that the change from deposition of the "Errol Beds" to accumulation of the "Clyde Beds" may mark the rapid temperature amelioration that, on the basis of evidence from terrestrial stratigraphy (Coope 1975:164), is thought to have occurred around 13500 to 13000 B.P. (Peacock 1974:66). Because the "Errol Beds" and to a lesser extent the "Clyde Beds" occur within and to the west of the confines of the present estuaries of the Rivers Forth and Tay (Peacock 1975:fig. 1), it may be concluded that by 13000 B.P. not only had the icefront retreated westwards to a line well to the west of Stirling and Perth, but the contemporaneous eastern shoreline of Scotland also was established in the vicinity of these cities, being linked (as at present) between the valleys of the Rivers Forth and Tay around the peninsula of Fife (Fig. IV-48).

Remnants of coastal deposits that accumulated towards the end of the Weichselian, together with traces of former shorelines of similar age, occur at various localities along the eastern seaboard of Britain, especially in southern Scotland. The positions of the shorelines, however, have not always been published because interest has been concentrated on the altitude and the degree of tilting of the shorelines rather than on showing their exact locations. In northern England, near Hartlepool, a planation surface occurring at approximately +25 m O.D., and covered by sediments that include marine shells, is attributed to a Late Weichselian sea, although the only positive indication of its age is that is pre-dates antlers (in a peat deposit) dated 8100 ± 180 to 8700 ± 180 B.P. (Smith & Francis 1967:246−247; cf. King 1976:152).

Along the southern shore of the Moray Firth near Elgin (Fig. IV-48), remnants of Late Weichselian shorelines and associated coastal sediments are present at altitudes ranging between about +15 and +26 m O.D. (Peacock et al. 1968:107−109). The evidence of the sediments and morphological features suggests that ground ice occurred in this area at the time when the highest of the shorelines were formed.

In southeast Scotland, broadly the area from the Firth of Tay to Berwick-upon-Tweed, including the Firth of Forth and its former westward extensions, Late Weichselian shoreline altitudes and tilting have been studied in great detail (cf. Sissons 1976:120–136). The detailed stratigraphy of the raised beach and estuarine deposits is given in publications of the Institute of Geological Sciences (e.g. Francis et al. 1970:274–282).

In summary, the results of these studies as indicative of the former position and nature of Late Weichselian shorelines are:

1. The six oldest identifiable Late Weichselian shorelines occur in eastern Fife at altitudes ranging from about +15 to +30 m O.D. The older (more easterly) shorelines are more steeply tilted than the younger, and the more westerly extension of the younger shorelines indicates that westward penetration of the sea kept pace with ice-front recession. The age of these shorelines is uncertain.

2. Westwards from the Fife shorelines are to be found traces of later shore features the most conspicuous of which is that termed (by Sissons) the Main Perth Shoreline. This feature was formed when the receding ice front stood slightly east of Stirling in the Forth Valley and near Perth in the Tay Valley. Near Stirling the shoreline is at about +37 m O.D. but declines to the east (along a line oriented E 17° S) at an average gradient of 0.43 m/km. Formerly the shoreline was thought to have been formed at about 13000 B.P. but on the basis of recent indications of rapid temperature rises around that time it must now be concluded that formation of the shoreline at the ice front was somewhat earlier than 13000 B.P. (cf. Armstrong et al. 1975:40).

After formation of the features that identify the "Main Perth" and later, minor, shorelines, a period of land uplift resulted in an episode of widespread marine erosion. The earlier Late Weichselian inshore marine deposits, e.g. the "Errol Beds", were planed by the sea to produce a gently-sloping layer of gravel. The layer is best developed in the vicinity of Grangemouth where it occurs at about O.D., but the shoreline associated with it can be traced (mainly in boreholes) from about +6 m O.D. at locations to the west of Grangemouth eastwards in the Forth Valley to −4 m O.D. near Rosyth, and to −9 or −10 m O.D. slightly west of Edinburgh. Much farther east, 9 km north of Berwick-upon-Tweed, a submarine platform, 600 m wide, rises gently landwards from about −27 m O.D. to a shoreline at −18 m O.D. The shoreline at −18 m O.D. is equated by Sissons (1974a:44) with the Buried Gravel Shoreline of the Grangemouth area. The age of this shoreline was for long uncertain. Recently it was suggested (Sissons 1974a:46) that it was formed in severe periglacial conditions immediately prior to 10500 to 10300 B.P., i.e. about the time of the cold episode during which the Loch Lomond ice readvance occurred in western Scotland. The coastal feature has been termed the Main Lateglacial Shoreline, and has been equated with the "Main line" (P_{12} shoreline) of northern Norway (Sissons 1974a:47).

Holocene shoreline evolution

About the beginning of the Holocene epoch (10000 B.P.), global sea level was still several tens of metres below its present level (at −45 m O.D., West 1972:fig. 7; cf. −45 m O.D. at 9600 B.P., Greensmith & Tucker 1973:197), but in southeast Scotland and adjacent areas the *relative* level of the sea was higher than Ordnance Datum. The Holocene coastal history of the western part of the North Sea is largely concerned with the effects produced in two contrasting areas by the continued rise of ocean level until approximately 5000 B.P.: the southern part where glacial rebound was negligible; the part adjacent to southeast Scotland where glacial rebound was still in progress until after 5000 B.P. The effects in the northernmost part of the Scottish mainland and in Orkney and Shetland may have been somewhat similar to those in the southern part of the North Sea, but known details are few. The history of Holocene shoreline evolution may best be examined first in the area of the Forth Valley because there, shortly before the beginning of the Holocene epoch, the shoreline was located many kilometres west of its present limit and therefore its position thereafter is recorded in readily-accessible sediments.

At the maximum of the Loch Lomond readvance (approximately 10500 B.P.) a lobe of glacier ice extended into the western part of the Forth Valley to the vicinity of the Lake of Menteith. Contemporaneously or slightly later the Late Weichselian sea occupied the Forth Valley east of the ice front (Sissons 1966: 25—27, 1974b:329). Melting of the ice lobe in the Forth Valley allowed brief penetration of the sea to a position west of the end moraine, followed by a series of pulsatory marine regressions and minor transgressions between approximately 9600 and 8800 B.P. (Sissons 1966:27, 1974b:329). Shortly before 8500 B.P. the rate of land uplift in southeast Scotland was greater than the rate of rise of global sea level and the *relative* level of the sea in the Forth area dropped to an altitude of around +6.20 m O.D. (Sissons & Brooks 1971, fig. 2).

Meantime, in the southern part of the North Sea the marine surface had risen and by around 8700 B.P. its level (if the same as global sea level) may have been approximately −36 m O.D. (Lacaille 1954:fig. 20, cf. the coloured chart; −34m O.D. in the outer Thames estuary, Greensmith & Tucker 1973:197) or rather higher (about −24 m O.D. at 8000 to 9000 B.P.) if the altitude is extrapolated from the well-known sea-level curves of Shepard, Jelgersma and Mörner (cf. curve in Kidson & Heyworth 1973:fig. 6). Certainly, at 8425 ± 170 B.P. terrestrial peat was still forming on the Leman and Ower bank, off the coast of Norfolk, on a surface that is now at about −37 m O.D. (Godwin 1960:316). Relative sea level in that area must have been at least one or two metres lower than −37 m O.D. then; global sea level probably was even lower.

The rate of rise of sea level shortly after 8400 B.P. was more rapid than the rate of terrestrial rebound in southeast Scotland and adjacent areas. As a result, not only did the rising sea begin to flood the southern part of the North Sea area about this time in the course of the main Holocene marine transgression, but it also invaded once more parts of the Firth of Forth for many kilometres beyond the present coast, and the borders of the Firth of Tay to a lesser extent. Simultaneously much of the remainder of the coast of eastern Scotland was inundated to a few tens of metres beyond its present limit, and in the Moray Firth and adjacent inlets the shoreline at the maximum of the transgression was located several kilometres inland from its present position.

The progress of the marine transgression along various parts of the western edge of the North Sea may be judged by the data given in Fig. IV-51, keeping in mind that almost all of the altitudes quoted are for the *relative* level of the sea at the given time. Although not specified in each publication cited, the level quoted in most cases must be approximately that of local high water mark (which in some places may be as much as 4 or 5 m higher than the corresponding mean sea level). From the data given in Fig. IV-51 it should be noted also that the *relative* level of the sea at about 8500 to 8700 B.P. was progressively higher the farther an area was located northwards from the Thames estuary (chosen here arbitrarily as the standard). This suggests that a map showing the position of the coastline of the North Sea at about 8700 B.P. (cf. the coloured chart) should no be based solely on the position of the (present) −36 m contour line. Similar arguments apply (but in progressively lesser degree) to the construction of maps of the shoreline of the North Sea at about 7500, 5000 and 2000 B.P.

Some of the areas listed in Fig. IV-51 require special comment. In the Forth Valley, the main Holocene marine transgression began later than 8500 B.P. and by about two thousand years later marine waters had penetrated into the area west of the relict end-moraine in the vicinity of the Lake of Menteith. According to Sissons & Brooks (1971:126) the culmination of the marine transgression occurred slightly before 6490 ± 125 B.P. and the corresponding shoreline, traces of which occur both to the west and to the east of the Menteith moraine, dates from about the same time.

Fig. IV-51. Position of sea level during the Holocene epoch at locations between the Thames estuary, in the south, and the Shetland Islands, in the north. In most cases the altitudes refer to the level of high water mark at the given location. The positions of the locations are indicated in Figs. IV-48, 49.

Sea level	Approximate date	Reference in literature
Thames estuary		
lower than − 4 m O.D.	4450 B.P.	Akeroyd 1972: 159
− 4.9 m O.D.	3900 B.P.	Greensmith and Tucker 1971: 317
− 8.3 to − 5.5 m O.D.	between 3936 ± 110 and 3580 ± 75 B.P.	Greensmith and Tucker 1971: 317, 1973: 199
− 10.7 m O.D.	7000 B.P.	Greensmith and Tucker 1971: 317
− 12 m O.D.	5650 ± 240 B.P.	Greensmith and Tucker 1973: 199
− 17 m O.D.	8000 B.P.	d'Olier 1972: 128
− 18.3 m O.D.	7516 ± 250 B.P.	Greensmith and Tucker 1971: 316
− 22 m O.D.	8300 B.P.	d'Olier 1972: 127
− 28 m O.D.	8600 B.P.	d'Olier 1972: 127
− 34 m O.D.	8900 B.P.	Greensmith and Tucker 1973: 197
− 34 m O.D.	9000 B.P.	d'Olier 1972: 126
− 40 m O.D.	9300 B.P.	d'Olier 1972: 126
− 45 m O.D.	9600 B.P.	d'Olier 1972: 125
Orford, Suffolk		
− 3.4 m O.D.	3460 ± 100 B.P.	Carr and Baker 1968: 114
− 12.7 to − 13.7 m O.D. (mean sea level at − 14.2 to − 15.2 m O.D.)	8640 ± 145 to 8460 ± 145 B.P.	Carr and Baker 1968: 117
English Fenlands		
lower than − 4 m O.D.	4450 B.P.	Akeroyd 1972: 159
Humber area		
− 2.35 m O.D.	6170 ± 180 B.P.	Gaunt and Tooley 1974: 36 and table 1
− 4.00 m O.D.	5240 ± 100 B.P.	Gaunt and Tooley 1974: 36 and table 1
− 9.14 m O.D.	6681 ± 130 B.P.	Gaunt and Tooley 1974: table 1
− 9.75 m O.D.	later than 6890 ± 100	Gaunt and Tooley 1974: 29
mean sea level at − 18 m O.D.	8000 B.P.	Gaunt and Tooley 1974: 39
West Hartlepool		
between −1.95 and − 0.34 m O.D.	between 5285 ± 120 and 5240 ± 70 B.P.	Gaunt and Tooley 1974: 36
Aberlady Bay (between Edinburgh and Dunbar)		
mean sea level at + 8.0 to + 9.0 m O.D.	5535 ± 160 B.P.	S.M. Smith 1972: 42
Forth Valley (west of Stirling)		
maximum of transgression	5492 ± 130 B.P.	Godwin and Willis 1962: 67
+ 14.4 m O.D.	6490 ± 125 B.P.	Sissons and Brooks 1971: 125
+ 13.3 m O.D.	7480 ± 125 B.P.	Sissons and Brooks 1971: 125
+ 9.5 m O.D.	8010 ± 130 B.P.	Sissons and Brooks 1971: 125
+ 7.70 m O.D.	8270 ± 160 B.P.	Sissons and Brooks 1971: 125

Fig. IV-51 cont.

Sea level	Approximate date	Reference in literature
Eastern Fife		
+7.22 m O.D.	5830 ± 110 B.P.	Chisholm 1971:104
+5.87 m O.D.	7605 ± 130 B.P.	Chisholm 1971:103
lower than +4.47 m O.D.	9945 ± 160 B.P.	Chisholm 1971:91 and 103
Firth of Tay		
local commencement of main Holocene trangression	8150 ± 50 to 7778 ± 55 B.P.	Harkness and Wilson 1974: 240–241
Lossiemouth (Moray Firth coast north of Elgin)		
marine maximum at +5.49 m O.D.	later than 7450 B.P.	Peacock et al. 1968:114
Cromarty Firth		
a few metres below O.D.	8500 B.P.	Peacock 1974:64 and 66
Shetland Islands		
lower than 9 m below present high water mark	6970 ± 100 to 6670 ± 100 B.P.	Hoppe et al. 1965:201

Formerly the shoreline of the main Holocene marine transgression in the Forth Valley was considered to date from about 5500 B.P. on the evidence of a date of 5492 ± 130 B.P. (Godwin & Willis 1962:67) for a sample from 12 cm above the base of a peat layer that covers the estuarine sediments at a location east of the Menteith moraine. The evidence cited by Sissons & Brooks is no more convincing than that of the earlier data. Access by the sea to the area to the west of the Menteith moraine was through a narrow gap. There is no reported evidence of the presence of low bars across or partially across the gap, but the situation is similar to that in the former Lochar Gulf near Dumfries in southwest Scotland where it has been shown that exclusion of the sea from the gulf was caused by local building of gravel and sand bars long before the termination of the main Holocene marine transgression in the adjacent area of the Solway Firth (Jardine 1975b:183–184 and fig. 6). Clearly it is possible that the sea was excluded from the area to the west of the Menteith moraine around 6490 ± 125 B.P. because of local modifications in coastal morphology, and that marine waters were still present a short distance to the east of the Menteith moraine for many hundreds of years after that event.

From the above it may be seen that the age of the main Holocene shoreline in the Forth Valley is still an open question. Broadly, however, the main Holocene transgression had ceased in southeast Scotland by around 5500 B.P. and thereafter the rate of land uplift exceeded that of sea-level rise so that marine regression rather than transgression occurred.

This was a situation rather different from that in some areas on the borders of the southern part of the North Sea. In the English Fenlands, for example, fluctuations in the position of the shoreline between about 4700 and 1800 B.P. are recorded. About 4450 B.P. relative sea level in that area may have stood lower than −4.0 m O.D. (Akeroyd 1972:159), possibly because of regional downwarping in East Anglia rather than because of global sea level being lower than at present. It was approximately at that time (4680 to 4185 B.P.; Godwin 1975:24; Willis 1961:372) that the sea penetrated extensively into the area to the south of The Wash where previously peat had been forming. As a result, a thin wedge of marine sediment, the Fen Clay, accumulated until approximately 4150 to 3950 B.P. (Willis 1962:373) when again the shoreline of the sea shifted to near its present position. A further, brief, marine transgression commenced in the Fenlands around 2035 to 1840 B.P. (Willis 1961:374). The transgression, sometimes termed the Romano-British marine transgression, is identifiable also, but to a lesser degree, in the Humber area of the North Sea coast (A. G. Smith 1958:44—46 and fig. 13). It is interesting to note that in Romano-British times there probably was a pause in regression of the sea in southern Scotland (Jardine 1975b:187, cf. Sissons 1976:126) whereas contemporaneously a marine regression may have been in progress in the area of the Thames estuary (Greensmith & Tucker 1973:199—200).

The situation in the Thames estuary and other parts of the southern North Sea area since the end of the Weichselian glaciation perhaps was somewhat similar to that in the northernmost part of the North Sea in the vicinity of the Orkney and Shetland Islands although the causes may not have been precisely the same. In the latter area, during periods of glaciation there may have been peripheral terrestrial uplift in association with glacial loading of northern Britain and Scandinavia (cf. Flinn 1964:339). Correspondingly, when terrestrial uplift in much of northern Britain occurred simultaeously with global rise of sea level in Late Weichselian and Holocene times, terrestrial depression of the Orkney and Shetland area may have accompanied global sea-level rise. As a result, marine transgression rather than regression has been dominant in that area since the melting of the Weichselian ice sheets began to affect sea level and terrestrial rebound. As stated above, detailed information is minimal, but the location of submerged peat, dated 6970 ± 100 B.P. and 6670 ± 100 B.P. (Hoppe et al. 1965:201), about 9 m below present high water mark of ordinary spring tides on the coast of Whalsay in the Shetland Islands indicates that marine transgression has been in progress since about 6600 B.P. This is in marked contrast with marine regression associated with uplift of the same order of magnitude in central and southern Scotland since approximately the same time.

Conclusion

During the Eemian (Ipswichian) and Holocene time intervals the position of the western shoreline of the North Sea about the times of the highest marine levels did not differ greatly from its present position. Exceptions occurred in the vicinity of the Humber estuary and The Wash (in Eemian times), and on the borders of the Moray Firth and adjacent areas, the Firth of Tay, Firth of Forth and The Wash (in Holocene times). The position of the coast varied greatly in the course of the Weichselian (Devensian) time interval. At the maximum of the last glaciation the shoreline lay on the floor of the North Sea area far to the east or northeast of its present position, and earlier in the Weichselian (about 40000 B.P.) the shoreline may have been located still farther from its present position when global sea level was even lower than during the maximum of the British and Scandinavian glaciations around 18000 B.P.

Towards the end of the Weichselian interval and in the early part of the Holocene epoch the shoreline lay west of its present position in the valley of the River Forth and on the borders of the Firth of Tay in southeast Scotland. In contrast, contemporaneously in the southern North Sea area the shoreline lay far to the east and northeast of its present position. In the extreme north, around the Orkney and Shetland Islands, the Late Weichselian and Early Holocene shorelines were located farther offshore than at present. In areas not greatly affected by terrestrial uplift, the present position of the western shoreline of the North Sea was attained at approximately 5500 B.P., but even since then there have been some fluctuations in the position of the shoreline, especially in the Fenland area south of the Wash.

Acknowledgements. — Professor T. Neville George read the text and made helpful suggestions for its improvement. Figs. IV-48 and IV-49 were drawn by Miss Sylvia Leek.

REFERENCES

Akeroyd, A. V., 1972: Archaeological and historical evidence for subsidence in southern Britain. *Philos. Trans. R. Soc. Lond. A* 272 151—169.

Armstrong, M., Paterson, I. B. & Browne, M. E. A., 1975: Late-glacial ice limits and raised shorelines in east central Scotland. In A. M. D. Gemmell (editor), *Quaternary studies in North East Scotland*, 39–44, Department of Geography, University of Aberdeen.

Baden-Powell, D. F. W., 1934: On the marine gravels at March, Cambridgeshire. *Geol. Mag. 71*, 193–219.

Carr, A. P. & Baker, R. E., 1968: Orford, Suffolk: Evidence for the evolution of the area during the Quaternary. *Trans. Inst. Br. Geogr. 45*, 107–123.

Catt, J. A. & Penny, L. F., 1966: The Pleistocene deposits of Holderness, East Yorkshire. *Proc. Yorks. geol. Soc. 35*, 375–420.

Chisholm, J. I., 1971: The stratigraphy of the post-glacial marine transgression in N.E. Fife. *Bull. geol. Surv. G. B. 37*, 91–107.

Coope, G. R., 1975: Climatic fluctuations in northwest Europe since the Last Interglacial, indicated by fossil assemblages of Coleoptera. In A. E. Wright and F. Moseley (editors), *Ice Ages: Ancient and Modern*, 153–168, *Geol. J. Spec. Issue 6*.

Curray, J. R. 1965: Late Quaternary history, continental shelves of the United States. In H. E. Wright and D. G. Frey (editors), *The Quaternary of the United States*, 723–735, Princeton University Press.

d'Olier, B., 1972: Subsidence and sea-level rise in the Thames Estuary. *Philos. Trans. R. Soc. Lond. A 272*, 121–130.

Fannin, N. G. T., 1975a: A regional assessment of the Quaternary succession in the central North Sea. Unpublished contribution, *Geol. Soc. Lond.* 22 October 1975.

Fannin, N. G. T., 1975b: A regional assessment of the Quaternary succession in the north central North Sea. Unpublished contribution, *Quaternary Newsl.*, November 1975, 5–7.

Flinn, D., 1964: Coastal and submarine features around the Shetland Islands. *Proc. Geol. Assoc. 75*, 321–339.

Francis, E. H., Forsyth, I. H., Read, W. A. & Armstrong, M., 1970: The geology of the Stirling district. *Mem. geol. Surv. U. K.*

Gaunt, G. D., Bartley, D. D. & Harland, R., 1974: Two interglacial deposits proved in boreholes in the southern part of the Vale of York and their bearing on contemporaneous sea levels. *Bull. geol. Surv. G. B. 48*, 1–23.

Gaunt, G. D. & Tooley, M. J., 1974: Evidence for Flandrian sea-level changes in the Humber estuary and adjacent areas. *Bull. geol. Surv. G. B. 48*, 25–41.

Godwin, H., 1960: Radiocarbon dating and Quaternary history in Britain. *Proc. R. Soc. Lond. B 153*, 287–320.

Godwin, H., 1975: *The history of the British flora* (2nd edition). Cambridge.

Godwin, H. & Willis, E. H., 1962: Cambridge University natural radiocarbon measurements V. *Radiocarbon 4*, 57–70.

Greensmith, J. T. & Tucker, E. V., 1971: The effects of late Pleistocene and Holocene sea-level changes in the vicinity of the River Crouch, East Essex. *Proc. Geol. Assoc. 82*, 301–322.

Greensmith, J. T. & Tucker, E. V., 1973: Holocene transgressions and regressions on the Essex coast outer Thames estuary. *Geol. Mijnbouw 52*, 193–202.

Harkness, D. D. & Wilson, H. W., 1974: Scottish Universities Research and Reactor Centre radiocarbon measurements II. *Radiocarbon 16*, 238–251.

Hoppe, G., Fries, M. & Quennerstedt, N., 1965: Submarine peat in the Shetland Islands. *Geogr. Annlr. 47A*, 195–203.

Jardine, W. G., 1975a: The determination of former sea levels in areas of large tidal range. In R.P. Suggate and M. M. Cresswell (editors), *Quaternary Studies. Selected papers from 9th INQUA Congress, Christchurch, N.Z.*, 163–168. Wellington.

Jardine, W. G., 1975b: Chronology of Holocene marine transgression and regression in southwestern Scotland. *Boreas 4*, 173–196.

Jardine, W. G., 1976: Some problems in plotting the mean surface level of the North Sea and the Irish Sea during the last 15000 years. *Geol. Fören. Stockh. Förh. 98*, 78–82.

Jardine, W. G., 1977: The Quaternary marine record in southwest Scotland and the Scottish Hebrides. In C. Kidson and M. J. Tooley (editors), *The Quaternary history of the Irish Sea*, 99–118, *Geol. J. Spec. Issue 7*.

Kidson, C. & Heyworth, A., 1973: The Flandrian sea-level rise in the Bristol Channel. *Proc. Ussher Soc. 2*, 565–584.

King, C. A. M., 1976: *The geomorphology of the British Islands: Northern England*. London.

Lacaille, A. D., 1954: *The Stone Age in Scotland*. London.

Milliman, J. D. & Emery, K. O., 1968: Sea levels during the past 35000 years. *Science 162*, 1121–1123.

Oele, E. & Kirby, R., 1976: Effects of the Pleistocene glaciations in the Southern Bight of the North Sea. Unpublished contribution, *Quaternary Newsl.*, February 1976, 14–15.

Peacock, J. D., 1974: Borehole evidence for late- and post-glacial events in the Cromarty Firth, Scotland. *Bull. geol. Surv. G. B. 48*, 55–67.

Peacock, J. D., 1975: Scottish late- and post-glacial marine deposits. In A. M. D. Gemmell (editor), *Quaternary studies in North East Scotland*, 45–48, Department of Geography, University of Aberdeen.

Peacock, J. D., Berridge, N. G., Harris, A. L. & May, F., 1968: The geology of the Elgin district. *Mem. geol. Surv. U.K.*

Sissons, J. B., 1966: Relative sea-level changes between 10300 and 8300 B.P. in part of the

Carse of Stirling. *Trans. Inst. Br. Geogr. 39,* 19—29.
Sissons, J. B., 1974a: Late-glacial marine erosion in Scotland. *Boreas 3,* 41—48.
Sissons, J. B., 1974b: The Quaternary in Scotland: a review. *Scott. J. Geol. 10,* 311—337.
Sissons, J. B., 1976: *The geomorphology of the British Islands: Scotland.* London.
Sissons, J. B. & Brooks, C. L., 1971: Dating of early postglacial land and sea-level changes in the western Forth Valley. *Nat. Phys. Sci. 234,* 124—127.
Smith, A. G., 1958: Post-glacial deposits in South Yorkshire and North Lincolnshire. *New. Phytol. 57,* 19—49.
Smith, S. M., 1972: Palaeoecology of post-glacial beaches in East Lothian. *Scott. J. Geol. 8,* 31—49.
Smith, D. B. & Francis, E. A., 1967: Geology of the country between Durham and West Hartlepool. *Mem. geol. Surv. U.K.*
Sparks, B. W. & West, R. G., 1963: The interglacial deposits at Stutton, Suffolk. *Proc. Geol. Assoc. 74,* 419—432.
Sparks, B. W. & West, R. G., 1970: Late Pleistocene deposits at Wretton, Norfolk. *Philos. Trans. R. Soc. Lond. B 258,* 1—30.
Sparks, B. W. & West, R. G., 1972: *The Ice Age in Britain.* London.
Straw, A., 1961: Drifts, meltwater channels and ice-margins in the Lincolnshire Wolds. *Trans. Inst. Br. Geogr. 29,* 115—128.
van der Heide, S., 1957: Correlations of marine horizons in the Middle and Upper Pleistocene of the Netherlands. *Geol. Mijnbouw 19,* 272—276.
Veenstra, H. J., 1970: Quaternary North Sea coasts. *Quaternaria 12,* 169—184.
West, R. G., 1969: Pollen analyses from interglacial deposits at Aveley and Grays, Essex. *Proc. Geol. Assoc. 80,* 271—282.
West, R. G., 1972: Relative land-sea-level changes in southeastern England during the Pleistocene. *Philos. Trans. R. Soc. Lond. A 272,* 87—98.
West, R. G., 1975: Evidence for transgression and regression around the British Isles in the Pleistocene. Unpublished contribution, *Geol. Soc. Lond.,* 22 October 1975.
Willis, E. H., 1961: Marine transgression sequences in the English Fenlands. *Ann. N. Y. Acad. Sci. 951,* 368—376.

Late Quaternary sedimentation in the North Sea

J.H. FRED JANSEN, TJEERD C.E. van WEERING & DOEKE EISMA

Jansen, J. H. F., van Weering, Tj. C. E. & Eisma, D., 1979: Late Quaternary sedimentation in the North Sea. In E. Oele, R. T. E. Schüttenhelm & A. J. Wiggers (editors), The Quaternary History of the North Sea, 175–187. *Acta Univ. Ups. Symp. Univ. Ups. Annum Quingentesimum Celebrantis: 2*, Uppsala. ISBN 91-554-0495-2.

In the North Sea sedimentation in the Quaternary was greatly influenced by the alternation of glacial and interglacial stages. Late Pleistocene glacial sediments are present in a large part of the northern North Sea. In the northwestern part a succession of glacial and glaciomarine deposits can be recognised in the Swatchway Beds, the Hills Deposits and the Fladen Deposits. These are locally covered by the marine Witch Deposits of which the sedimentation ended between 9000 B.P. and 8500 B.P., when the connection with the open sea was still incomplete. Morainic deposits, present on the floor and along the margins of the Norwegian Channel and the Skagerrak, are mostly covered by locally very thick, well-stratified glaciomarine sediments, gradually passing into a fine-grained, homogeneous toplayer of Holocene age.
   A belt of tunnel valleys is associated with the Fladen Deposits, marking the last maximum extension of the Weichselian ice front in the northern part, whereas further south a system of buried and infilled channels indicates that the British and Norwegian glaciers did not meet there.
   In the southern North Sea periglacial circumstances led to the formation of fluvial and fluvio-glacial deposits covered with aeolian sands. During the Holocene transgression peat was formed, followed by the development of tidal flat deposits containing a mollusc fauna which has been dated by radiocarbon measurements. Subsequently the so called Young Sea-sands were deposited, probably in late or post-Medieval time, by repeated reworking and by removal of the fine fraction by wave action.

Drs. J. H. F. Jansen, Drs. Tj. C. E. van Weering and Dr. D. Eisma, Nederlands Instituut voor Onderzoek der Zee, Postbus 59, Den Burg, Texel, the Netherlands.

Introduction

The development of the present North Sea started as a secondary rift system simultaneously with the Mesozoic rifting in the North Atlantic (Ziegler 1975, Kent 1975, Ziegler & Louwerens 1977). When in the Early Tertiary, Europe had become separated from North America and Greenland, secondary subsidence caused the development of a sedimentary basin, with its deepest parts along the former rift zone. The floor of this basin, which had already about the same extension as the present North Sea, is presently found at a maximum depth of 3500 m. The Tertiary infill mainly consists of clastic shelf sediments with intercalations of tuffs; the Tertiary strata wedge out towards the East and West; locally they are absent (Ronnevik et al. 1975). In the centre of the basin the thickness of the Quaternary glacial and interglacial deposits exceeds 1000 m (Caston 1977 b), but decreases towards the periphery.

The waterdepth in the North Sea increases gradually from less than 50 m in the southern North Sea to about 200 m at the shelf edge in the North; in the Northeast the Norwegian Channel is generally more than 300 m deep and in the Skagerrak a depth of 700 m is reached.

The present sea-floor consists almost entirely of sandy deposits, locally with admixtures of gravel and, chiefly in the deeper parts, fine-grained material (Fig. IV-56). The distribution of sediments is patchy. Hardrock locally crops out off England and Scotland, the Orkneys, the Shetlands, Norway, at Helgoland and in the Straits of Dover-Calais.

Quaternary sedimentation

During the Quaternary, sedimentation was controlled by alternating glacial and interglacial stages with low and high sea levels respectively. Ice sheets developed several times during the Pleistocene; the mountain areas of Scandinavia and Britain were the glacial growth centres that mainly influenced conditions in the North Sea. The extension of the ice sheets on land is rather well known, but contrasting views have been put forward, e.g. on ice-free areas along the Atlantic coast of Norway during several glaciations (Dahl 1972, Lindroth 1972). The extension of ice sheets in the North Sea, however, is still problematical. Reviews of paleogeographic reconstructions were given by Veenstra (1965, 1970) and Maisey (1972). As during the Elsterian the inland ice reached the Netherlands (Zagwijn 1974) and possibly also East Anglia (Zagwijn 1977, this vol.) it may be assumed that glaciers also covered part of the North Sea during that time. From the presence of boulders and boulder clay on the present sea-floor it can be concluded that Saalian and Weichselian glaciers covered large parts of the present North Sea (Pratje 1951, Veenstra 1970, Oele 1971a, 1971b). Large amounts of fluvial sands were supplied to the North Sea area by the Rhine, the Thames, the Meuse and the North German rivers during the glacial stages.

During interglacial stages, the extension of the North Sea was about the same as the present one, marine deposits of Tiglian age being known from the Netherlands and East Anglia, and of Cromerian age from the Netherlands (Doppert et al. 1975, Zagwijn 1977, this vol.). Marine deposits of Holsteinian and Eemian age have been found on the British, Dutch, North German, Danish and Norwegian mainlands as well as in the Dutch sector of the North Sea (Oele 1971a, 1971b); marine deposits only of Eemian age have been found also in Belgium and in the German sector.

Late Pleistocene glacial sediments

Northwestern North Sea. — In the northern North Sea a Weichselian age has been assigned to a large part of the Quaternary infill at the Witch Ground (Fig. IV-52). Here a 200 m thick succession of glaciomarine sands and clays (the Aberdeen Ground Beds) was dated as Weichselian on the basis of radiocarbon datings of 33000 and 48000 B.P. (Fannin 1976, Holmes, pers. comm.). This means, however, that during the first 2.4 million years of the Pleistocene roughly the same quantity of sediment was formed as during the last 75000 years, or that enormous quantities of sediment must

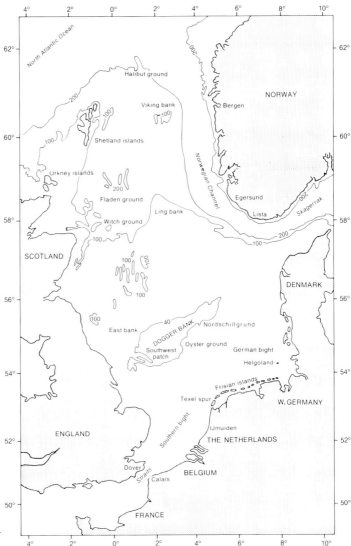

Fig. IV-52. Locations mentioned in the text.

have been eroded during the Quaternary. As it is hazardous to draw definite conclusions from radiocarbon datings of 30000 years or more, it is possible that at least parts of these deposits were formed during one or more of the previous glaciations. The Aberdeen Ground Beds are overlain by the 40 m thick Swatchway Beds (Holmes et al. 1975) containing glaciomarine and glacial sediments (Fannin 1976), which are thought to have been formed approximately between 27000 and 20000 B.P. (Holmes, pers. comm.). The development of the Swatchway Beds was preceded and followed by periods of erosion and channeling. A detailed contour map of the base of the older valley system in a small area at the Witch Ground (by Holmes et al.

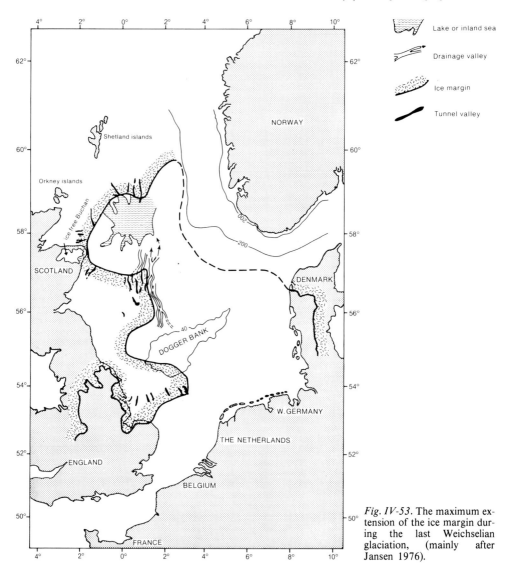

Fig. IV-53. The maximum extension of the ice margin during the last Weichselian glaciation, (mainly after Jansen 1976).

1975) shows local overdeepening, probably of glacial origin. A probably related system of buried channels, from Dogger Bank towards the north (Figs. IV-53, 54), was charted by Jansen (1976). Here at least three periods of channeling during periods of low sea level can be distinguished, alternating with periods of sedimentation in deeper water. Dingle (1971) interpreted buried channels in the East Bank area northwest of Dogger Bank as Weichselian tunnel valleys.

Northeastern North Sea. — In the Norwegian Channel morainic deposits are widely distributed as far south as Stavanger, off Lista and

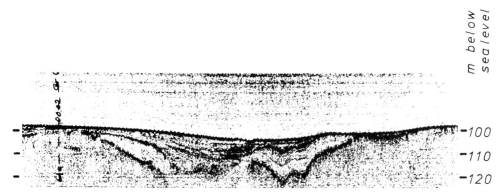

Fig. IV-54. Acoustic reflection record of a buried channel at 56°36'N, 1°28'E, vertical exaggeration is 17x (from Jansen 1976).

along both sides of the Skagerrak (van Weering et al. 1973, van Weering 1975 and unpubl. data). Close to the Norwegian coast, ice marginal accumulations can be traced as ridges over large distances (Andersen 1954, 1960, Klemsdal 1969). The thickness of these deposits varies from more than 40 m off Lista to less than a few metres south of Bergen (Braithwaite et al. 1974). As the Norwegian glaciers are known to have reached the sea during the Weichselian and most of the glacial deposits are covered by postglacial sediments, the moraines are considered to be of Weichselian age. When the Weichselian ice sheet reached its maximum extension, the Norwegian Channel was probably completely covered with ice, locally floating along the Norwegian west coast (Sellevol and Sundvor 1974). In the Skagerrak the ice must have eroded the trough considerably, as can be concluded from the great depth and the thin Quaternary deposits. Further southwest, the ice probably floated during part of the last glacial period whereas off Lista the glacier probably was grounded as indicated by the presence of glacially eroded deposits (van Weering et al. 1973).

Weichselian ice front. — In the northern North Sea the largest extension of the ice front (Fig. IV-53) is marked by a belt of tunnel valleys which are relatively straight U-shaped depressions, 1—3 km wide, 25—60 km long and at least 100 m deeper than the surrounding seafloor (Flinn 1967, Jansen 1976, Eisma et al. 1977, this vol.). Athought there is no consensus of opinion about the origin, tunnel valleys generally are supposed to have been formed subglacially. Their outflow-ends generally coincide with the ice front and their orientation is roughly parallel to the ice current (Woodland 1970, Wright 1973). Further south the ice front was probably situated between the tunnel valleys and the buried channels, as the channels do not show glacial influence and the channel fill, which is assigned to this period, points to a sediment supply from the West (Jansen 1976). Consequently, the British ice sheet did not meet the Scandinavian glacier in this area. West and southwest of the Dogger Bank, the extension of the ice in the North Sea is drawn according to Valentin (1957). The exact extension of the Weichselian ice sheet is not known. West of the Norwegian Channel the gravel deposits on the banks reaching from Ling Bank in the south to the Viking Bank and the Halibut Ground in the north, were interpreted by Pratje (1951) as reworked terminal moraines and glacial outwash deposits. The most likely maximum extension of the Weichselian ice is indicated in Fig. IV-53.

The extension of the ice probably represents several non-simultaneous advances in different areas between 20000 and 15000 B.P. The main source areas were the Scandinavian and British mountains and to a lesser degree Shetland. Glaciation of Shetland ended with a local ice cap which probably disappeared 13000–12000 B.P., but also erratics from an earlier Norwegian glacier were found (Flinn 1967, Hoppe 1974). It is not known whether this glacier was present during an early phase of the last glacial maximum or during a previous (sub)stage (Flinn, pers. comm.).

Simultaneously with the maximum glaciation, glaciomarine deposits originated in the North Sea. The Swatchway Beds mentioned above are thought to belong to this period (Holmes et al. 1975). Jansen (1976) mapped the overlying Hills Deposits and the subsequently formed Fladen Deposits, which are both ascribed to the last major ice advance. The Hills Deposits have been divided into a morainic and a glaciomarine facies. The glaciomarine Fladen Deposits are somewhat more than 15 m thick at most and consist of sandy clay with some gravel; in some places they are finely laminated (Jansen, et al., 1978). Mollusc fragments and Foraminifera point to an arctic glaciomarine environment; this indicates a connection with the open sea at least during part of the maximum glaciation.

Southern North Sea. – In the southern North Sea, marine sediments from the Weichselian have not been found. This was a land area with a periglacial climate and with rivers and lakes, as indicated by the presence of glaciolacustrine deposits (clays and fine sands), river sands and fresh-water clays (Oele 1969, 1971a, 1971b, Kolp 1974, 1976, Jelgersma et al., this vol.). There are no indications of a large inland lake at the Oyster Ground, as suggested by Valentin (1957), but during the Early Weichselian a large fresh-water lake was present in the Southern Bight. The Brown Bank Bed (Oele 1971b), fresh-water clays deposited in this lake, was cut during the Weichselian by streams and subsequently partly covered with fluvial sands (the Kreftenheye Formation) in the South. Further north, towards Dogger Bank, there are but few river- and lake deposits; most deposits are of fluvioglacial origin. During the Late Weichselian aeolian sands (the so-called cover sands) were deposited on top of these sediments. The aeolian sands are absent, however, at Dogger Bank (Jelgersma et al. 1977, this vol.) and at least locally north of this area (Jansen, unpubl. data). The reason is uncertain. Data of Kolp (1974, 1976) and Oele (1969) indicate that at least locally postglacial peat rests directly on boulder clay from the Saalian glaciation, and that aeolian sands were not deposited there. In the southern North Sea, however, considerable reworking has taken place during the Holocene and therefore the aeolian sands must have been more widely distributed than known at present.

Postglacial sediments

Northwestern North Sea. – The maximum fall of sea level in the North Sea is placed at the transition of the Fladen Deposits into the overlying Witch Deposits (Jansen 1976). The deepest mark of coastal erosion points to a relative sea level of about 110 m below the present level. The subsequent transgression coincided with the formation of the marine Witch Deposits. These are mainly restricted to the Witch and the Fladen Grounds; the maximum thickness is 15 m and they consist of finely laminated clay and very fine sand (Jansen 1976, Jansen et al. 1978). The fauna and pollen indicate arctic to cool marine conditions. Sedimentation of the Witch Deposits ended between 9000 and 8500 B.P., when an incomplete connection remained with the open sea, as indicated by Foraminifera and pollen (Jansen et al. 1978, de Jong & Jansen in prep.). The Witch Deposits and the above-mentioned Hills and Fladen deposits were combined into one postglacial unit, the Witch Ground Beds by Holmes et al. (1975) and Fannin (1976).

Northeastern North Sea, – In the deeper part of the Norwegian Channel a well-stratified

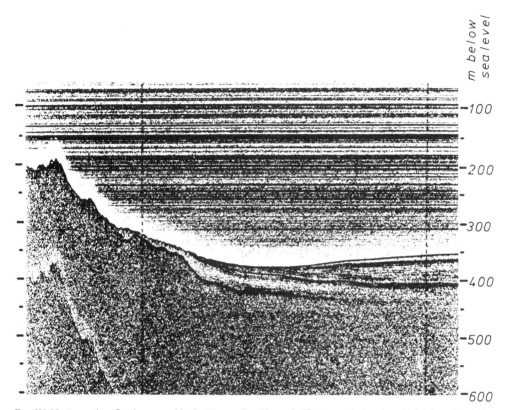

Fig. IV-55. Acoustic reflection record in the Norwegian Channel off Egersund, showing glacial deposits, well-stratified sedimentary cover and homogeneous toplayer. Norway at the left. Vertical exaggeration is 20 x (from van Weering et al. 1973).

sedimentary layer covers the glacial drift deposits (Fig. IV-55) (van Weering et al. 1973). Below the layer, however, sediments locally present off Stavanger wedge out towards the East, the stratification reflecting the topographic irregularities of the underlying glacial surface. These sediments probably were formed during an early stage of deglaciation. In the deeper parts of the Skagerrak, a deposit with a flat erosional surface, underlying the postglacial sediments, probably dates from before the major Weichselian glaciation. The well-stratified postglacial sediments have been interpreted on the basis of acousttical profiles as glaciomarine deposits containing more coarse components in the lower part (van Weering et al. 1973). The toplayer of these sediments has been dated with Foraminifera as Late Weichselian (Doppert 1976a) and contains a large amount of reworked Tertiairy and Mesozoic pollen (Zagwijn 1976). The stratified deposits are covered by an acoustically homogeneous toplayer consisting of fine-grained sediment (Lange 1956, van Weering unpubl. data). In the eastern Skagerrak the transition from stratified sediments to fine-grained toplayer is gradual, further north the transition is sharp.

The postglacial sediments reach a maximum thickness of 127 m along the southern border of the Skagerrak and wedge out towards the N and NE where the thickness becomes less than 1.5 m near the Norwegian and Swedish coast. Only where nearshore glacial deposits have been eroded and winnowed out, the thickness increases again (Holtedahl and Bjerkli 1975). Along the Norwegian Channel towards the North the postglacial deposits become gradually thinner and locally the glacial drift forms the sea-floor. The thickest postglacial deposits outside the Skagerrak are found in depressions off Egersund (70 m), off Bergen (50 m) and on the west side of the Norwegian Channel (43 m).

The differences in thickness are mainly attributed to the topographic undulations of the underlying glacial drift, especially on the east side of the Norwegian Channel where irregular strips of postglacial infill of 5 m or less are present. A local thickening of the postglacial deposits off Viking Bank is probably related to erosion of the flanks. The decrease in thickness towards the NE in the Skagerrak and along the Norwegian Channel towards the N suggests a mainly southern source area and a current pattern similar to the present one, but with higher velocities. Such a situation may have existed already when during the postglacial sea level rise the North Sea north of Dogger Bank attained roughly its present shape when sea level became higher than minus 60 m (van Weering et al. 1973). During that time the influence of surface waves on the sea-floor was much stronger than now and it may be assumed that the glacial deposits south of the Norwegian Channel were partly eroded and coarse material was deposited in the deeper parts. With rising sea level and increasing depth the deposits gradually became finer grained and there are strong indications that the deposition of finegrained material still continues at present (Lange 1956, van Weering unpubl. data).

Central North Sea. — In the East Bank area northwest of Dogger Bank a system of tidal sand ridges is present, similar to the linear ridge systems in the Southern Bight (Dingle 1970, Jansen 1976, Eisma et al. 1977, this vol.). These ridges probably were formed in the mouth of an embayment within the present −40 to 45 m isobath and are roughly parallel to the former shore, thus reflecting the direction of the main tidal currents. The Late Weichselian − Holocene sea level rise reached a standstill at approximately −45 m, a world-wide registered level of stagnation within the period 12000−9000 B.P. (Fairbridge 1961, Curray 1965, Mörner 1971). Kolp (1974, 1976) described a submarine terrace at −45 m around the Oyster Ground dated about 9000 B.P.; therefore the ridges are thought to have been formed at that time. During the formation of the tidal ridges a considerable amount of sand must have been mobile and a superficial layer of clayey sand, the East Bank Deposit, was formed in the northern North Sea. The thickness varies from 10 m in the East Bank area to a few dm north of 57° N (Eden 1975, Jansen 1976, Jansen et al. 1978, Caston 1977a) and, as indicated by Foraminifera and molluscs, the age is Holocene. At the Witch Ground it has been dated with pollen as Early Boreal (de Jong Jansen et al. 1978, Caston 1977a). At Dogger Bank and in the area just to the North this deposit contains also Late Holocene molluscs (Stride 1959, Spaink 1976, 1977).

Southern North Sea. — With the amelioration of the climate and the decrease in sea level rise the formation of peat started in the shallow area south of Dogger Bank. The peat was found from Dogger Bank up to the Straits of Dover-Calais (Jelgersma 1961, Oele 1969, 1971a, 1971b, Kolp 1974, 1976, Kirby & Oele 1975), and is associated with fresh-water clays; the overlying deposits are slightly brackish at the top (Oele 1969, Behre & Menke 1969). Pollen analysis and radiocarbon dating of the peat indicate a Preboreal and Boreal age. The oldest peats near Dogger Bank are dated 9900−9000 B.P. and those near the Straits of Dover-Calais 9900−9400 B.P. In the central Southern Bight the peat is dated 9000 B.P. to 8300 B.P. near the Dutch coast at IJmuiden. Younger peat is found at lesser depths; this corresponds very well to the curve of rising sea level (Jelgersma 1961). She concluded from the peat data that flooding of the southern North Sea began before 9300 B.P. through the Straits of Dover-Calais as well

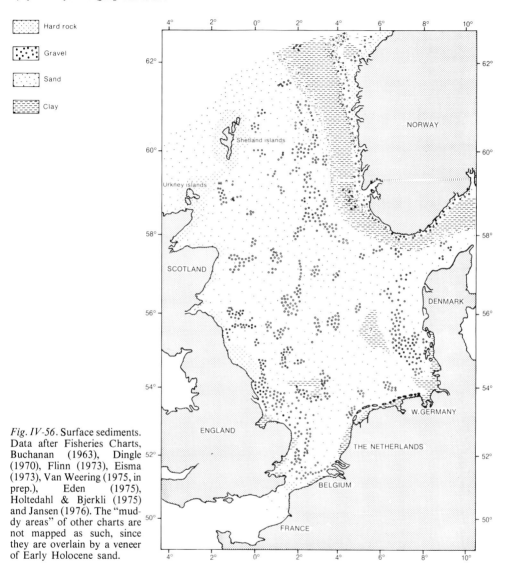

Fig. IV-56. Surface sediments. Data after Fisheries Charts, Buchanan (1963), Dingle (1970), Flinn (1973), Eisma (1973), Van Weering (1975, in prep.), Eden (1975), Holtedahl & Bjerkli (1975) and Jansen (1976). The "muddy areas" of other charts are not mapped as such, since they are overlain by a veneer of Early Holocene sand.

as near Dogger Bank. The area south of Dogger Bank was flooded between 8900 B.P. and 8600 B.P. (Kolp 1974), and the connection between the Oyster Ground and the Southern Bight was effected at about 8300 B.P. (Eisma & Mook, unpubl. data).

South of Dogger Bank peat formation was followed by the formation of tidal flat deposits (the Elbow Deposits, Oele 1969), which have been found in the Oyster Ground and in the Southern Bight off the Dutch coast (Oele 1969, 1971a, 1971b, Kolp 1974, Jelgersma et al. 1977, this vol.) A reworked tidal-flat mollusc-fauna, however, is more widely distributed and has been found on Dogger Bank (Pratje 1929, Stride 1959, Veenstra 1965), in the Oyster Ground north of the Frisian islands (Pratje 1929), and in the Southern Bight (Eisma and Spaink, unpubl. data.). Radiocarbon dating of this fauna indicated that in the deepest parts of the Southern Bight a tidal flat was already formed around 10000 B.P. when sea level stood about −55 m (Jelgersma pers. comm. 1977). When sea level rose, almost the entire Southern Bight became a tidal flat area between 8600 B.P. and 8400 B.P., prior to the complete flooding about 8300 B.P. (Jelgersma 1961, 1966, Eisma and Mook, unpubl. data). In 8000 B.P. when sea level was still at −20 m, the southern North Sea nearly reached its present extension. At 7250 B.P. when sea level stood at −12 m, deposits in the Dutch coastal area were already fully marine (Eisma and Mook, unpubl. data).

The tidal flat deposits are covered by the so-called Young Sea-sands (Oele 1971, Jelgersma et al. 1977, this volume). They form the present sea-floor of almost the entire southern North Sea except the places with (reworked) gravel beds (Jarke 1956, see Fig. IV-56). In the southern part of the Southern Bight the thickness of the Young Sea-sands is more than 10 m, but the thickness is less than 50 cm on the Zeeland ridges, off northern Holland, on the Texel Spur and in the Oyster Ground. The Young Sea-sands are partly a recent deposit of late or post-Medieval age (Oele 1971b). South of Dogger Bank the deposition of the uppermost 40 cm has been dated as Medieval or younger by Behre and Menke (1969) on the basis of pollen analysis. In the southern North Sea the median grainsize of the Young Sea-sands decreases from south to north from more than 400 μm to 120 μm with exceptions on Dogger Bank (200 μm) and in a shallow area like the Southwest Patch (500 μm). The size distribution seems to be related principally to the tidal current pattern, while on shallow banks like Dogger Bank wave action is more important. On the basis of the size distribution Kruit (1963) and Stride (1965) suggested that the Young Sea-sands must have been transported unidirectionally from the Southern Bight towards the Oyster Ground and the German Bight. Oele (1969, 1971b), however, considered this unlikely in view of the variations in thickness and the mineralogical difference between the sands in the Southern Bight and those in the Oyster Ground and on Dogger Bank.

A possible explanation for the deposition of the Young Sea-sands is repeated reworking and removal of fine-grained material during the Holocene. Kolp (1974, 1976), Diebel & Pietrzeniuk (1971), Behre & Menke (1969) and Pazotka von Lipinski & Wiegank (1969) stressed the reworked nature of the upper 75 cm of the sandy deposits south of Dogger Bank. In the Southern Bight the Young Sea-sands contain Tertiary, Pleistocene and Early Holocene molluscs, the distribution of which is roughly related to the known distribution of the older deposits (Eisma and Spaink, unpubl. data).

Already during the flooding of the southern North Sea, Pleistocene and probably also Tertiary deposits were reworked into tidal flat sands, estuarine sand banks and, presumably, beach deposits. These were later reshaped into sheet sands, megaripples, sand waves and linear sand ridges (Swift 1975, Nio 1976, Eisma et al. 1977, this vol.). An example is Brown Bank, which is a remnant of a former beach barrier (Oele 1971b). Wave action on the sea floor diminished gradually during the rise of sea level but must have increased again during the periods of increased storm frequency since ca. 3000 B.P. Present storm waves can remove the upper two metres of sand waves at a water depth of 18−20 m (Terwindt 1971, McCave 1971) and can move sand on the sea-floor in the entire southern North Sea (Draper 1967), but the effect is probably small below 30 m. Increased reworking of the sea-floor sediments by wave action and the churning up of bottom sands will facilitate a gradual removal of the fine sand fractions northward by tidal currents. An admixture of sand smaller than 160 μm from the South would probably hardly affect the mineralogy of the fine sands in the Oyster Ground as mineralogical differences between sands of different origin along the Dutch coast are only apparent in the coarser fractions (Eisma 1968) Stride (1970) explained the differences in heavy mineral composition by selective transport. The last period of increased reworking evidently is the late to post-Medieval transgression period and some reworking probably is going on at present.

REFERENCES

Andersen, B. G., 1954: Randmorener i Sørvest-Norge. *Nor. Geogr. Tidsskr.* 14, 5/6, 273—342.
Andersen, B. G., 1960: Sørlandet i sen- og postglacial tid. *Nor. Geol. Unders.* 210, 1—142.
Behre, K.-E. & Menke, B., 1969: Pollenanalytische Unterschungen an einem Bohrkern der südlichen Doggerbank. *Beitr. Meereskd.* 24—25, 123—129, Deutsche Akademie der Wissenschaften zu Berlin.
Braithwaite, P., Fritzner, H. E., Löfaldli, M., Maisey, G. H. & Myhre, L. A., 1974: Profiling and sampling in the North Sea in an area outside Bergen. NTNF's Kontinentalsokkelkontor, Survey Rep. Survey 7311.
Buchanan, J. B., 1963: The bottom fauna communities and their sediment relationships off the coast of Northumberland. *Oikos* 14, 2, 154—175.
Caston, V. N. D. 1977a: The Quaternary deposits of the Forties field, northern North Sea. *Inst. Geol. Sci., Rep. No 77/11*, 9—22.
Caston, V. N. D., 1977b: A new isopachyte map of the Quaternary of the North Sea. This volume.
Curray, J. R., 1965: Late Quaternary history, continental shelves of the United States. In H. E. Wright & D. G. Frey (editors), The Quaternary of the United States, 723—735, Princeton Univ. Press.
Dahl, R., 1972: The question of glacial survival in western Scandinavia in relation to the modern view of the Late Quaternary climate history. *Ambio Spec. Rep.* 2, 45—49.
Diebel, K. & Pietrzeniuk, E., 1971: Holozäne Ostrakoden von der Doggerbank, Nordsee. *Bull. Centre Rech. Pau-SNPA suppl.* 5, 377—390.
Dingle, R. V., 1970: Quaternary sediments and erosional features off the north Yorkshire coast, western North Sea. *Mar. Geol.* 9, 3, M17—M22.
Dingle, R. V., 1971: Buried tunnel valleys off the Northumberland coast, western North Sea. *Geol. Mijnbouw* 50, 5, 679—686.
Doppert, J. W. Chr. 1976a: Micropaleontologisch onderzoek boring G75—31 in het Skagerrak. *Rijks Geol. Dienst, Afd. Microfauna Kaenozoicum, Intern. Rep. 1236*, 2 pp.
Doppert, J. W. Chr., Ruegg, G. H. J., van Staalduinen, C. J., Zagwijn, W. H. & Zandstra, J. G., 1975: Formaties van het Kwartair en Boven-Tertiair in Nederland. In W. H. Zagwijn & C. J. van Staalduinen (editors), Toelichting bij geologische overzichtskaarten van Nederland, 11—56, Rijks Geologische Dienst, Haarlem.
Draper, L., 1967: Wave activity at the sea bed around north western Europe. *Mar. Geol.* 5, 2, 133—140.
Eden, R. A., 1975: North Sea environmental geology in relation to pipelines and structures. Oceanology International 75, Conf. Pap., 302—309.
Eisma, D., 1968: Composition, origin and distribution of Dutch coastal sands between Hoek van Holland and the island of Vlieland. *Neth. J. Sea Res.* 4, 2, 123—267.
Eisma, D., 1973: Sediment distribution in the North Sea in relation to marine pollution. In E D Goldberg (editor), North Sea Science, 131—150, MIT Press, Cambridge, Mass.
Eisma, D., Jansen, J. H. F. & Weering, Tj. C. E. van, 1977: this vol.
Fairbridge, R. W., 1961: Eustatic changes in sea level. In L. H. Ahrens, F. Press & S. K. Runcorn (editors), Physics and Chemistry of the Earth 4, 99—185, Pergamon, London.
Fannin, N. G. T., 1976: The Quaternary geology of the British sector of the North Sea. Joint Oceanogr. Ass. 1976, Book of Abstr., 81, FAO, Rome.
Flinn, D., 1967: Ice front in the North Sea. *Nature* 215, 5106, 1151—1154.
Flinn, D., 1973: The topography of the seafloor around Orkney and Shetland and in the northern North Sea. *J. geol. Soc.* 129, 1, 39—59.
Holmes, R., Fannin, N. G. T. & Tully, M. C., 1975: Geological report on the Forties pockmark detailed survey area, M. V. "Sea-lab" drill sites and Forties to Piper, Forties to engineering study area reconnaissance lines. *Inst. Geol. Sci. CSU II Rep. 75/13*, 1—12.
Holtedahl, H. & Bjerkli, K., 1975: Pleistocene and recent sediments of the Norwegian continental shelf ($62°$ N—$71°$ N) and the Norwegian Channel area. *Norg. Geol. Unders,* 316, 241—252.
Hoppe, G., 1974: The glacial history of the Shetland Islands. *Inst. Geogr., Spec. Publ.* 7, 197—210.
Jansen, J. H. F. 1976: Late Pleistocene and Holocene history of the northern North Sea, based on acoustic reflection records. *Neth. J. Sea Res.* 10, 1, 1—43.
Jansen, J.H.F., Doppert, W.W. Chr., Hoogendoorn-Toering, K., Jong, J. de & Spaink, G., 1978: Late Pleistocene and Holocene deposits in the Witch and Fladenground area, northern North Sea. *Neth. J. Sea Res.* 12, 4, in press.
Jarke, J., 1956: Eine neue Bodenkarte der südlichen Nordsee. *Dtsch. Hydrogr. Z.* 9, 1, 1—9.
Jelgersma, S., 1961: Holocene sealevel changes in the Netherlands. *Meded. Geol Stichting,* C, VI, 7, 1—100.
Jelgersma, S., 1966: Sea-level changes during the last 10,000 years. *R, Meteor. Soc. Proc. Int. Symp. World Climate 8000—0 B.C.,* 54—71.

Jelgersma, S., 1977: this volume.
Jelgersma, S., Oele, E. & Wiggers, A. J. 1977: this volume.
Jong, J. de & Jansen, J. H. F., in prep.: A palynological study of Late Quaternary deposits in the Witch and Fladen Ground area, northern North Sea. *Neth, J. Sea Res.*
Kent, P. E., 1975: Review of North Sea basin development. *J. Geol. Soc. 131*, 435–468.
Kirby, R. & Oele, E., 1975: The geological history of the Sandettie-Fairy Bank area, southern North Sea. *Phil. Trans. R. Soc. Lond. A. 279*, 257–267.
Klemsdal, K., 1969: A lista-stage moraine on Jaeren. *Nor. Geogr. Tidsskr. 23*, 4, 193–199.
Kolp, O., 1974: Submarine Uferterrassen in der südlichen Ost- und Nordsee als Marken eines stufenweise erfolgten holozänen Meeresanstiegs. *Baltica 5*, 11–40, Vilnius.
Kolp, O., 1976: Submarine Uferterrassen in der südlichen Ost- und Nordsee als Marken des holozänen Meeresanstiegs und der Überflutungsphasen der Ostsee. *Petermanns Geogr. Mitt. 120*, 1, 1–23, Gotha.
Kruit, C., 1963: Is the Rhine delta a delta? *Verh. K. Ned. Geol. Mijnbouwkd. Genoot. 21*, 2, 259–266.
Lange, W., 1956: Grundproben aus Skagerrak und Kattegatt, mikro-faunistisch und sedimentpetrografisch untersucht. *Meyniana 5*, 51–86.
Lindroth, C. H., 1972: Reflections on glacial refugia. *Ambio Spec. Rep. 2*, 51–54.
Maisey, G. H., 1972: Summary of previous works on Quaternary geology of the North Sea. NTNF's Kontinentalsokkelkontor Oslo, 1–60.
McCave, I. N., 1971: Sandwaves in the North Sea off the coast of Holland. *Mar. Geol. 10*, 3, 199–225.
Mörner, N.-A., 1971: Eustatic changes during the last 20.000 years and a method of separating the isostatic and eustatic factors in an uplifted area. *Palaeogeogr., Palaeoclim., Palaeoecol. 9*, 3, 153–181.
Nio Swie-dijn, 1976: Marine transgressions as a factor in the formation of sandwave complexes, *Geol. Mijnbouw 55*, 1/2, 18–40.
Oele, E., 1969: The Quaternary geology of the Dutch part of the North Sea, north of the Frisian Isles. *Geol. Mijnbouw 48*, 5, 467–480.
Oele, E., 1971a: Late Quaternary geology of the North Sea south-east of the Dogger Bank. *Inst. Geol. Sci., Rep. No. 70/15*, 25–34.
Oele, E., 1971b: The Quaternary geology of the southern area of the Dutch part of the North Sea. *Geol. Mijnbouw 50*, 3, 461–474.
Pazotka von Lipinski, G. & Wiegank, F., 1969: Foraminiferen aus dem Holozän der Doggerbank. *Beitr. Meereskd. 24–25*, 130–174, Deutsche Akademie der Wissenschaften zu Berlin.
Pratje, O., 1929: Subfossile Seichtwassermuscheln auf der Doggerbank und in der südlichen Nordsee? *Zentralbl. Miner. Geol. Paläontol. Abt. B*, 56–61.
Pratje, O., 1951: Die Deutung der Steingründe in der Nordsee als Endmoränen. *Dtsch. Hydrogr. Z. 4*, 3, 106–114.
Ronnevik, H., Bergsager, E. I., Moe, A., Ovrebø, V., Navrestad, T. and Stangenes, J., 1975: The geology of the Norwegian continental shelf. In A.W. Woodland (editor), Petroleum and the continental shelf of North-Western Europe, 1, Geology, 117–129, Applied Sci. Publ., Barking.
Sellevoll, M. A. & Sundvor, E., 1974: The origin of the Norwegian Channel a discussion based on seismic measurements. *Can. J. Earth Sci. 11*, 2, 224–231.
Spaink, G., 1977: Mollusken-onderzoek van de Noordzee-boringen 72J33 en 72J34. *Rijks Geol. Dienst, Afd. Macro-palaentologie, Intern. Rep. 789*, 4 pp. and *790*, 1 p. (Faunal lists only).
Stride, A. H., 1959: On the origin of the Dogger Bank, in the North Sea. *Geol. Mag. 96*, 1, 33–44.
Stride, A. H., 1965: Periodic and occasional sand transport in the North Sea. In *Revue Pétrolière I*, Abstr. 1st Int. Congr. "Le pétrole et la mer", Monaco, 111, 1–4.
Stride, A. H., 1970: Shape and size trends for sand waves in a depositional zone of the North Sea. *Geol. Mag. 107*, 469–477.
Swift, D. J. P., 1975: Tidal sand ridges and shoal-retreat massifs. *Mar. Geol. 18*, 2, 105–134.
Terwindt, J. H. J., 1971: Sandwaves in the Southern Bight of the North Sea. *Mar. Geol. 10*, 1, 51–67.
Valentin, H., 1957: Die Grenze der letzten Vereisung im Nordseeraum. *Tagungsber. und wissensch. Abh. Dtsch. Geografentag, Hamburg 1955, 30*, 359–366, Steiner, Wiesbaden.
Veenstra, H. J., 1965: Geology of the Dogger Bank area, North Sea. *Mar. Geol. 3*, 234–262.
Veenstra, H. J., 1970: Quaternary North Sea coasts. *Quaternaria XII*, 169–184.
Weering, Tj. C. E. van, 1975: Late Quaternary history of the Skagerrak; an interpretation of acoustical profiles. *Geol. Mijnbouw 54*, 3/4, 130–145.
Weering, Tj. C. E. van, in prep.: On the distribution and character of late- and postglacial sediments in the Skagerrak and the Norwegian Channel.
Weering, Tj. C. E. van, Jansen, J. H. F. & Eisma, D., 1973: Acoustic reflection profiles of the Norwegian Channel between Oslo and Bergen. *Neth. J. Sea Res. 6*, 1/2, 241–263.
Woodland, A. W., 1970: The buried tunnel-valleys of East Anglia. *Proc. Yorks. geol. Soc. 37*, 4, 521–578.
Wright, H. E. 1973: Tunnel valleys, glacial surges, and subglacial hydrology of the Superior Lobe, Minnesota, *Mem. geol. Soc. Am. 136* 251–276.
Zagwijn, W. H., 1974: Palaeogeographic evolution

of the Netherlands during the Quaternary. *Geol. Mijnbouw 53*, 6, 369–385.

Zagwijn, W. H., 1976: Pollenanalytisch onderzoek van boring G75-31 in het Skagerrak. *Rijks Geol. Dienst, Afd. Palaeobotanie, Intern. Rep. 760*, 2 pp.

Zagwijn, W. H., 1977: this volume.

Ziegler, P. A., 1975: North Sea basin history in the tectonic framework of North Western Europe. In A. W. Woodland (editor), Petroleum and the continental shelf of North-Western Europe, 1, Geology, 131–148, Applied Sci. Publ. Barking.

Ziegler, P. A. & Louwerens, C. J., 1977: this volume.

Chapter V

Summary of present knowledge on the Late Quaternary History of the North Sea

Introduction

The nine papers in the preceding chapter describing regional aspects of ancient shorelines and coastal development in the North Sea area since the beginning of the Eemian, necessitate a summary of the essential data in a few review papers. These papers include recently released information, data and theories on certain topics of phenomena not discussed in the regional papers of chapter IV. The three review papers in this chapter offer a summary of knowledge and problems with regard to the Late Quaternary history of the North Sea area.

They discuss the above-mentioned subject from three different points of view. The paper by Oele & Schüttenhelm summarizes the development of the North Sea from the Saalian glaciation to present time. Amongst other things it presents a line of maximum advance of the Saalian ice across the southern North Sea. The course and effects of the Eemian and Holocene transgressions are dealt with. The transgressions are apparently similar. The problems in the understanding of the Weichselian cold climatic period are briefly discussed.

The paper by Eisma et al. summarizes the data on sea-floor morphology and recent sediment movement in the North Sea. Morphological relicts and the recent mud transport are emphasized.

The last paper of this chapter consists of a summary of the present knowledge on sea level changes in the North Sea area during the Late Quaternary. The North Sea has been and still is the working area of many students interested in sea level changes. It has appeared, however, that various phenomena are more complicated than previously assumed. Tectonic movements, compaction, changes in the shape of the geoid and glacio-isostatic movements are not yet completely understood. Neglect of these factors has adversely affected the reliability or even the sense of a great number of sea level studies.

V-*a*

Development of the North Sea after the Saalian glaciation

ERNO OELE & RUUD T.E. SCHÜTTENHELM

Oele, E, & Schüttenhelm, R. T. E., 1979: Development of the North Sea after the Saalian glaciation. In E. Oele, R.T.E. Schüttenhelm & A.J. Wiggers (editors), The Quaternary History of the North Sea, 191–215. *Acta Univ. Ups., Symp. Univ. Ups. Annum Quingentesimum Celbrantis: 2*, Uppsala. ISBN 91-554-0495-2.

A broad outline is presented of the development of the North Sea area since the Saalian glacial maximum. Use has been made of the data and conclusions of the authors who contributed to this volume.
The known extent and nature of land ice advances in Saalian and Weichselian times, as far as relevant to the history of the North Sea, is described. A few remarks on sea level stands and glacio-isostatic movements have been included. A tentative line of maximum advance of the Saalian land ice across the southern North Sea is shown in the attached map. The development of Eemian and Holocene marine transgressions and so of the position of ancient shorelines is emphasized, and the progress of these transgressions is depicted on the colour map. A schematic section across the Holocene deposits in the southern and northwestern North Sea is presented.

Dr. E. Oele and Dr. R.T.E. Schüttenhelm, Rijks Geologische Dienst, Spaarne 17, Haarlem, the Netherlands.

Introduction

During the Saalian, inland ice for the last time covered large parts of the southern North Sea region. Apparently, two ice sheets, one from Scandinavia and one from the British Isles, met as far south as the Oyster Ground area. In contrast to the Elsterian glaciation, the traces of which are well covered by the deposits of the Holsteinian transgression and the subsequent

Saalian glaciation, the Saalian deposits are found in some places at or near the present sea floor.

The Eemian is the last complete interglacial interval characterized by a climatic amelioration and a related marine transgression after the Saalian cold climatic period. This makes a study of the Eemian particularly rewarding because this interval serves as an example for the history and probably also for the future course of the present Holocene interglacial. In some places, particularly outside subsiding areas, however, it proves difficult to distinguish Eemian from Holsteinian deposits and even more from intercalated temperate interstadial deposits.

Sutcliffe (1976), on the basis of British mammalian evidence, suggested that the Eemian does not show a continuous climatic amelioration and subsequent deterioration or transgression/regression, but that it includes a colder period that would lead to a fluctuation in sea level. Other authors expressed their doubt to this view (Mayhew 1976). Allen (1977) and Hollin (1977) studied the interglacial estuarine deposits along the north side of the river Thames that provide substantial evidence of an interglacial rise in sea level and a drop in sea level during the mid-interglacial as mentioned above. Even then, some doubt remains as to the Eemian age of the deposits. The alternative would be a Holsteinian age or an interglacial between Holsteinian and Eemian. Menke (1968) proved the existence of an interglacial following the Holsteinian and preceding the advance of the Saalian land ice in Wacken, Holstein. Erd (1970) described a similar Dömnitzian interglacial in the G.D.R., and Zagwijn (1973) discussed the vegetational history of the Hoogeveen interstadial, which strongly resembles the succession of a true interglacial.

There is evidence of an Early Weichselian (Early Devensian) glaciation that was widespread in North America, limited in Scandinavia and minimal in the British Isles. Sea level seems to have been low. The Middle Weichselian (the period between 55000 B.P. and 13000 B.P.) witnessed a long succession of mainly cold-climate regimes.

The most extensive glaciation occurred after 26000 B.P. The continental European terminology that places this glaciation in the Middle Weichselian, the so-called "Pleniglacial", is followed in this paper. The English-speaking countries, however, include this glaciation in the Late Devensian (or Late Weichselian), in order to distinguish it from earlier Weichselian glacial advances. Both views are used in the preceding contributions. As a consequence, the Late Weichselian in this paper begins 13000 B.P., somewhere during the climatic amelioration and the corresponding sea level rise (see Hafsten, this vol.). In Anglo-Saxon usage the Late Devensian starts with the advance of the land ice 26000 B.P. (Mitchell 1977).

After 26000 B.P. during the late Middle Weichselian glaciation most parts of the North Sea must have become dry land and in the remaining shallow marine waters ice masses were grounded (Mitchell 1977). This picture must be viewed with caution, as the extent of glacio-isostatic compensation movements is poorly understood. The late Middle Weichselian glaciation was less extensive than the Saalian glacial maximum. Virtually the whole of the southern North Sea and probably also the central North Sea were ice-free. After the glacial maximum at 18000 B.P. the climate improved and the sea level rose, at first slowly but after 15000 B.P. more rapidly but not without fluctuations.

At the start of the Holocene, the southern and central parts of the North Sea area had witnessed for several thousands of years the combined processes of climatic amelioration and sea level rise. The shorelines were located somewhere north of Dogger Bank (see Jelgersma, this vol.). These shore features were gradually submerging and became inaccessible until a few years ago when their outlines were unveiled by modern seismic methods and improved coring equipment.

In places the Pleistocene-Holocene boundary can be mapped rather easily, e.g. in the southern North Sea, where marine Holocene sediments overlie terrestrial Pleistocene deposits, especially where a Holocene basal Peat is intercalated. Holocene deposits are difficult to distinguish in the more northern parts of the North Sea where the marine sedimentation started already in the Weichselian. In many places, a distinction cannot be registered geophysically. Jansen (1976a), who was unable to make a further subdivision, combined all postglacial (marine) sediments in the Witch deposits.

The distribution of Holocene deposits is fairly well known, but a map showing the depths and thicknesses in the whole of the North Sea area cannot be drawn at present. For data and opinions about the sea level stands during the Holocene, the reader is referred to Jelgersma (this vol.).

In former heavily glaciated areas as Scandinavia and Scotland, however, the depressed land surface continued to rise. The uplift after the last glaciation of the northern areas bordering the North Sea is readily visible and was recognized already long ago. This uplift proceeded at different rates: at the head of fjords over 200 m (Hafsten, this vol.), but in nearshore areas a few tens of metres only.

In this respect Hafsten (this vol.) mentions the presence of Boreal peat off southernmost Norway. He estimated the local Boreal sea level at 4 m below the present one. Generally it is accepted that at that time sea level had reached a position of about 35 m below present sea level. This means that a local rebound of about 30 m must have taken place. Jansen (1976a), on the other hand, concluded a postglacial isostatic subsidence in the central North Sea of 25 m.

In the northwestern North Sea the isostatic movements apparently must be measured in tens of metres (Jardine, this vol.), a figure that greatly surpasses the total thickness of the Holocene deposits in the North Sea. At the moment an overall picture of the isostatic movement of the area, covering the present North Sea, cannot be given due to lack of coordinated information. The Holocene sedimentation as a result of tectonic warping of the North Sea basin during that period is masked in places by crustal reaction to the withdrawal of the Weichselian land ice.

Holmes (1977) calculated a subsidence of the Central Trough in the northern part of the North Sea since the late Middle Weichselian of about 30 m. This tectonic subsidence, however, had no influence on the sedimentation.

The southern North Sea basin during the Saalian glaciation

The Netherlands and northern Germany

The Saalian glaciers reached the central parts of the Netherlands and extended even further south in the adjoining German area (Duphorn et al. 1973). It was the most extensive glaciation in the onshore part of the Netherlands, reaching its southernmost position in the second of its five phases (Jelgersma & Breeuwer 1975). The land ice advancing over thick series of unconsolidated deltaic and fluvial sediments of the Rhine-Meuse and the North German rivers excavated glacier-tongue basins and formed ice-pushed ridges during glacier surges. The resulting relief exceeds 200 m in some places. Lodgement tills were deposited, each of which is associated with a series of elongated basins produced by ice lobes and reaching 8 km in width and -100 m below present sea level. Around the margins of some of those basins, deformed, ice-pushed Holsteinian deposits are present, which prove the Saalian age of the glaciation. Glacial basins and ice-pushed ridges are far less conspicuous in areas with a thin Quaternary cover overlying older indurated rocks, as for example in Eastern England.

The southern North Sea

The continuation of the Saalian maximum glacial advance to the West of the Netherlands is rather uncertain (Fig. V-1). A large glacial valley has been observed to the West of and parallel to the Dutch coast. Ice-pushed structures are present at the margins of this valley. These may be related to one or several of the Saalian recession and/or readvance stages that can be recognized in the Netherlands (Maarleveld 1953, ter Wee 1962). It is also possible that the glacial valley was formed during an earlier (Elsterian) glaciation.

Farther west the geology is poorly known. Glacigene sediments of Saalian age are absent in deeper borings in a number of Dutch K, L and P blocks in the central area between Holland and

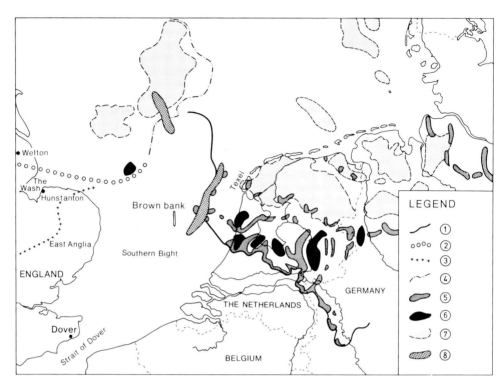

Fig. V-1. The Saalian glaciation in the southern North Sea. 1. Maximum extension of the Scandinavian land ice, in part assumed. 2. Tentative maximum extension of British land ice based on the presence of tills and warved clays. 3. Tentative maximum extension of British land ice based on the presence of sandy drift (outwash) and constructional features (after West 1977a). 4. Maximum eastward extension of British erratics, assumed to be deposited by Saalian land ice. 5. Ice-pushed ridges. &. Glacier tongue basins. 7. Till plateaus. 8. Galcial valley of Saalian age or older.

East Anglia. Saalian glacial deposits are known from the northernmost K and the eastern L blocks and are widespread farther to the North and East. This information makes it unlikely that Saalian ice covered the central part of the Dutch sector. A somewhat similar situation is found in parts of East Anglia, where some authors advocate a Saalian land ice cover although there are no tills and probably no other glacigene sediments of firmly established Saalian age.

Ice-pushed structures are known from the Brown Bank area. The sediments are largely or exclusively of Waalian (Early Pleistocene) or "Cromerian Complex" (Middle Pleistocene) age. Formerly they were regarded as Saalian glacial products (Oele 1971). Since the views on the glaciations in East Anglia and their age are strongly under discussion after the work of Bristow & Cox (1973), it must be admitted that the Brown Bank glacial features might be Saalian or even older. Given the absence of Saalian glacial deposits in the area and the position halfway between the Anglian (Elsterian) ice limit in East Anglia and an Elsterian lacustroglacial deposit west of the island of Texel (Zagwijn, this vol.), we are in favour of an older, i.e. an Elsterian age.

Northwest of the Brown Bank in block J 14, there seems to be a glacial basin of Saalian age

(Zagwijn 1977a), with a similar warved-clay fill as the Saalian basins in the western Netherlands at Ijmuiden and Amsterdam. In the area south of Dogger Bank, part of these Saalian deposits, amongst others lacustroglacial clays, were formerly regarded as of Elsterian age (Oele 1969). Pollen-analytical investigations of deeper borings that have been made since prove that these deposits of Scandinavian origin are partly younger than considered previously, and that they should be assigned to the Saalian glaciation.

The British part of the North Sea was at least partly covered by land ice. This follows already from the presence on the eastern (Dutch) flank of Dogger Bank of glacigene deposits including tills, which contain typical British gravels on the southern flank of the Dogger Bank. He assumed that they had been transported by Weichselian land ice. Although there may be doubt about his conclusions, especially as to the age, the idea of British land ice reaching the Dogger Bank is supported by later investigations (Zagwijn & Veenstra 1966, Zandstra 1971, 1972, 1973). More recently, Zagwijn succeeded in establishing the Saalian age of deposits. In Fig. V−1, the tentative extension of the influence of the British land ice is shown.

The glaciated Strait of Dover

Recently, the idea has been propagated that land ice, probably of Saalian age, had penetrated into the North Sea through both the English Channel and the Strait of Dover (Kellaway et al. 1975). In the view of other authors, however, all the facts mentioned in favour of this glaciation can be interpreted otherwise. None of them is a definite indication of glaciation. Kidson & Bowen (1976) especially caution against concluding to processes based on morphological evidence without supporting sedimentological data, which is a notorious source of potential error. See for instance Reinhard (1974). Their views have been challenged by other authors (Briggs 1976).

D'Olier (1975) described deep, infilled channels east of Suffolk and Essex, which have several of the known or supposed characteristics of tunnel valleys. The channels have not been dated; according to d'Olier they may be Wolstonian (Saalian) or Devensian (Weichselian). In our opinion tunnel valleys of both ages mentioned above are unlikely to be present in this part of the North Sea. It must be doubted whether they are all of the same origin or even whether they are glacial at all. Similar depressions have been found on non-glaciated shelves with strong tidal currents and also in some tidal inlets between barrier islands (McCave et al. 1977). Infilled valleys in the Strait of Dover are considered by some authors and on the same kind of arguments to represent tunnel valleys (Destombes et al. 1975). The southern Dutch and the Belgian sectors of the North Sea do not provide any unambiguous evidence in support of the hypothesis of a land-ice cover in these areas.

East Anglia

The extent, if any, of the Saalian (Wolstonian) glaciation in East Anglia remains a problem. Bristow & Cox (1973), after some fieldwork in the Norwich and Chelmsford areas, concluded that there is evidence of only one glaciation (the Anglian) in East Anglia, which they correlated at that time with the Saalian. However, the pollen-analytical data, especially of the overlying Hoxnian interglacial deposits, point to a correlation with the Elsterian glacial period of continental Europe. See discussiom by Francis and Turner in Bristow & Cox (1973) and Zagwijn (this vol.).

In other areas of East Anglia there is evidence of constructional features, such as river-terraces (Waveney valley) and outwash deposits (amongst others Kelling Sandur, Weybourne), which are post-Anglian (post-Elsterian) and pre-Devensian (pre-Weichselian). To the West there are several Wolstonian (Saalian) tills known in the Midlands (Shotton in Mitchell et al. 1973). In East Anglia proper, there are no tills that can be unambiguously assigned to the Wolstonian. One of the dubious cases that merits additional investigation is situated in the Nar Valley, NW

Norfolk. Gravel interpreted as of glacial origin overlies fresh-water and marine beds of proven Holsteinian age (Stevens 1960). The gravel is reported to pass laterally in glacial outwash interlayered with till. This till differs from the Hunstanton Till farther north that might be Weichselian in age (see below). The large quantity of outwash without firm evidence of related tills is puzzling. See Shotton et al. (1977) for more details.

The southernmost limit of the Saalian glaciation in East Anglia is drawn in Fig. V–1 after West (1977a). It is based on constructional features and outwash deposits (sandy drift). A more conservative Saalian limit, based on the presence of tills and warved clays, runs from the J 14 area (warved clays in a glacial basin) to the West, north of Norfolk to Lincolnshire, where Saalian (Wolstonian) tills have been recognized by Alabaster & Straw (1976) at Welton-le-Wold. These Welton and Calcethorpe Tills are overlying gravel beds with Acheulean handaxes and horse teeth, reported to be of Hoxnian (Holsteinian) age. Recently, some doubts have been expressed (Straw in Catt 1977a) whether the Calcethorpe Till at Welton is in situ or soliflucted in Late Devensian time.

The Calcethorpe Till is lithologically similar to the Marly Drift (till), situated on the other (eastern) side of the Wash. The Marly Drift is considered to be Elsterian (see West 1977b). But lithological similarities are a poor argument to prove similarities in age (see Alabaster & Straw 1976). This very argument, however, is used to prove that the post-Elsterian Hunstanton Till, a local feature in the coastal area around Hunstanton NW Norfolk, should be of Weichselian (Devensian) age. Its lithology is similar to that of the Purple, Drab and Hassle Tills of Holderness, Yorkshire. Moss fragments from a silt basin at Dimlington below these tills were dated with ^{14}C at 18240 ± 250 B.P. (Catt & Penny 1966, Shotton et al. 1969). If the age of the Hunstanton Till would be incorrect, the age can only be Saalian.

Research and more fieldwork especially at sea may clarify the issues mentioned above and will present us with a more reliable Saalian limit across the southern North Sea than tentatively drawn in Fig. V–1.

Morphological impact

The maximum extension of the Saalian glaciation and the successive recession stages strongly influenced the topography in the Netherlands and in northern Germany (ter Wee 1962, Duphorn et al. 1973). In the Netherlands several very deep glacial basins were formed (Jelgersma & Breeuwer 1975). They are flanked by ice-pushed ridges continuing far to the East (Duphorn et al. 1973). This topography later determined the extent of the Eemian transgression (Behre et al., this vol. and Jelgersma et al., this vol.). Moreover, glacigene sediments like tills and proglacial clays, both resistent, suppressed the erosive impact of the Eemian transgression and subsequent regression.

The North Sea basin during the Eemian interglacial

General remarks

Eemian deposits have been recorded from a great number of places in the marine and coastal areas of the North Sea. Most offshore recordings are located in Dutch waters and in the German Bight (Behre et al., this vol., Jelgersma et al., this vol.). In the northern areas, however, Eemian deposits are practically unknown. Part of the Eemian deposits may have been removed by the subsequent Weichselian land ice, especially in the coastal areas of Scandinavia. The waterdepth and the thick cover of Weichselian sediments are other factors that prevent Eemian deposits offshore to be reached by most marine surveys. In the southernmost part of the North Sea, including the Strait of Dover, Eemian deposits are also scarce to absent; this cannot be attributed to glacial activity. The Eemian deposits in this area consist to a large extent of reworked fluvial sand originally deposited by the Rhine riversystem during the preceding cold low sea level period. During this period the Rhine discharged into the Enlish Channel. The Eemian sediments may be difficult to distinguish, as part of them was reworked again by Weichselian rivers and during the Holocene marine transgression.

Sea level stands

After Eemian pollen zones had been established for the Netherlands (Zagwijn 1961, 1975) and surrounding areas, data became available on the bathymetric position of sediments in certain zones. On the basis of these data Zagwijn (1977b) constructed a tentative Eemian relative sea level curve – the first for the southern North Sea area. A more detailed study on Eemian sea levels by the same author is in preparation. See also Jelgersma (this vol.) for information on this subject.

In the southern North Sea the Eemian transgression was not noticed below pollen zone E3 (Jelgersma et al., this vol.), probably because the southern North Sea was still dry land.

The Eemian transgression in the Netherlands culminated during pollen zone E5 (Jelgersma et al., this vol.). The maximum Eemian relative sea level in more northern areas may have been far earlier, as this region witnessed a glacio-isostatic rebound. In Scandinavia this might have coincided with an Eemian postglacial transgression maximum, following the melting of most of the Saalian land ice.

At the transgression maximum, the relative sea level in the Netherlands was about 8 to 9 m below N.A.P. The difference between this figure and the maximum Eemian sea level of 4 to 5 m below N.N., in Schleswig-Holstein (Behre et al., this vol.), may be explained by tectonic downwarping in the Netherlands and/or isostatic readjustment in Schleswig-Holstein. In the eastern part of the Belgian coastal plain marine deposits of Eemian age are generally present at a depth of 5 m below the present surface (Paepe & Baeteman, this vol.), which is slightly above sea level. Paepe et al. (1972) observed Eemian beach deposits at −1 m O.P. (this is between −3 and −4 m N.A.P.). Paepe & Vanhoorne (1972) and Paepe (1971) described Eemian deposits at + 1 m O.P. and + 2 m O.P. respectively.

This suggests a rise of the top of the marine Eemian from the Netherlands towards Belgium. In several of the data just mentioned, the role of compaction seems to have been neglected. See Jelgersma (this vol.) for more details on this subject. In the English Channel, Eemian raised-beach deposits are known up to + 15 m O.D.,
but the sea level had risen to + 10 m O.D. (Mottershead 1977). The regional configuration supported by the differences in depth between the top of the Eemian deposits in the Amsterdam area and in the Dutch L blocks (Jelgersma et al., this vol.) and by the thicknesses of the Quaternary (Caston, this vol.), seems to point to epeirogenetic movements.

The northern North Sea

As mentioned before, Eemian deposits are hardly known from this area. One of the few well-studied deposits is that at Fjøsanger near Bergen, discovered by Mangerud (Hafsten, this vol., Mangerud et al. 1977). It shows a rather complete marine interglacial sequence between two (Saalian? and Weichselian) tills, not displaced by glacial forces. Marine clays, believed to be of Eemian age, are known from Kroken in the Sognefjord farther north (Vorren 1972). Marine sediments and high strand marks, probably (Vorren 1972). Marine sediments and high strand marks, probably of pre-Weichselian age, were found in Jaeren (Hafsten, this vol.). This scarce evidence indicates that the maximum Eemian sea level was higher than the present sea level. Notwithstanding differential isostatic movements, marine conditions must have been present in the present offshore area during the Eemian. Information on the Eemian (Ipswichian) deposits in Scotland is practically absent. The till-covered remnants of presumed interglacial rock platforms at + 18 to + 25 m O.D. near Dunbar, SE Scotland, might represent Eemian shorelines (Jardine, this vol.).

The eastern North Sea

In Denmark, marine Eemian deposits are known from SW Jylland, the Kattegat area and the Western Baltic (Krog, this vol.). It seems that a North Sea area west of Jylland was dry land in Eemian times, with only one connection of unknown extent between the North Sea and the Baltic Sea by way of the Skagerrak, Kattegat

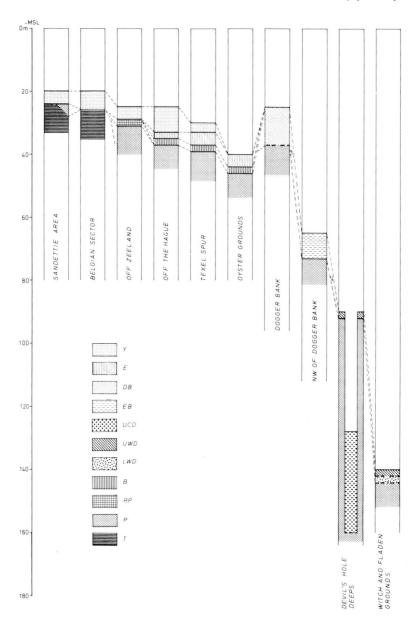

Fig. P-2. Schematic section across the Holocene deposits of the North Sea. Not to scale. Y — Young Sea sands, E — Elbow deposits (Oele 1969), DB — Dogger Bank sands, EB — East Bank deposits (Jansen 1976a) UCD — part of Upper Channel deposits (Holmes 1977), UWD — Upper Witch deposits (Jansen 1976a Jansen in press), LWD — uppermost part of Lower Witch deposits (Jansen in press), B — Basal Pea (Jelgersma 1961), RP — reworked Pleistocene fluviatile sands, P — Pleistocene deposits, T — Tertiary deposits Note: the Holocene deposits in the Witch and Fladen Ground areas are according to Holmes (1977) part of the Witch Ground beds.

and the Sont/Belt straits. Eemian deposits seem to have been eroded or displaced by Weichselian glaciers. A few noticeable exceptions are Eemian marine deposits in situ, e.g. at Skärumhede and Ejby (Krog, this vol.). Also in the Baltic area the marine deposits were disturbed by the Weichselian land ice. The course of the Eemian transgression in these areas is not yet fully known.

Eemian deposits predominantly consisting of marine sands are widely distributed in the German North Sea sector (Sindowski 1970). The Eemian coastline had a certain similarity to the present coast. A presently marine and tidal flat area around the North Frisian islands, however, seems to have been dry land. In Germany and in the Netherlands the Eemian coastline was characterized by numerous embayments, which are related to Saalian features. In the present coastal areas tidal flat and other coastal sediments were deposited in Eemian times also (Behre et al., this vol.). The Eemian transgression in the present coastal region started in the zones f and g from Jessen & Milthers (1928), equivalent to the zones E3–E5 of Zagwijn (1961). It was a single-phase Eemian transgression, possibly interrupted by two phases of standstill. Already in the upper Eemian h zone, equivalent to E6 of Zagwijn (1961), a regression set in (Behre et al., this vol.).

The level of the top of marine Eemian deposits slopes from -4 to -5 m N.N. in Schleswig-Holstein to -6.5 to -9 m N.N. in Niedersachsen (Behre et al., this vol.).

The Southern North Sea

Eemian deposits have been recorded from various places in the southern North Sea. They are found in the Dutch sector as a more or less continuous layer in the South and as infill of glacial depressions of Saalian age in the North Oele in Jelgersma et al., this vol.). From the present information it seems doubtful that Eemian sediments are present in the southernmost part of the North Sea. Subbottom profiler records from the Westerschelde to the Sandettié Bank indicate the mere presence of Young Sea-sand overlying Tertiary strata (Fig. V–2).

Eemian sediments are reported to be present as a basin infill in the Sandettié area (Kirby & Oele 1975). The deposit does not contain a typical Eemian fauna and is overlain by dated Weichselian material. Kirby and Oele suggested a Late Eemian age in view of the mollusc species indicating cool to cold climatic conditions. In our opinion it could also be late glacial Saalian.

Also on land firm evidence is lacking for Eemian deposits in the southernmost part of the North Sea area. Eemian deposits, contiguous to similar deposits in the southwest Netherlands (Jelgersma et al., this vol.), are known from northwestern Belgium (Paepe & Baeteman, this vol.). South of Oostende, however, they seem to be absent. The available evidence for Eemian in northernmost France (Sommé, this vol.) is not convincing either. On the western side of the North Sea, Eemian deposits are unknown from the coastal area south of the Thames estuary (Jardine, this vol.). Pollen-analytical evidence is, however, in support of an open Strait of Dover. During zone E4 (Zagwijn 1961, 1975) *Picea* spread along the southern shores of the North Sea but did not reach the British Isles. This suggests that the British Isles were isolated from continental Europe during zones E4/E6a, i.e. the range of *Picea* during the Eemian (West 1970).

If the Strait of Dover was open, the question arises how to explain the different effects of the Eemian and the Holocene transgressions; the latter caused a blanket of Young Sea-sands throughout the Southern Bight, up to 10 m. thick. The tidal movement must have been roughly the same at the maximum sea level stands. The cause, therefore, must be sought in the availability of material and/or the transgressive movement. It is possible that the Rhine during the Saalian supplied relatively small amounts of material only, whereas in the foregoing period large quantities of sediment formed the delta in a more NW direction (Zagwijn 1974). Consequently, much material cannot have been available. In the Eemian sediments in the Dutch sector, elements with a southern origin, i.e. from the English Channel, are unknown. Such elements are commonly present in the Holocene Young Sea-sand cover at the same latitude. The absence may indicate that the transgression from the South must have passed very quietly, if at all.

The Eemian deposits in the southern North Sea roughly consist of a transgressive series followed by a regressive series, which is closed by clay and peat in the Eemian type-locality in the onshore part of the Netherlands (Zagwijn 1961, Jelgersma et al., this vol.). The general depth of the offshore marine sandy Eemian series in the Dutch and German sectors is approximately 40 to 60 m below mean sea level (Oele in Jelgersma et al., this vol. and Behre et al., this vol.). Brackish deposits are described from the coastal areas at shallower depths, i.e. from less than 10 to about 20 m below sea level. In the Netherlands the brackish, generally clayey deposits are dated as Late Eemian (pollen zone E5–E6 of Zagwijn 1961). Sandy deposits are known from shallower depths as well, but part of them has been subjected to intensive reworking during the Holocene transgression (Oele in Jelgersma et al., this vol.). In the Southern Bight the Eemian deposits are overlain by a fresh-water clay of Early Weichselian age at c. 40 m below mean sea level. This is additional proof for the rapid lowering of the sea level at the end of the Eemian interglacial.

The western North Sea

In the English North Sea, Eemian deposits are practically unknown. North of the Dogger Bank they must be covered by thick Weichselian deposits (Jansen 1976a, McCave et al. 1977). Jansen (1976b) interpreted one of his geophysically delineated units NW of Dogger Bank, the Middle Channel Fill, as Eemian to Middle Weichselian. South of Dogger Bank and west of the Dutch sector hardly any work has been done. In block J 14 there are indications of marine Eemian sands in a glacial basin of Saalian age (see the paragraph on the Saalian glaciation). West of the southern part of the Dutch sector Tertiary beds seem to be covered by Young Sea-sands; Pleistocene strata, possibly including Eemian, are found closer to the shore, e.g. off Norfolk (Oele 1974). McCave et al. (1977) correctly stated that much work remains to be done in the British half of the southern area of the North Sea.

In the past, the study of Eemian (Ipswichian) deposits in onshore Eastern England was impeded by the difficulty to distinguish Holsteinian (Hoxnian) from Eemian interglacial deposits by means of mammals and handaxes, the more so by the apparent lack of firmly dated Saalian (Wolstonian) glacial deposits (see Shotton et al. 1977). In recent years, however, pollen analysis proved a succesful method to distinguish Holsteinian from Eemian floras (Turner 1970).

Along the English North Sea coast, the Easington beach gravel in county Durham was suggested to represent Ipswichian, but its height of + 27.5 m O.D. raises doubt about the age (Mitchell 1977, Jardine, this vol.). The alternative might be Holsteinian. A shell bed at Speeton (see Jardine, this vol.) near a recent beach level could be of Eemian age, as it is overlain by a Weichselian till. A nearby sandy deposit at 27 to 30 m O.D., reportedly overlain by a Saalian till, is likely to be of Holsteinian age, although the pollen flora bears some resemblance to Ipswichian zone II (West 1969, Gaunt et al. 1974). This zone appears to be equivalent to the zones E2–E3 of Zagwijn (1961, 1975).

In the Vale of York – Humber region estuarine deposits of Eemian age are present at 7–12 m below O.D. near Goole and at 2.5 m O.D. near Doncaster. The first mentioned (Langham) deposits date from the early part of Ipswichian zone IIb (West 1968), about equivalent to early zone E3 of Zagwijn (1961) The other (Westfield Farm) deposits are from the Ipswichian zone III/IV (Gaunt et al. 1974) probably equivalent to a level in zone E6 Together with evidence from an Eemian beach at Sewerby (Catt & Penny 1966, Gaunt et al 1974), this indicates that sea level rose by about 13 m from a relatively low position in the early zone IIb (early E3) to a maximum of about 1.5 m O.D. during zone III or E4–E5 (Gaunt et al 1974, Jardine, this vol.). The base of the buried cliff at Sewerby, dating from the Eemian climatic optimum or immediately thereafter (Catt 1977a), is at 2 to 3 m O.D., whereas th sea level is thought to have been at 0.9 to 1.5 m O.D. (Catt & Penny 1966). It will be noted that the above-mentioned estuarine deposits are located on the landward side of the Eemian shoreline as drawn on the colour chart. Lithological evidence, as well as pollen, dinoflagellates and foraminifera present in the above-mentioned Langham and Westfield Farm deposits indicate fluvial sedimentation within the tidal reaches of a river, at no great height above contemporary sea level (Gaunt et al. 1974). Apparently, the Humber estuary existed already in Eemian times. The Eemian shoreline on the colour map indicates the approximate position of a buried coastal cliff, in places associated with a shore platform (Jardine, this vol.). It seems that for instance at Sewerby the cliff and platform are older than Eemian but there is little doubt that these features formed the shoreline in Eemian times (Jardine, this vol.).

Farther south, there are traces of a poorly defined Eemian transgression in the Wash area (Jardine, this vol.). The only other evidence of an Eemian transgression in East Anglia is the presence of brackish-water deposits at Stutton near Ipswich. As compared with Eemian deposits in the English Channel area there are indications of downwarping in both eastern East Anglia and the Humber region (Jardine, this vol.). As mentioned already, Allen (1977) and Hollin (1977) worked in estuarine deposits along the river Thames, probably of Eemian age.

The information about Eemian deposits on the English east coast permits a forecast of the nature and distribution of Eemian deposits offshore. Given the sea level stands, about the whole of the British sector must have been covered during some part of the Eemian by marine waters. The marine deposits in the South must have consisted of transgression sands that were subsequently moved back and forth in a non-depositional shelf environment when sea level continued to rise and the shoreline moved in a landward direction. Farther north the Eemian transgression sands, if present, might be covered by a veneer of fine-grained material reflecting quiet deeper water conditions. The transgressive series were reworked and overlain by a sandy regressive series that in depressed areas terminates with brackish to fresh-water clay layers. Part of the Eemian deposits were removed during the Weichselian and Holocene (Fig. V–2) by rivers, land ice, ablation and tidal currents.

The North Sea basin during the Weichselian glacial period

General remarks and sea level stands

At the Eemian–Weichselian boundary between 75000 and 80000 B.P. the oxygene-isotope curve (Shackleton & Opdyke 1973) drops steeply, suggesting the rapid build-up of ice. Tills considered to be deposited during this stadial are wide-spread in eastern Canada (Dreimanis & Goldthwait 1973) and are probably present in Ireland (Colhoun et al. 1972). Most of Scandinavia was ice-covered during an Early Weichselian cold period that ended before the Brørup interstadial, i.e. before 66000 B.P. (Worsley 1977). In Britain, no Weichselian till has been recognized, that is not late Middle Weichselian (Mitchell 1977). During the Early Weichselian cold period, just mentioned, sea level must have been considerably lower. If glaciation of the British Isles was minimal, as it probably was, then also the glacio-isostatic depression was minimal, and the shorelines of eastern Scotland and England were located far to the East of their present position (Jardine, this vol.). The land area of northern Britain probably was even larger than during the late Middle Weichselian glacial maximum.

In Early Weichselian time there are at least two interstadials, i.e. Amersfoort and Brørup or Chelford (Zagwijn 1961, Lamb 1977), or three interstadials including the Odderade interstadial (Averdieck 1967) when forests covered the shores of the North Sea. However, the type of vegetation indicates that winters were severe (Lamb 1977).

Oxygene-isotopic records and climatic considerations suggest that during parts of the Middle Weichselian the sea level was higher (Mitchell 1977, Jardine, this vol.). Only very briefly, around 43000 B.P. (the Upton Warren interstadial), interglacial conditions seem to have been approached. This interstadial did not last long enough for a forest vegetation to spread from southern Europe (Lamb 1977). The idea of a Weichselian sea level only a little below that of the present one was proposed by Curry (1961). Especially in the British Isles this idea has found many followers (Kidson 1977). It should be pointed out, however, that tectonic and isostatic movements are easily confused with world-wide eustatic movements. It is known that there is no evidence of a marine Weichselian transgression in the southern North Sea. This means that sea level at any time during the Weichselian must have been lower than 50 m below present sea level. Isostatic movements were negligible in this area, and tectonic downwarping cannot have been more than a few metres.

Jansen (1976b) described for the Middle Channel Fill in the NW North Sea evidence of a cycle covering a low sea level, a high sea level and a low sea level again. The low sea levels are considered Saalian and late Middle Weichselian. The high sea level deposits in his view probably originated during the Eemian and Early-Middle Weichselian.

During the late Middle Weichselian glacial interval, here considered to cover the period of 26000 to 13000 B.P., sea level according to many authors fell more than − 100 m below the present level, e.g. −130 m in Scotland (Mitchell 1977) and −110 m in the northwestern North Sea (Jansen, in press). According to Jardine (this vol.), however, the relative sea level was −60 to −70 m O.D. in the more inshore western North Sea during the late Middle Weichselian glacial maximum. In northern Britain and adjacent parts of the shelf, ice loading and unloading resulted in a regional depression and a rebound of several decametres and possibly of more than 100 m (Jardine, this vol.). This resulted in a much higher relative sea level for heavily glaciated areas during the glacial maximum than for unglaciated areas (Jardine, this vol.).

In the northern North Sea sea level studies must deal also with the apparently anomalous high Late Quaternary sedimentation and subsidence rates as compared to those in the southern areas, mentioned amongst others by Holmes (1977). If the subsidence rate for the area between 56° and 58° N and 1° and 3° E is extrapolated from the trend established during the preceding 100 million years into the 2 million years of the Quaternary (Clarke 1973), then a subsidence of approximately 160 m has taken place since the beginning of the Middle Weichselian more than 50000 years ago. According to Holmes (1977), the alleged rapid Weichselian sedimentation may well be related to subsidence associated with glaciation. Jansen (in press), however, has expressed doubts as to the reliability of some of the starting points of this reasoning (see below). The reader is referred to Jelgersma (this vol.) for more information on Weichselian sea levels.

The northern North Sea

The Weichselian glacial period witnessed at least two major glaciations. The first one directly followed the Eemian interglacial and ended before the Brørup interstadial (Worsley 1977). The land ice that covered most of the Scandinavian peninsula seems to have originated in the mountains. Glaciation of Northern Britain might have been minimal, thus keeping glacio-isostatic depression values low. This might have resulted in large parts of the northwestern North Sea being dry land (Jardine, this vol.).

At the Witch Ground in the northwestern North Sea, a 200 m thick glaciomarine series is considered to be Weichselian on the basis of radiocarbon datings of 33000 and 48000 years B.P. (see Jansen et al., this vol.). These so-called Aberdeen Ground Beds are overlain by the Swatchway Beds (40 m acc. to Fannin 1975; 50 m acc. to Jansen, in press; about 100 m acc. to McCave et al. 1977). These beds include both glaciomarine and temperate marine deposits and contain erosion surfaces and channel-fill sequences. They are thought to be of Weichselian age (Fannin 1975) or Early to Middle Weichselian possibly also Eemian age (Jansen, in press). A core from the Swatchway beds near Old Devil's Hole was radiocarbon dated: 33800

B.P. for a sample from 1.5 m below sea bed and 37500 B.P. for 2.5 m below sea bed (Jansen, in press). In the inshore area other Weichselian sediments including tills have been recognized. The Swatchway Beds are preceded and followed by erosion and channeling. A probably related system of channels was charted in this area by Jansen (Jansen 1976a). At least three fases of channeling coinciding with periods of low sea level have been distinguished (Jansen et al., this vol.).

According to Holmes (1977) Weichselian sediments in the Josephine Field in U.K. block 30/13 reach to at least 310 m below sea bed. Radiocarbon datings around 120 m below sea bed gave an age of $23170 +$ B.P., around 250 m below sea bed $32682 ^{+530}_{-500}$ B.P. and around 310 m below sea bed $47715 ^{+1940}_{-1560}$ B.P. (Holmes 1977). Jansen (in press), however, is doubtful about the reliability of the Josephine Field datings; these were obtained from wood of in part thermophilous trees, and remarkably produced ages indicating a very cold to polar climate. A possible explanation is that the trees drifted in from the North, from the Atlantic. The suggestion by McCave et al. (1977) of a southern origin by drifting is unlikely; the southern North Sea was dry land without large trees from 55000 B.P. to late glacial time. Reference is made to Kolstrup & Wijmstra (1977), who discussed the absence of large trees in the period 50000 to 30000 B.P. that included three interstadials.

It does not seem easy to relate the Josephine datings to the datings near Old Devil's Hole obtained slightly below the sea bed. According to Holmes (1977) the northern North Sea is characterized by anomalous high Late Quaternary sedimentation rates compared to those in the southern areas, but given Jansen's recent datings it is quite possible that sedimentation rates were considerably lower.

In Norway, marine beds of Early to Middle Weichselian age were observed at Fjøsanger, Bergen (Mangerud et al. 1977), SE of Stavanger in Foss-Eigeland and Sandnes (Feyling-Hanssen 1974) and at Reve, Jaeren (Hafsten, this vol.). Datings of the deposits in Jaeren range from 28000 to 48000 B.P. (Mangerud 1972). The marine clays around Foss-Eigeland suggest a sea level of about 200 m above the present one (Rose et al. 1977). The height of the interstadial marine sediments in Jaeren is much greater than that of the Late Weichselian postglacial transgression maximum. Consequently, the ice in Jaeren, which caused the isostatic depression, must have been thicker during the Early Weichselian than during the late Middle Weichselian glacial maximum.

During the main (late) Middle Weichselian glaciation, the ice shed east of the Scandinavian mountain range shifted eastward (Worsley 1977). The extent of this glaciation, especially in the offshore areas, is only partly known. Boulton et al. (1977) assumed a zone of confluence of British ice with Scandinavian ice in the northern North Sea. This is based on the apparent deflection of the ice both to the North and to the South in east Scotland and northeast England, and on the suggestion by Hoppe (1974) that ice moved over Shetland in a westerly direction during the last glaciation. The glaciation of Shetland ended around 13000—12000 B.P. (see Jansen et al., this vol.). The tilt of about 1 m/km of the 10000—13000 B.P. shorelines on the west coast of Norway indicates that the ice load was small or that the North Sea grounded ice was broken up early (Andersen 1977). Jansen (1976a) postulated an ice-free central North Sea and an inland sea or proglacial lake in the northern North Sea.

A steady-state model of the British late Middle Weichselian ice sheet (Boulton et al. 1977) presents the following data: summit height 1800 m; velocities in marginal areas 150 to 500 m/year; the basal ice cold in the centre and temperate near the margins; about 15000 years required for its build-up. To explain all features of the model, Boulton et al. suggested that the British ice sheet never reached a steady state, but was halted by a rapid climatic amelioration. The central parts of the ice sheet were relatively inactive, e.g. in low-lying northeast Scotland (Buchan) where erosional features of glacial origin are absent, not because this area was ice-free (e.g. Jansen 1976b) but lack of ice-movement (Boulton et al. 1977). The intensive glacial erosion in highland Britain was not primarily caused during glacial maxima but during periods of lesser glacierization (Boulton et al. 1977). In the Moray Firth area, ice moving out of Cromarty Firth and the Great Glen carved a series of east-west basins. The fill consists of basal till interbedded with glacio-lacustrine deposits and is overlain by marine beds, all supposed to be of (late) Middle Weichselian age (McCave et al. 1977).

The uppermost 50 m of the Quaternary cover in the west-central North Sea consists of tills and glaciomarine sediments separated by levels with channel formation, all of Weichselian age (Thomson & Eden 1977). In the central U.K. sector between 56° and 58°N, only thin Weichselian sediments are found in the nearshore areas; according to Holmes (1977) and McCave et al. (1977), farther offshore, Weichselian series are known, several hundreds of metres thick and predominantly consisting of Middle Weichselian glacial and marine sediments. Jansen (1976a) studied an area between 57°30′ and 59°30′ N and 1°30′ W to 2° E and described the late Middle Weichselian morainic and glaciomarine Hills and glaciomarine Fladen deposits. A belt of tunnel valleys marks the maximum extension of the late Middle Weichselian ice, which coincided about with the formation of the Hills deposits (Jansen, in press). The area between 55° and 57° N and 0° and 3° E contains a number of channel fills of which the Upper Channel Fill is Late Weichselian (Jansen 1976a). Unfortunately in some publications the same lithostratigraphic terms have a different meaning.

At Frigg Field there is no firm evidence of late Middle Weichselian glacial deposits. This suggests that this area was ice-free during the late Middle Weichselian glacial maximum (Løfaldli 1973). During the late Middle Weichselian glacial maximum (20000–18000 B.P.), southern Norway was covered by land ice. The Norwegian Channel was subjected to renewed glacial scour. Morainic deposits are widely distributed in the Norwegian Channel. Close to the Norwegian coast ice-marginal ridges can be traced over large distances. In some places, the ice that covered the Norwegian Channel must have been floating (Jansen et al., this vol.). The Skagerrak was eroded to a great extent during the late Middle Weichselian glacial maximum. Several substages with more or less distinct morainic ridges can be recognized, e.g. Lista and Ra. The Lista substage dates from Older Dryas time (13500–13000 B.P.), the Ra moraine from Younger Dryas time (11000–10000 B.P.). Holtedahl & Bjerkli (1975) mentioned the presence of a submarine ridge off southern Norway. Apparently, it represents a Younger Dryas moraine. A well-stratified probably glaciomarine layer covers the late Middle Weichselian morainic deposits in the Norwegian Channel (Jansen et al., this vol.).

Jardine (this vol.) summarizes the evidence for late Middle and Late Weichselian shorelines in Eastern Scotland. Near Elgin on the Moray Firth the shorelines are found between +15 and +26 m O.D. In eastern Fife the oldest Weichselian shorelines near the present coast reach similar heights. Farther inland shore features are younger and higher. The highest postglacial marine levels in Norway are related to local deglaciation and, consequently, not synchronous (Hafsten, this vol.). Some of the highest shoremarks seem to belong to Late Weichselian time. Most of the coastal area of southern Norway was below sea level at that time. In western Norway, however, a rather broad deglaciated strip of land rose above sea level already in Younger Dryas time. Sea level curves for outermost SW Norway show a relative maximum in the period 13000–12000 B.P. (Hafsten, this vol.).

The eastern North Sea

Both the Skagerrak and the Kattegat were ice-covered from the time when the ice started to readvance (after 27000 B.P.) until about 13000 B.P. when the Lista moraine was deposited (van Weering 1975). At the time of deposition of the Kristiansand moraine (12500 B.P.) the eastern part of the Skagerrak was probably still covered by ice (van Weering 1975). According to Eriksson (this vol.) it was still covered by Norwegian ice during the Ågård interstadial (12700–12350 B.P.) The Swedish West coast became free of ice betwee the late-glacial Ågård and Alleröd interstadials (11800—11000 B.P., Mörner 1969). The area somewhat more inland was deglaciated between 11000 and 10000 B.P. (Eriksson, this vol.). This deglaciation process caused extensive terminal moraines, deltas and ice-marginal deposits. Large parts of the present land areas were covered by marine waters at the Pleistocene — Holocene transition (see the colour map). As from the Alleröd interstadial, a connection existed between the Skagerrak and the Baltic Sea through the Vänern basin. Immediately after the deglaciation the Swedish west coast was subjected to isostatic land upheaval that was strongest in the North. In Denmark, Late Weichselian marine deposits are only known from northernmost Jylland (Krog, this vol.). Information on Late Weichselian shorelines in the Danish part of the North Sea is not available.

The Scandinavian ice sheet did not extend beyond the river Elbe during the late Middle Weichselian glacial maximum. It covered the eastern half of Schleswig-Holstein and east and north Jylland. The glaciated areas have tills and ice-pushed ridges, the ice front is marked by a more or less continuous belt of sandur deposits (Duphorn et al. 1973), notably in the present North Frisian coastal and tidal flat area (see Behre et al., this vol.). The sandar have been subjected to fluvial erosion in late glacial times. The relief was moulded by aeolian erosion or smoothed by cover sands (Behre et al., this vol.).

The German Bight and Niedersachsen were areas with periglacial conditions. Typical deposits are cover sands with frost wedges, loess and fluvial deposits (Duphorn et al. 1973). Bertelsen (1972) considered similar deposits in the southern part of the Danish sector to be of Weichselian age, on the basis of a correlation with the nearby Dutch sector. Sindowski (1970) observed up to 17 m of sandy sediments of Weichselian age in the German Bight, probably representing cover sands and fluvio-periglacial deposits. The Elbe valley is one of the most significant morphological elements in the German sector (see Behre et al., this vol.). The continuation of the Elbe is less clear. It is quite possible that the river Elbe was smaller than at present, because precipitation in the drainage area in front of the land-ice sheet was probably less than at present. There is no proof for late Middle Weichselian ice-dammed lakes in the German Bight nor in the Dutch sector south of Dogger Bank, as suggested by Valentin (1957).

The southern North Sea

The Weichselian cold period in the Netherlands and in the Dutch sector was a period of considerable erosion and sedimentation, causing a further levelling of the Saalian and Eemian relief. Subaerial conditions prevailed in the Dutch sector during the whole of the Weichselian. Climatic conditions were polar or subarctic for most of the time. Forests were present only during the interstadials in the Early Weichselian (Jelgersma et al., this vol.).

Weichselian deposits are included in the Twenthe Formation (mainly cover sands and fluvio-periglacial deposits), the Kreftenheye Formation (fluvial deposits), and the Brown Bank Bed consisting of lacustrine clays (Jelgersma et al., this vol.). The Early Weichselian Brown Bank lake (Oele 1971) probably covered a large part of the Southern Bight. It is possible that other lakes were formed when sea level fell at the end of the Eemian; Kirby & Oele (1975) found an Early Weichselian lacustrine clay in the Sandettié bank area.

A few local peat horizons were dated as Middle and Late Weichselian. Part of the datings are supported by radiocarbon dates (see Jelgersma et al., this vol.). Thin Weichselian cover sands and fluvio-periglacial deposits are scattered in most of the Dutch sector. Fluvial deposits of Rhine origin (the Kreftenheye Formation) were recorded from the area in front of the present Rhine delta and more to the South (Oele in Jelgersma et al., this vol.). The Rhine discharged through the English Channel. The topmost part of the Kreftenheye Formation contains a few intercalated pumice levels from Late Weichselian postglacial time. The Kreftenheye Formation contains also a large quantity of reworked Eemian material (see Jelgersma et al., this vol.).

Sediments dating from the Weichselian "Pleniglacial", i.e. the Middle Weichselian, are well developed in NW and Central Belgium. The basal laminated sands and silts originated in a humid subarctic to cold temperate climatic period. A covering loam layer gave radiocarbon ages of 30700 ± 350 B.P. This suggests an age equivalent to the Denekamp interstadial in the Netherlands, but the vegetational contents indicate a forest cover (Vandenberghe & Gullentops 1977). The overlying cover sands with several ice-wedge cast levels were deposited under severe climatic conditions. Radiocarbon datings from the base of the cover sands range from 28100 ± 300 B.P. to 35760 ± 590 B.P. (Vandenberghe & Gullentops 1977). It is doubtful whether the deposits mentioned above continue under the Young Sea-sand cover in the Belgian sector.

Similar aeolian silty sands with paleosols, stone pavements, loess, ice-wedge casts and solifluction deposits are known in the French North Sea coastal area. An intercalated peat layer at Gravelines with a pollen flora indicative of an open park landscape has been radiocarbon dated: 33110 ± 740 B.P. (Sommé, this vol.). The known Weichselian deposits in the French sector are an Early Weichselian lacustrine clay, Weichselian sands with a fresh-water fauna and a Late Weichselian non-marine clay, all observed in the Sandettié Bank area on the French-British median line (Kirby & Oele 1975). Weichselian sea level in this area seems to have been below −45 m (Jardine, this vol.).

The western North Sea

In eastern England Early and Middle Weichselian deposits indicative of very cold to somewhat more temperate climatic conditions are best preserved in the Fenlands, in the terrace deposits of modern streams (Shotton et al. 1977). The short-lived and temperate Middle Weichselian Upton Warren interstadial was dated about 43000 B.P. (Shotton 1977). It has been recognized e.g. at Tattershall (Lincolnshire) and Earith (Huntingdonshire). Some of the deposits, e.g. at Tattershall, are famous for their rich mammal fauna (Shotton et al. 1977).

Little is known about Early to Middle Weichselian sediments in the English sector, i.e. south of 56° N. Some part of the Middle Channel Fill, NW of Dogger Bank (Jansen 1976a), is considered to belong to this period. The Early Weichselian Brown Bank Lake (Oele 1971) probably covered parts of the U.K. blocks 49 and 53.

Loess from the late part of the Middle Weichselian must have been wide-spread in a belt running from Belgium to south and east England, but a large part was removed by later Weichselian and Holocene processes. It has been suggested that the English loess originated from Weichselian glacial outwash in the North Sea basin (Catt 1977b).

Cover sands of Weichselian age are widespread in a belt running from northern Germany, the Netherlands and Belgium across the southern North Sea to East Anglia, Lincolnshire and Yorkshire. Part of the cover sands in East Anglia is thought to be of Early Weichselian age (West et al. 1974). Most of the cover sands situated to the North were deposited at the end of the Weichselian or later (Catt 1977b).

Eastern England was partly covered by late Middle Weichselian land ice. It seems that a Weichselian ice lobe extended far south along the east coast, while interior parts from the North York Moors to the South remained ice-free. Another ice sheet moved south along the Vale of York, farther to the West (Catt 1977a). This configuration blocked the pre-existing drainage pattern and resulted in a number of ice-marginal lakes (Catt 1977a). There is no evidence of similar ice-marginal lakes in the North Sea proper.

Two distinct tills are present in Holderness, Yorkshire and Lincolnshire, i.e. the Withernsea (or Purpe) Till and the lower Skipsea (or Drab) Till (Madgett 1975). The latter is reported to extend as far south as North Norfolk. Donovan (1973) traced these tills south of Dogger Bank as far as 2° E. Late Weichselian Withernsea (Purple) Till was reported from the Silver Pit (Donovan 1973). This till contains Triassic material from the Tees Valley, whereas the

Skipsea Till was formed mainly by the North Sea ice (Catt 1977a). The late Middle Weichselian age of both tills was proved by a number of radiocarbon datings e.g. 18240 ± 250 B.P. (Shotton et al. 1969). However, the lobe of late Middle Weichselian ice that moved down the east coast of England is suggested to have been caused by a surge starting on the location of the river Tees (Boulton et al. 1977). This implies that the radiocarbon datings related to this surge may pre- or postdate the date of the maximum extension of Weichselian ice in Britain, i.e. 18000 B.P.

Gravel of British origin is found between Lincolnshire and Dogger Bank in reportedly Weichselian glacial deposits (Veenstra 1969). Similar deposits of Saalian age, however, have been reported from the Dogger Bank area (see the paragraph on the Saalian). The Weichselian ice sheet appears to have stopped at the northern and western side of Dogger Bank. The bank contains fluvioglacial outwash and cover sands (Oele 1971). Numerous valleys are present in the North Sea, especially on the western side (see Jansen et al., this vol. and Eisma et al., this vol.). Some of the valleys are thought to represent tunnel valleys formed during the late Middle Weichselian glaciation (Jansen et al., this vol.). D'Olier (1975) postulated a possible Weichselian age for supposed tunnel valleys off Essex and Suffolk.

After 15000 B.P., a more or less continuous climatic amelioration set in, combined with a rise of sea level. Data from the North Sea area in support of this view are scarce. Jardine (this vol.) summarizes the evidence for a Late Weichselian shoreline that possibly coincided with the Loch Lomond readvance at −18 m O.D. near Berwick on Tweed. This shoreline has been recognized as far as the Firth of Forth.

The North Sea basin during the Holocene

General remarks

Holocene sediments in the North Sea proper consist of terrestrial, brackish-marine and marine sediments. Terrestrial deposits are found close to the present coasts and in the central part of the southern North Sea, south of Dogger Bank. The greatest depth recorded so far is 55 metres. Oele in Jelgersma et al. (this volume) mentions Holocene peat at this depth in the Dutch sector of the North Sea. The terrestrial deposits consist of thin layers of peat and freshwater clays, the latter normally on top.

Generally speaking, the brackish-marine sediments are found in the same areas as the terrestrial deposits. The brackish sediments in the central southern North Sea are thinner than in the coastal zones and measure a few metres only. The brackish sediments are fine sands, silts and clays. In the central southern North Sea parallel bedding indicates quiet sedimentation; in the coastal areas cross-bedding and channeling are observed on a large scale, indicating higher energy depositional conditions.

Marine deposits are present in the entire North Sea. In the northern part they generally consist of silts and clays, but south of the Dogger Bank sands are known that tend to coarsen in the direction of the coasts and the Strait of Dover. Veenstra (1969), Owens (1977) and others described the presence of marine gravelly sands north of the Dogger Bank, but these are supposed to represent a lag deposit.

The section, presented in Fig. V-2, gives a schematic picture of the different Holocene sediments in the southern and northwestern North Sea.

The northern North Sea

In the southern part of the North Sea basin the Holocene deposits on land form the marginal infill, which extends into the North Sea proper. In these deposits, and in the older Holocene marine series, a number of lithostratigraphic units can be distinguished (Jelgersma et al., this vol. and Fig. V-2). The position of the coastline must have been shifting continuously, as a result of interacting eustatic sea level rise, sedimentation rate and tectonic movements. In the northern North Sea the situation is completely different. In nearshore areas of Scandinavia and Scotland the isostatic uplift reduced the effect of the eustatic sea level rise. The isostatic movement farther inland strongly exceeded the movement in the coastal area. Farther offshore, the isostatic compensation movements seem to be decreasing. Consequently, the position of the coastline there must have been governed by the interaction between the eustatic sea level rise, the sedimentation rate, tectonic movements, if any, and the isostatic rebound of the land masses.

Shortly after the beginning of the deglaciation, some of the land was covered by marine waters, e.g. in the Oslo region (Hafsten, this vol.); in other places this happened somewhat later (Firth of Forth; Jardine, this vol.). Subsequently, the unloading caused the land to rise; soon this movement greatly exceeded the rate of eustatic sea level rise, and the effect was similar to that of a regression.

Depending on the momentum and the time of the deglaciation and on the height of the land in relation to the sea level, the eventual submergence and subsequent rebound with its effect differed from place to place and in time. Thus the marine transgression on the Swedish west coast was recorded as Late Weichselian (Eriksson, this vol.), whereas the transgression in the Forth Valley is younger (10500 B.P. or somewhat later) (Jardine, this vol.). For the beginning of the just-mentioned regressional movement, which is related to the shoreline of the maximum marine ingression after the deglaciation, different ages are found for this region. According to Eriksson (this vol.), the regression began in the period of 13000 to 9500 B.P. in the Swedish west coast area. Hafsten (this vol.) dates the maximum ingression in the Oslofjord area younger (Early Preboreal); the regression started later. On the Norwegian west coast the regression started about 12000 B.P. (Hafsten, this vol.).

In Scandinavia, a general trend in the tilting appears to be caused by unloading. The uplift increases from the coast to the interior and from south to north. The highest shoreline of the marine ingression after the deglaciation is found at a height of over 220 metres in the Oslo area, and it slopes to about 120 metres in the South. Hafsten (this vol.) estimates the lowering at 0.8 m/km; Eriksson (this vol.) mentions a figure of 0.47 m/km for the Swedish west coast. As said before, the age of the post-glacial shoreline changes: following the receding ice the shoreline is becoming younger towards the North. This explains the differences in age mentioned by Hafsten and Eriksson.

Still during the Early Holocene the rate of the isostatic uplift decreased, even so much that the rate of the sea level rise exceeded the rebound. This resulted in a transgression; the turning point was called the Boreal regression minimum by Hafsten and the regression maximum by Eriksson. In Norway this transgression is known as the Tapes transgression, in other areas as Litorina transgression (Eriksson, this vol.). The transgression affected Sweden as early as 9600 B.P., whereas it reached Norway somewhat later (after 8770 B.P.). In southeast Scotland the transgressive movement is dated 8400 B.P. (Jardine, this vol.).

Along the Swedish west coast and in the Oslo area the coastline related to this last transgression never reached so high as the coastline belonging to the maximum transgression directly after the deglaciation (Eriksson, this vol., Hafsten, this vol.). For these areas, the line drawn on the coloured map depicting the maximum Holocene marine ingression represents some stage during this first ingression after the deglaciation. In the Oslo area the transgression after the Boreal regression never caught up with the isostatic uplift (Hafsten, this vol.).

Local conditions, superimposed on the global trend, actually governed the course of the Flandrian transgression. Four active phases with some stationary or even regressional phases can be recognized in the Scandinavian part of the area. These phases cannot be distinguished in east Scotland.

The transgression is reported to have lasted from 9500—7000 B.P. in Sweden, whereas its influence became noticeable at about 8500 B.P. in southeast Scotland. Its culmination is estimated at 7300—7000 B.P. for Sweden, and at 6500 B.P. for southeast Scotland. As a result of a decrease in the rate of sea level rise and an increase in the rate of uplift, the transgression turned once more into a regression in both areas just mentioned. The Middle and Late Flandrian regression (Eriksson, this vol.) still continues.

Only little mention has been made of sediments that originated as a result of Holocene processes. The northern North Sea, in contrast to the southern North Sea basin, does not contain characteristic sedimentary series. Shell beds are the most conspicuous markers of the Holocene coastline. In addition, various types of sediments are mentioned in relation to the shoreline displacements, e.g. marine and lagoonal deposits and beach ridges.

Jansen et al. (this vol.) describe the postglacial deposits of the northwestern North Sea as clays and fine sand. The deposits consist of the Upper Witch deposits, the uppermost part of the Lower Witch deposits (Jansen, in press) and part of the Upper Channel deposits (Holmes 1977; see Fig. V-2). According to Jansen et al. (this vol.) deposition came to a halt already in the Early Holocene (9000—8500 B.P.). Holmes (1977) supposed that since the Late Weichselian, especially during the lower sea level stands, erosion prevailed. In the deeper parts of the North Sea sediments settled, which have not yet been dated. See Eisma et al. (this vol.) for more information on this subject. In the west-central North Sea Thomson & Eden (1977) distinguished an unconformity that is said to be situated on the Weichselian — Holocene boundary and divides the two marine units. They concluded that the upper Forth beds in the Firth of Forth and its approaches belong to the Flandrian (Postglacial) transgression. The transgression in this area was dated as younger than 10300 B.P. by Sissons (1976), who assigned the upper Forth beds to the Holocene. The upper Forth and middle Forth beds are separated by an erosion surface that can be correlated with the lowest postglacial sea level recorded in the area (Thomson & Eden 1977). Also in the northeastern part, Holocene sedimentation is restricted; the datings mentioned amongst others by Jansen et al. (this vol.) indicate Late Weichselian ages for the younger sediments. In the Norwegian trough and other deep parts, mud continues to be deposited (Eisma et al., this vol.).

In the Early Holocene, sands were deposited on a large scale in the East Bank area northwest of Dogger Bank (Fig. V-2). The transgressive movement caused the formation of a series of bars along an embayment. Jansen (1976a) concluded to an age of 9000 B.P., when sea level was approximately 45 metres below the present level. Kolp (1974) mentioned the presence of a submarine terrace (—46 metres) near the Dogger Bank and concluded that the marine transgression stagnated in the Preboreal.

The southeastern and southern North Sea

The southeastern and southern North Sea basin, covering the area south of Dogger Bank and the adjoining area between south Jylland in Denmark and the coastal plain in France, can be considered as one depositional region, where the Holocene processes were roughly similar.

According to the various authors in this volume the Holocene series reaches a thickness of about 30 metres in the coastal areas. In the Thames estuary it may be up to 36 metres thick (Greensmith & Tucker 1976). In the North Sea the thickness appears to be a few metres only, especially in the direction of Dogger Bank. This bank consist of more than 10 metres of sandy material, the Dogger Bank sands (Fig. V-2), which is in strong contrast to its immediate surroundings.

Behre et al. (this vol.) describe the earliest Holocene period as strongly erosive, causing deep V-shaped valleys. In the Netherlands, in Belgium and in France, this is unknown. In the greater part of these areas, a stratigraphic hiatus seems to be present between Weichselian and Early Holocene sediments, but indications of severe erosion have not been found. In the North Sea, where the strata are geophysically registered erosion as mentioned above has not been noticed as yet. The Rhine may even have deposited some sediment where crossing the present coastal plain.

In large parts of the area, Holocene sedimentation started with peat formation. The age of this Lower Peat or Basal Peat (Fig. V-2) varies from place to place; the peat is not a continuous layer. Where peat growth was not interrupted by the invading sea, it continued until historic time, when man interfered (see Behre et al., this vol.), e.g. in northwestern Germany and in the northeastern and western Netherlands. Various authors reported the peat from the area southeast of Dogger Bank, but the Basal Peat there may have been patchily developed. As mentioned already, the continuous sea level rise caused a flooding of the Basal Peat. In some places in the Netherlands and in the North Sea, a fresh-water clay but more generally brackish marine sediments were deposited on top of the Basal Peat. In Germany the oldest transgressive sediments are found in deep channels. In other parts of the region, where such channels are absent, the land simply drowned, and tidal flat and lagoonal conditions developed.

In the older (brackish)-marine sediments, which are part of the Elbow deposits (Fig. V-2) in the North Sea, regressive phases have not yet been observed as in the younger sediments forming the Calais deposits. The Elbow series indicates quiet sedimentary conditions, despite the rapid sea level rise in the Early Holocene. The earliest Calais transgressive phases have neither been very erosive in the Netherlands or in the Dutch sector of the North Sea. Deep channeling (to 40 metres below present mean sea level) accompanies the later Calais III and especially the Calais IV transgression (see Jelgersma et al., this vol.).

In the period between the onset of the Holocene and the Late Atlantic transgressions, the shoreline moved across the North Sea area. Already in the Preboreal the northeast part of the North Sea must have been flooded, as can be concluded from the marine character of the sediments of that age, but the position of the successive shorelines is not known. In the central North Sea the shoreline was still far away from the present one, but in France the sea reached the present coast at the end of the Boreal (Sommé, this vol.); in the Netherlands this happened in the Early Atlantic (Jelgersma et al., this vol.). By then the North Sea had flooded the Southern Bight from the South through the Strait of Dover as well as from the North. On the previous land and coastal areas another series of tidal flat deposits was formed, similar to the Elbow and older Calais deposits in the present North Sea area.

In the Atlantic a coastal barrier system began to develop. In France (Sommé, this vol.) this happened already in the Early Atlantic, whereas in the Netherlands the oldest barriers are dated as Middle Atlantic (5300—4700 B.P.). In Germany, where beach barriers are scarce, the barrier system is still somewhat younger (Atlantic—Subboreal). See Behre et al., (this vol.).

There has been some speculation about the presence of beach barriers in the earlier Holocene (Baak 1936, Jelgersma 1961). Swift (1975) considered them as former estuarine banks reshaped in an open shelf environment (see Eisma et al., this vol.). Jelgersma (this vol.) described the relation between successive shoreline and offshore bars. The presence of linear sand ridges inspired other authors to consider the barriers as old beach ridges (amongst others Jelgersma 1961). E.g. Oele (1971) reached the conclusion that the Brown Bank is a former beach barrier. The Sandettié Bank, too, may have originated as such a barrier (Kirby & Oele 1975). If this supposition is correct, both banks must be Boreal to Atlantic in age, in view of their depth (top at 20 metres, base at 35 metres below sea level). Zagwijn (1974) presented some paleogeographic maps suggesting the presence of Atlantic coastal barriers west of the present coastline.

If the sands banks in the North Sea represent beach barrier systems, their origin must be different from that of the younger coastal ridges onshore. To explain the formation, Hageman (1969) assumed a landward migration of barriers; the "bulldozer effect" caused new beach barriers while the older ones were destroyed. In the onshore part of the Netherlands the oldest barriers are found in the East, and the barriers become younger in the direction of the coast. This is attributed to the Subboreal retardation in sea level rise that resulted in a seaward movement of the coastline (Jelgersma et al., this vol.). Nothing is known about the ages of the various banks in a complex of sand banks in the North Sea. It is hardly conceivable that a retardation of sea level rise caused them to be younger in a seaward direction. If they become younger in the opposite direction, the question is how they survived the aforementioned bulldozer effect by the transgressing sea.

The formation of the coastal barriers was completed with the deposition of windblown sands, known as "Older Dunes" (Jelgersma et al., this vol.). The Holland Peat, still expanding along the eastern margin of the Atlantic lagoonal/tidal flat zone, could extend again in a westward direction under the protection of the barriers during a retarded sea level rise. The outstanding stratigraphic unit, the Holland Peat or surface peat, is found in the whole coastal zone from France to the northern Netherlands, and separates the Calais and Dunkerque Members (Jelgersma et al. this vol.). Its absence in the northeastern Netherlands and northwestern Germany is due to a weakly developed or an even absent barrier system.

The sea, regaining its influence, once more covered the coastal area. As a result the (Dunkerque) tidal flat deposits were formed from France to Jylland (Denmark); at Ringköbing these deposits are known to have settled behind a barrier (Krog, this vol.). Retardations in the sea level movement are reflected by intercalations of minor peat layers.

The various Dunkerque transgression phases were different in intensity and effect. In the western Netherlands especially the younger phase, the Dunkerque III, was attended with intensive erosion. In the northern Netherlands the Dunkerque I was strongly erosive (ter Wee, 1976). During the Dunkerque II, the sea apparently reached its deepest penetration in the whole area from France to Germany (Sommé, this vol., Paepe & Baeteman, this vol., Jelgersma et al., this vol.). As mentioned before, the coasts around the German Bight had only a weakly developed barrier system, which explains the absence of the surface peat. As a result, a distinction between Calais and Dunkerque deposits in this area cannot be made. Of course, in the offshore area the surface peat is absent. Therefore, a distinction between Calais and Dunkerque deposits in the North Sea proper is neither possible. A Dunkerque age is assigned to all oxydized sands. Near the Dutch southwest coast, a grey sandy layer is intercalated between this Young Sea-sand (Fig. V-2) and the Calais dated deposits. The layer has been dated by pollen analyses as younger than 2000 B.P., but this age does not necessarily apply to all places (Jelgersma et al., this vol.).

At the beginning of the Dunkerque period (about 3700 B.P.) the coastline from northern France to the northwestern Netherlands was more or less fixed. The sea still had entrance to the tidal flat area behind the beach barriers through tidal inlets and river outlets. The formation of the Older Dunes on top of the beach barriers continued until Roman time (Jelgersma et al., this vol.).

The coastline continues to the Northeast in the chain of Frisian islands, extending to Denmark. They seem to be of Dunkerque age, but reliable datings are scarce (van Staalduinen 1977). The origin of the islands is not definitely known. In the Netherlands and in Germany some of the islands have a core of Pleistocene deposits. These islands must have developed by acuumulation of material around the cores. The islands consisting of Holocene material, however, are considered by Behre et al. (this vol.) as the result of sand from the foreshore accumulated on a tidal flat shoal. In Germany some of the islands are unstable. In the Netherlands the delineation of the islands is changing, but none of them can be considered as actually unstable.

The coastline, from France to Denmark, moved at first from the North, West and Southwest in easterly directions. Shortly before the end of the Atlantic it reached its maximum extension, and then it retreated westward in the period from 4500–3500 B.P. At the beginning of the Dunkerque period the position of the coastline was mainly fixed, with minor changes taking place in the outlines of the (tidal flat) area behind the barriers.

The southwestern North Sea basin

The northern and southern parts of the North Sea basin each had a distinct development during the Holocene; in the northern part it was mainly governed by an interaction of eustasy and isostasy, in the southern part by the interaction of eustasy and sedimentation. In the southwestern North Sea basin, along the British southeast coast, the Holocene history has its own characteristics also; tectonism seems to be a more prominent factor. Some parts along the British east coast subsided, whereas other areas have been affected by a local rise. Subsiding areas were the Thames estuary and, at certain times, East Anglia and the Wash.

At the beginning of the Holocene (10000 B.P.), climatic conditions and sea level stands had markedly changed since the late Middle Weichselian 'glacial maximum, but in the geological record of the British eastern regions this change was still not expressed (Jelgersma, this vol.). It took some time before the rising sea level took effect. Formation of peat started in the western part of the North Sea basin in the same way as in the eastern part. The peat on the Leman/Ower Bank, mentioned by Godwin (1960) was dated 8425 ± 170 B.P. (Jardine, this vol.). The sea reached the area of the present-day coastline already in the beginning of the Boreal. Greensmith & Tucker (1973) estimated the age of the first marine influence in the Thames estuary to be about 8700 B.P.; this coincided with a sea level stand of 34 m below O.D. Whether this estimate is correct or not, it is clear that, compared to the eastern margin of the North Sea basin, the sea reached the western margin of the basin relatively fast. In the eastern part of the North Sea the transgression reached the present coastal area not until the beginning of the Atlantic.

The further Holocene history of the offshore area is rather uncertain, owing to lack of data. The transgression caused marine sands to be deposited on top of the older sediments; in part these sands are equivalent to the Young Seasands (Fig. V-2) of the Dutch sector (see Jansen et al., this vol.). Groups of linear sand ridges have a common origin as compared to their Belgian and Dutch equivalents (see Eisma et al., this vol.). It has not yet been proved whether outside the present estuaries and their approaches tidal flat sediments were ever deposited. A local reworking of older sediments may be assumed. Gravels, representing lag deposits and/or tidal bars, are now present as surface sediments, e.g. off Norfolk. A layer of relatively coarse sand covers almost the whole area. The sand has probably been reworked and redeposited since the area submerged.

The transgression was interrupted by regressive movements; as in the east part of the North Sea basin, these do not always reflect a global lowering of the sea level but may have been caused by local conditions also. Gaunt & Tooley (1974) presented evidence for a sea level rise with interruptions in the Humber area. A close relationship was established with sea level movements in northwestern England. They also concluded that the sea level curve fits the one drawn by Greensmith & Tucker (1973) for the Thames estuary. Greensmith & Tucker (1973) distinguished six transgressive phases; the second one corresponds to a global event and could be correlated with the Calais I transgression of the continent. The rising sea then invaded the Essex area. According to the mentioned authors the transgression in the Thames estuary came to a halt about 5000 B.P., when sea level was only a few metres below the present level. See Jardine (this vol.) for more details on shore level changes on the British east coast. In the Thames estuary the transgression caused cheniers, i.e. ridges of shells and sand. These are found on the seaward side of the salt marshes, and the origin is directly related to the transgression (Greensmith & Tucker 1975).

After 5000 B.P. a regression set in, lasting until 4350 B.P. In the Fenlands and in the Wash area already as early as 4700 B.P. the sea penetrated deeply, which is atributed to regional warping (Jardine, this vol.). The invading sea deposited the Fen Clay.

Although there were more transgressive phases than mentioned above, the resulting shorelines coincide about with the position of the present shoreline. As shown on the coloured map, the shorelines of 5000 and 2000 B.P. coincide more or less, except for a deeper penetration in the Wash area around 5000 B.P. Where the present coast consists of Pleistocene deposits, erosion may be strong (N and E Norfolk, Holderness). The coastline may retreat as much as 3 metres a year (Wilson 1948, Chatwin 1961).

Acknowledgements. The authors are grateful to Dr. S. Jelgersma and Dr. W.H. Zagwijn for their critical remarks.

REFERENCES

Alabaster, C. & Straw, A., 1976: The Pleistocene context of faunal remains and artefacts discovered at Welton-le-Wold, Lincolnshire. *Proc. Yorks. Geol. Soc. 41*, 75—94.

Allen, T., 1977: Interglacial sea-level change: evidence for brackish water sedimentation at Purfleet, Essex. *Quaternary Newslett. 22,* 1—3.

Andersen, B.G., 1977: Problems in Constructing the Weichselian Ice-front lines in Northern Europe for the CLIMAP project. *Quaternary Newslett. 22,* 9—10.

Averdieck, F.R., 1967: Die Vegetationsentwicklung des Eem-Interglazials und der Frühwürm Interstadiale von Odderade — Schleswig Holstein. *Fundamenta 2,* 101—125.

Baak, J.A., 1936: Regional petrology of the southern North Sea. Veenman & Zonen, Wageningen.

Behre, K.-E., Menke, B. & Streif, H., 1979: The Quaternary geological development of the German part of the North Sea, this volume.

Bertelsen, D., 1972: Azolla species from the Pleistocene of the Central North Sea area. *Grana Palynologica 12,* 131—145.

Boulton, G.S., Jones, A.S., Clayton, K.M. & Kenning, M.J., 1977: A British ice-sheet model and patterns of glacial erosion and deposition in Britain. In F.W. Shotton (editor), British Quaternary studies — recent advances, 231—246, Clarendon Press, Oxford.

Briggs, C.S., 1976: Cargoes and field clearance in the history of the English Channel. *Quaternary Newslett. 17,* 5.

Bristow, C.R. & Cox, F.C., 1973: The Gipping Till: a reappraisal of East Anglian stratigraphy. *J. Geol. Soc. 129,* 1—37.

Caston, V.N.D., 1979: A new isopachyte map of the Quaternary of the North Sea, this volume.

Catt, J.A., 1977a: Guidebook for excursion C7 (Yorkshire and Lincolnshire). Xth INQUA Congress, Birmingham 1977, 56 pp.

Catt, J.A., 1977b: Loess and coversands. In F.W. Shotton (editor), British Quaternary studies — recent advances, 221—229, Clarendon Press, Oxford.

Catt, J.A. & Penny, L.F., 1966: The Pleistocene deposits of Holderness, East Yorkshire. *Proc. Yorks. Geol. Soc. 35,* 3, 375—420.

Chatwin, C.P. 1961: British Regional Geology — East Anglia and adjoining areas. Fourth edition. Inst. Geol. Sci. London, 101 pp.

Clarke, R.H., 1973: Cainozoic subsidence in the North Sea. *Earth and Planet. Sci. Lett. 18,* 329—332.

Colhoun, E.A., Dickson, J.H., McCabe, A.M., & Shotton, F.W., 1972: A Middle Midlandian freshwater series at Derryvree, Maguiresbridge, County Fermanagh, Northern Ireland, *Proc. R. Soc. Lond.* B 180, 273—292.

Curry, J.R., 1961: Late Quaternary sea levels: a discussion. *Geol. Soc. Am. Bull.* 73, 159—162.

Destombes, J.-P., Shepard-Thorn, E.R. & Redding, J.H., 1975: A buried valley system in the Strait of Dover. *Philos. Trans. R. Soc. London A 270,* 243—256.

Donovan, D.T., 1973: The geology and origin of Silver Pit and other closed basins in the North Sea. *Proc. Yorks. Geol. Soc. 39,* 267—293.

Dreimanis, A. & Goldthwait, R.P., 1973: Wisconsin glaciation in the Huron, Erie and Ontario Lobes. *Mem. Geol. Soc. Am. 136,* 71—105.

Duphorn, K., Grube, F., Meyer, K.-D., Streif H. & Vinken, R., 1973: A. Area of the Scandinavian glaciation, 1. Pleistocene and Holocene. *Eiszeitalter und Gegenwart 23/24,* 222—250.

Eisma, D., Jansen, J.H.F. & van Weering, Tj. C.E., 1979: Sea-floor morphology and recent sediment movement in the North Sea, this volume.

Erd, K., 1970: Pollen-analytical classification of the Middle Pleistocene in the German Democratic Republic. *Palaeogeogr. Palaeoclimat. Palaeoecol. 8,* 129—145.

Eriksson, K. Gösta, 1979: Late Pleistocene and Holocene shorelines on the Swedish West Coast, this volume.

Fannin, N.G.T., 1975: A Regional Assessment of the Quaternary Succession in the north central North Sea. *Quaternary Newslett. 17,* 5—7.

Fannin, N.G.T., 1976: The Quaternary Geology of the British sector of the North Sea. Joint Oceanogr. Ass. 1976, Book of Abstr., 81 FAO, Rome.

Feyling-Hanssen, R.W., 1974: The Weichselian section of Foss-Eigeland, south-west Norway. *Geol. Fören. Stockh. Förh. 96,* 341—353.

Gaunt, G.D., Bartley, D.D. & Garland, R., 1974: Two interglacial deposits proved in boreholes in the southern part of the Vale of York and their bearing on contemporaneous sea levels. *Bull. Geol. Survey G.B. 48,* 1—23.

Gaunt, G.D. & Tooley, M.J., 1974: Evidence for Flandrian sea-level changes in the Humber estuary and adjacent areas. *Bull. Geol. Surv. G.B. 48,* 25—41.

Godwin, H., 1960: Radiocarbon dating and Quaternary history in Britain. *Proc. R. Soc. Lond. B 153,* 287—320.

Greensmith, J.T. & Tucker, E.V., 1973: Holocene transgressions and regressions on the Essex Coast outer Thames estuary. *Geol. Mijnbouw 52,* 193—202.

Greensmith, J.T. & Tucker, E.V., 1975: Dynamic Structures in the Holocene Chenier Plain Setting of Essex, England. In J. Hails & A. Carr (editors), Nearshore Sediment Dynamics and Sedimentation, Wiley, London, 251–270.

Greensmith, J.T. & Tucker, E.V., 1976: Major Flandrian transgressive cycles, sedimentation and palaeography in the coastal zone of Essex, England. *Geol. Mijnbouw 55*, 131–146.

Hafsten, U., 1979: Late and post-Weichselian shore level changes in South Norway, this volume.

Hageman, B.P., 1969: Development of the western part of the Netherlands during the Holocene. *Geol. Mijnbouw 48*, 373–388.

Hollin, J.T., 1977: Thames interglacial sites, Ipswichian sea levels and Antarctic ice surges. *Boreas 6*, 33–52.

Holmes, R., 1977: Quaternary deposits of the Central North Sea, 5. The Quaternary Geology of the U.K. sector of the North Sea between 56° and 58 ° N. *Rep. Inst. Geol. Sci. 77/14*, 50 pp.

Holtedahl, H. & Bjerkli, K., 1975: Pleistocene and recent sediments of the Norwegian continental shelf (62° N – 71° N) and the Norwegian Channel area. *Norg. Geol. Unders. 316*, 241–252.

Hoppe, G., 1974: The glacial history of the Shetland Islands. *Inst. Geogr. Spec. Publ. 7*, 197–210.

Jansen, J.H.F., 1976a: Late Pleistocene and Holocene history of the northern North Sea, based on acoustic reflection records. *Neth. J. Sea Res. 10*, 1–43.

Jansen, J.H.F., 1976b: Late Pleistocene and Holocene history of the northern North Sea. Joint Oceanographic Assembly 1976, Edinburgh, Poster Session C12, abstract, 6 pp.

Jansen, J.H.F., in press: Late Pleistocene and Holocene deposits in the Witch and Fladen Ground area, northern North Sea. To be published in *Marine Geology*.

Jansen, J.H.F., van Weering, Tj.C.E. & Eisma, D., 1979: Late Quaternary sedimentation in the North Sea, this volume.

Jardine, W.G., 1979: The western (United Kingdom) shore of the North Sea in Late Pleistocene and Holocene times, this volume.

Jelgersma, S., 1961: Holocene sea level changes in the Netherlands. *Meded. Geol. Sticht., C, VI*, 7, 100 pp.

Jelgersma, S., 1979: Sea-level changes in the North Sea basin, this volume.

Jelgersma, S. & Breeuwer, J.B. 1975: Toelichting bij de geologische overzichtsprofielen van Nederland. In W.H. Zagwijn & C.J. van Staalduinen (editors), Toelichting bij de Geologische Overzichtskaarten van Nederland, 91–93, Rijks Geologische Dienst, Haarlem.

Jelgersma, S., Oele, E. & Wiggers, A.J., 1979: Depositional history and coastal development in the Netherlands and the adjacent North Sea since the Eemian, this volume.

Jessen, K. & Milthers, V., 1928: Stratigraphical and Paleontological Studies of Interglacial Freshwater Deposits in Jutland and Northwest Germany. *Danm. Geol. Unders. II. 48*, 1–379.

Kellaway, G.A., Redding, J.H., Shephard-Thorn, E.R. & Destombes, J.-P., 1975: The Quaternary history of the English Channel. *Philos. Trans. R. Soc. London A 279*, 189–218.

Kidson, C., 1977: Some problems of the Quaternary of the Irish Sea. In C. Kidson & M.J. Tooley (editors), The Quaternary History of the Irish Sea, *Geol. J. Spec. Issue 7*, 299–320.

Kidson, C. & Bowen, D. Q., 1976: Some comments on the history of the English Channel. *Quaternary Newslett. 18*, 8–9.

Kirby, R. & Oele, E., 1975: The geological history of the Sandettie-Fairy Bank area, southern North Sea. *Philos. Trans. R. Soc. London A 279*, 257–267.

Kolstrup, E. & Wymstra, T.A., 1977: A palynological investigation of the Moershoofd, Hengelo and Denekamp interstadials in the Netherlands. *Geol. Mijnbouw 56*, 85–102.

Krog, H., 1979: Late Pleistocene and Holocene shorelines in western Denmark, this volume.

Lamb, H.H., 1977: The Late Quaternary history of the climate of the British Isles. In F.W. Shotton (editor), British Quaternary Studies – recent advances, 283–298, Clarendon Press, Oxford.

Løfaldli, M., 1973: Foraminiferal biostratigraphy of late Quaternary from the Frigg Field and Booster Station. *Rep. NTNF Cont. Shelf Div. 18*, 83 pp.

Maarleveld, G.C., 1953: Standen van het landijs in Nederland. *Boor en Spade 6*, 95–105.

McCave, I.N., Caston, V.N.D. & Fannin, N.G.T., 1977: The Quaternary of the North Sea. In F.W. Shotton (editor), British Quaternary Studies – recent advances, 187–204, Clarendon Press, Oxford.

Madgett, P.A., 1975: Re-interpretation of Devensian till stratigraphy of eastern England. *Nature 253*, 105–107.

Mangerud, J., 1972: The Eemian Interglacial and the succession of glaciations during the Last Ice Age (Weichselian) in southern Norway. *Ambio Spec. Rep. 2*, 39–44.

Mangerud, J., Sønstegaard, E. & Sejrup, H.-P., 1977: Saalian-Eemian-Weichselian stratigraphy at Fjøsanger, Western Norway. Xth INQUA Congress, Birmingham 1977, abstracts, 286.

Mayhew, D.F., 1976: Comments on the British Glacial-Interglacial Sequence. *Quaternary Newslett. 19*, 8–9.

Menke, B., 1968: Beiträge zur Biostratigraphie des Mittelpleistozäns in Norddeutschland. *Meyniana 18*, 35–42.

Mitchell, G.F., 1977: Raised beaches and sea-levels. In F.W. Shotton (editor), British Quaternary Studies – recent advances, 169–186, Clarendon Press, Oxford.

Mitchell, G.F., Penny, L.F., Shotton, F.W. & West, R.G., 1973: A correlation of Quaternary deposits in the British Isles. *Geol. Soc. London, Spec. Rep. 4*, 99 pp.

Mörner, N.A., 1969: The Late Quaternary history of the Kattegatt Sea and the Swedish West Coast. Deglaciation, shorelevel displacement, chronology and eustacy. *Sver. Geol. Unders. C. 640*, 487 pp.

Mottershead, D.N., 1977: The Quaternary evolution of the south coast of England. In C. Kidson & M.J. Tooley (editors), The Quaternary History of the Irish Sea, *Geol. J. Spec. Issue 7*, 229–320.

Oele, E., 1969: The Quaternary geology of the Dutch part of the North Sea, north of the Frisian Islands. *Geol. Mijnbouw 48*, 467–480.

Oele, E., 1971: The Quaternary geology of the southern area of the Dutch part of the North Sea. *Geol. Mijnbouw 50*, 461–474.

Oele, E., 1974: Geologisch Onderzoek in de Grindconcessie Norfolk. *Rijks Geol. Dienst, Intern. Rep. 1136*, 15 pp.

D'Olier, B., 1975: Tunnel valleys and associated features of the Southern Bight of the North Sea. *Quaternary Newslett.* 17, 5.

Owens, R., 1977: Quaternary deposits of the central North Sea, 4. Preliminary report on the superficial sediments of the central North Sea. *Rep. Inst. Geol. Sci. 77/13*, 16 pp.

Paepe, R., 1971: Autosnelweg Brugge-Calais – Kb. Houtave, Bredene, Gistel. Prof. Paper 1971/9, *Belg. Geol. Survey: 59*, Brussel.

Paepe, R. & Baeteman, C., 1979: The Belgian coastal plain during the Quaternary, this volume.

Paepe, R. & Vanhoorne, R., 1972: An outcrop of Eemian Wadden Deposits at Meetkerke (Belgian Coastal Plain). Prof. Paper 1972/7, Belg. Geol. Survey, Brussel, 9 pp.

Paepe, R., Vanhoorne, R. & Deraymaker, D., 1972: Eemian Sediments near Bruges (Belgian Coastal Plain). *Prof. Paper 1972/9, Belg. Geol. Survey*, Brussel, 11 pp.

Reinhard, H., 1974: Genese des Nordseeraumes im Quartär. *Fennia 129*, Helsinki, 96 pp.

Rose, J., Pennington, W. & Catt, J.A., 1977: Quaternary research association study course S. Norway, August 1976. *Quaternary Newslett. 21*, 1–17.

Shackleton, N.J. & Opdyke, N.D., 1973: Oxygen isotope and palaeomagnetic stratigraphy of Equatorial Pacific core V28 – 238: oxygen isotope temperatures and ice volumes on a 10^5 and 10^6 year scale. *Quaternary Res. 3*, 39–55.

Shotton, F.W., 1977: British dating work with radioactive isotopes. In F.W. Shotton (editor), British Quaternary Studies – recent advances, 17–29, Clarendon Press, Oxford.

Shotton, F.W., Blundell, D.J. & Williams, R.E.G., 1969: Birmingham University Radiocarbon Dates III. *Radiocarbon 11*, 2, 263–270.

Shotton, F.W., Banham, P.H. & Bishop, W.W., 1977: Glacial-interglacial stratigraphy of the Quaternary in Midland and eastern England. In F.W. Shotton (editor), British Quaternary Studies – recent advances, 268–282, Clarendon Press, Oxford.

Sindowski, K.-H., 1970: Das Quartär im Untergrund der Deutschen Bucht (Nordsee). *Eiszeitalter und Gegenwart 21*, 33–46.

Sissons, J.B., 1976: The geomorphology of the British Islands: Scotland. London.

Sommé, J., 1979: Quaternary coastlines in Northern France, this volume.

Staalduinen, C.J. van, 1977: Geologisch onderzoek van het nederlandse Waddengebied. compiled by C.J. van Staalduinen, Rijks Geologische Dienst, Haarlem, 77 pp.

Stevens, L.A., 1960: The interglacial of the Nar Valley, Norfolk. *Quart. J. Geol. Soc. London 115*, 291–316.

Suttcliffe, A.J., 1976: The British Glacial – Interglacial sequence. *Quaternary Newslett. 18*, 1–7.

Swift, D.J.P., 1975: Tidal sand ridges and shoal-retreat massifs. *Mar. Geol. 18*, 105–134.

Thomson, M.E. & Eden, R.A., 1977: Quaternary deposits of the Central North Sea, 3. The Quaternary sequence in the west-central North Sea. *Rep. Inst. Geol. Sci. 77/12*, 18 pp.

Turner, C., 1970: The Middle Pleistocene deposits at Marks Tey, Essex. Philos. Trans. R. Soc. London B 257, 373–440.

Valentin, H., 1957: Glazialmorphologische Untersuchungen in Ostengland. *Abh. Geogr. Inst. F.U. Berlin 4*, 1–86.

Vandenberghe, J. & Gullentops, F., 1977: Contribution to the Stratigraphy of the Weichsel Pleniglacial in the Belgian cover sand area. Xth INQUA Congress, Birmingham 1977, abstracts, 474.

Veenstra, H.J., 1965: Geology of the Dogger Bank area, North Sea. *Mar. Geol. 3*, 245–262.

Veenstra, H.J., 1969: Gravels of the southern North Sea. *Mar. Geol. 7*, 443–464.

Vorren, T.O., 1972: Interstadial sediments with rebedded interglacial pollen from Inner Sogn, West Norway. *Nor. Geol. Tidsskr. 52*, 229–240.

Wee, M.W. ter, 1962: The Saalian glaciation in the Netherlands. *Meded. Geol. Sticht., Nieuwe Ser. 15*, 57–76.

Wee, M.W. ter, 1976: Toelichtingen bij de Geologische kaart van Nederland 1:50.000. Blad Sneek (10 W, 10 0), Rijks Geologische Dienst, Haarlem, 131 pp.

Weering, Tj.C.E. van, 1975: Late Quaternary history of the Skagerrak; an interpretation of acoustical profiles. *Geol. Mijnbouw 54*, 130–145.

West, R.G., 1968: Pleistocene geology and biology. XIII + 377 pp., Longmans, London.

West, R.G., 1969: A note on pollen analysis from the Speeton Shell Bed. In L.F. Penny & P.F. Rawson, Field Meeting in east Yorkshire and north Lincolnshire. Report by the directors, with an appendix. *Proc. Geol. Assoc. 80,* 193—218.

West, R.G., 1970: Pleistocene history of the British Flora. In D. Walker & R.G. West (editors), Studies in the Vegetational History of the British Isles, 1—11, Cambridge University Press, Cambridge.

West, R.G., 1977a: Pleistocene Geology and Biology. 2nd edition, Longmans, London.

West, R.G., 1977b: Guidbook for excursions A1 and C1 (East Anglia), Xth INQUA congress, Birmingham 1977, 65 pp.

West, R.G., Dickson, C.A., Catt, J.A., Weir, A.H. & Sparks, B.W., 1974: Late Pleistocene deposits of Wretton, Norfolk. II Devensian deposits. *Philos. Trans. R.Soc. London B 267,* 337—420.

Wilson, V., 1948: British Regional Geology — East Yorkshire and Lincolnshire. Inst. Geol. Sci., London, 1948, 94 pp.

Worsley, P., 1977: Problems of the Weichselian glaciation in Scandinavia. *Quaternary Newslett. 21,* 24—27.

Zagwijn, W.H., 1961: Vegetation, climate and radiocarbon datings in the Late Pleistocene of the Netherlands. Part I: Eemian and Early Weichselian. *Meded. Geol. Sticht., Nieuwe Ser. 14,* 15—45.

Zagwijn, W.H., 1973: Pollenanalytic studies of Holsteinian and Saalian Beds in the Northern Netherlands. *Meded. Rijks. Geol. Dienst, Nieuwe Ser. 24,* 139—156.

Zagwijn, W.H., 1974: The Palaeogeographic Evolution in the Netherlands during the Quaternary. *Geol. Mijnbouw 53,* 369—385.

Zagwijn, W.H., 1975: Indeling van het Kwartair op grond van veranderingen in vegetatie en klimaat. In W.H. Zagwijn & C.J. van Staalduinen (editors), Toelichting bij geologische overzichtskaarten van Nederland, 109—114, Rijks Geologische Dienst, Haarlem.

Zagwijn, W.H., 1977a: Stratigrafische interpretative van boringen tot ca. 100 m. onder zeeniveau in het Noordzeegebied. *Rijks Geologische Dienst, Afd. Palaeobot., Intern. Rep. 769,* 12 pp.

Zagwijn, W.H., 1977b: Sea level changes during the Eemian in the Netherlands. Xth INQUA Congress, Birmingham 1977, abstracts, 509.

Zagwijn, W.H., 1979: Early and Middle Pleistocene coastlines in the southern North Sea Basin, this volume.

Zagwijn, W.H. & Veenstra, H.J., 1966: A pollen-analytical study of cores from the Outer Silver Pit, North Sea. *Mar. Geol. 4,* 539—551.

Zandstra, J.G., 1971: Sedimentpetrologisch onderzoek van de Fugro-boring E12—B1 (Noordzee). *Rijks Geologische Dienst, Sedimentpetrologische Afd., Intern. Rep. 257,* 2 pp.

Zandstra, J.G., 1972: Grind uit spuitboring 72GS32 op de Oostbank. *Rijks Geologische Dienst, Sedimentpetrologische Afd., Intern. Rep. 343,* 1p.

Zandstra, J.G., 1973: Onderzoek van monsters uit spuitboringen in de zuidelijke Noordzee. *Rijks Geologische Dienst, Sedimentpetrologische Afd., Intern. Rep. 398,* 2 pp.

Sea-floor morphology and recent sediment movement in the North Sea

DOEKE EISMA, J.H. FRED JANSEN & TJEERD C.E. van WEERING

Eisma, D., Jansen, J. H. F. & van Weering, Tj. C. E. 1979: Sea-floor morphology and recent sediment movement in the North Sea. In E. Oele, R. T. E. Schüttenhelm & A. J. Wiggers (editors), The Quaternary History of the North Sea, 217–231. *Acta Univ. Ups. Symp. Univ. Ups. Annum Quingentesimum Celebrantis: 2*, Uppsala. ISBN 91-554-0495-2.

The morphology of the North Sea floor is dominated by relict features except in the Southern Bight and adjacent areas. The oldest are the glacially eroded Norwegian Channel and Skagerrak, the still unexplained Dogger Bank, and the channels and gravel deposits in the southern North Sea. Others originated during the last Weichselian glaciation: gravel banks bordering the Norwegian Channel, which are considered as terminal morianes and glacial outwash deposits, iceberg grooves and subglacial tunnel valleys. The tunnel valleys in the southern North Sea probably were reshaped later by tidal scour. During the postglacial transgression linear sand ridges were formed northwest of Dogger Bank as well as submarine terraces south of this area.

The Holocene Young Sea-sands in the southern North Sea are represented by nearly planar beds, megaripples, sand waves and linear ridges. The planar beds and the megaripples are the result of recent sediment movements. The sand waves have a certain relation to the present current regime, but at least partly originated during earlier Holocene times. The almost stationary linear sand ridges are possibly also older but adapted to present flow conditions.

Most of the fine-grained suspended matter, supplied from inflow and/or erosion or organic production is deposited in the Norwegian Channel and Skagerrak/Kattegat or transported into the North Atlantic. Deposition also takes place in the Outer Silver Pit area, in the Wadden Sea and in the German Bight.

Pockmarks in the northern North Sea are attributed to escaping gas.

Dr. D. Eisma, Drs. J. H. F. Jansen and Drs. Tj. C. E. van Weering, Nederlands Instituut voor Onderzoek der Zee, Postbus 59, Den Burg, Texel, the Netherlands.

Introduction

The North Sea is a shallow epicontinental sea. Depths in the southern North Sea increase from less than 30 m in the Southern Bight to 50 m south of Dogger Bank, and in the northern Sea from 80 m just north of Dogger Bank to 200 m at the shelf edge between Shetland and Norway (Figs. V-3 and V-4). At the eastern side a large depression, the Norwegian Channel, follows the Norwegian coast and continues in the Skagerrak, where a depth of 700 m is reached. Other but smaller depressions are found in the central part of the North Sea (Devil's Hole 263 m, Outer Silver Pit area 70–90 m,) and in the Fladen Ground (302 m). A conspicuous feature is the shallow Dogger Bank in the central North Sea with a depth of 18 m at the southwestern end.

The North Sea attained its present shape dur-

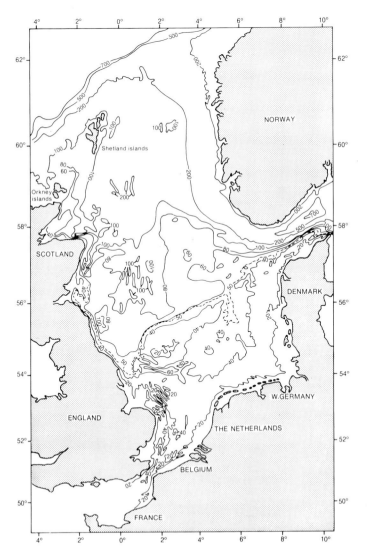

Fig. V-3. Bathemetry, mainly based on Fishery Charts Flinn (1973) and Jansen (1976).

ing Early Tertiary times (Ziegler 1975, Kent 1975, Ziegler & Louwerens 1977). During the Quaternary, glacial stages with lower sea level alternated with interglacial stages with higher sea level. Most morphological features are the result of the last Weichselian glaciation in the North Sea region, when relative sea level was about minus 110 m, and of the subsequent transgression (Jansen et al. 1977). Because of the great depths the Weichselian relief of the northern North Sea has been changed only to a minor degree by postglacial erosion and deposition, and the morphology is characterized by a predominance of relict features. These are less common in the southern North Sea: the Southern Bight and adjacent areas are predominated by subrecent and recent depositional features (Fig. V-5).

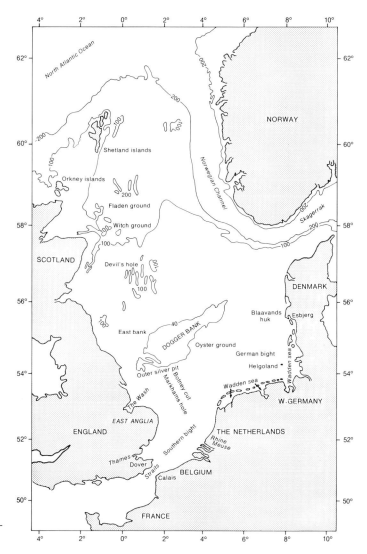

Fig. V-4. Locations mentioned in the text.

Relict features

Norwegian Channel and Skagerrak

The Norwegian Channel and the Skagerrak were shaped by large-scale glacial erosion during the Pleistocene (Sellevoll & Aalstad 1971, Flodén & Sellevoll 1972, Sellevoll & Sundvor 1974), but O. Holtedahl (1929, 1956, 1964) advocated a tectonic origin. Recent research indicates that at least along the southern border of the Skagerrak tectonic movements have been active (van Weering, in prep.).

The Norwegian Channel and the Skagerrak are only partly filled up with Quaternary sediments; also the Weichselian relief has not been obliterated by the postglacial deposition.

Iceberg grooves

On the continental slope between Norway and the Shetland Islands, as well as in the Skagerrak and the Norwegian Channel, large furrows are present in the sea floor, up to 10 m deep, several hundred metres wide and up to 3 km long (Belderson & Wilson 1973, Belderson, Kenyon & Wilson 1973; van Weering 1975). The origin of these furrows (Fig. V-6) has been ascribed to a ploughing action of grounded icebergs; they have been called iceberg grooves (Berkson & Clay 1973), iceberg plough marks (Belderson, Kenyon & Wilson 1973), ice gouges (Reimnitz & Barnes 1974) and iceberg furrow marks (Harris & Jollymore 1974). The relief of the grooves, which are found at depths of 140 to 500 m, increases with depth as the larger icebergs strand in deeper waters. Criss-cross patterns have been observed northeast of the Shetland Islands but subparallel trends following the depth contours are more common along the western slope of the Norwegian Channel, probably because upslope the drift of the icebergs was brought to a halt. Belderson and Wilson (1973), however, suggested a movement of the icebergs by longitudinally directed paleocurrents as an alternative explanation.

In the deeper parts of the Norwegian Channel and the Skagerrak the grooves may have existed as well, but they are now covered by recent mud deposits; off Stavanger remnants of ice grooves have been noted on slight rises in the seafloor (Belderson & Wilson 1973). In other parts of the North Sea, where iceberg grooves have not yet been reported, reworking may have destroyed them. Large icebergs possibly were present during the stage of deglaciation from about 18000 B.P. when the glaciers started to retreat, to about 12000 B.P. when the ice front had receded at least as far as the present coast line of Norway. The deepest grooves therefore probably date from this period.

Tunnel valleys

In the northern North Sea a series of deep valleys is present from Devil's Hole via northeast Scotland to about 59° N, 1° E (Figs. V-5 and V-7). These valleys are 1 to 3 km wide, 25–60 km long, and reach a depth of 100 m or more below the sea floor. They are U-shaped in cross section, have a characteristic undulating thalweg without a one-directional valley gradient, and terminate abruptly (Flinn 1967, Jansen 1976). According to Flinn (1967), Donovan (1973) and Jansen (1976), these valleys cannot have been formed by tidal scour, subaerial erosion, or tectonic movements. Valentin (1957) interpreted them as tunnel valleys comparable to those in northern Germany and Denmark. This view is now generally accepted. Tunnel valleys are considered to have been formed under a continental ice sheet, either by subglacial meltwater (Woodland 1970, Wright 1973) or by ice erosion (Hansen 1971); their point of outflow nearly always coincides with the ice front and their orientation is about parallel to the ice current. The formation of the tunnel valleys in the northern North Sea is correlated with the development of the glaciomarine Fladen Deposits during the Weichselian glacial maximum (Flinn 1967, Jansen 1976 Jansen et al. 1977).

The valleys between Dogger Bank and the Humber are less deep, from 30 m to 60 m lower than the surrounding sea floor. Subglacial erosion and also tidal scour can only partly explain their formation. Therefore, they are probably of composite origin, initiated as tunnel valleys and

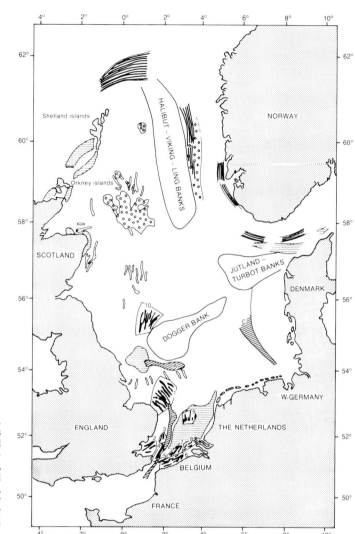

Fig. V-5. Morphological map. 1-Norfolk Banks, 2-Thames Estuary Banks, 3-Gabbard Banks, 4-The Falls, 5-Sandettie, 6-Flemish Banks, 7-Hinder Banks, 8-Zeeland Ridges, 9-Brown Bank ridges, 10-East Bank ridges. A-Outer Silver Pit, B-Deep Water Channel, C-Helgoland Channel.

reshaped by tidal scour during the Early Holocene, as suggested by Donovan (1973).

Other relict features

The most extensive change in the Weichselian relief of the northern North Sea was caused by the deposition of the East Bank Deposit together with the formation of the tidal sand ridges in the East Bank area. These relict ridges, which are similar to the linear ridges in the Southern Bight (see below), were formed probably in the mouth of an embayment southwest of Dogger Bank during the Early Holocene, when sea level was 40 to 45 m lower (Jansen 1976, Jansen et al. 1977). During this period also a terrace at minus

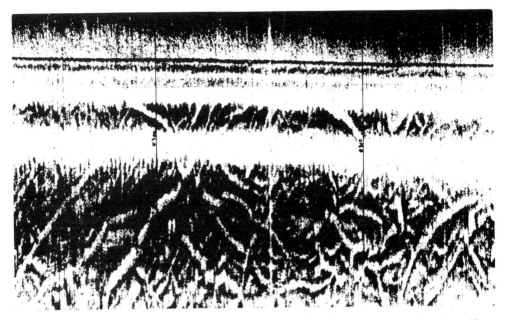

Fig. V-6. Side-scan sonar record of iceberg grooves, showing a random cirss-cross pattern of superimposition. Location off the Norwegian coast between Stavanger and Egersund (from Belderson & Wilson, 1973).

45 m was formed around the nearby Oyster Grounds (Kolp 1964, 1976). South of Dogger Bank, Kolp found terraces at −30 and −60 m as well. At present, deposition is hardly of importance; the East Bank Deposit forms the present sea-floor and is slightly reworked at the most (Jansen 1976 and unpubl. data).

The Dogger Bank, the origin of which is still unexplained (Oele 1969, 1971), is a large sand body covering Elsterian fluvioglacial clay and possibly some boulder clay, and can be considered as a relict feature. The upper part is regularly being reworked by stormwaves, and the steep slope on the northwest-side is most probably the result of coastal erosion during the Early Holocene when sea level was lower. Such a steep north-western slope is also found at the Turbot and Jutland banks further east. The gravel banks bordering the Norwegian Channel and the Oyster Grounds are considered to be reworked terminal moraines and glacial outwash deposits, although the original relief has partly been flattened (Pratje 1951).

Next to the Norwegian Channel and the Skagerrak, three smaller channels can be distinguished in the North Sea. The Outer Silver Pit probably was part of an old valley system, and was reshaped during the Weichselian by subglacial erosion (Donovan 1973). The Helgoland Channel is a Late Mesozoic fault zone (Sindowski 1970) which was traversed by rivers during the Pleistocene. The Deep Water Channel in the Southern Bight, continuing through the Straits of Dover-Calais into the Channel, is considered by some authors (Destombes et al. 1975) to be a former glacial valley.

The gravel deposits in the southern North Sea must also be considered relict features; they are probably not reworked any more, or only to a minor degree (Jansen et al. 1977): overgrowth and encrustations of Hydrozoa, Bryozoa and Serpulids, found on gravel south of Dogger Bank, off East Anglia and near the Hinder Banks, indicate that the gravel has been at rest for some time. Most gravel is present as an admixture in coarse sandy deposits, but gravel

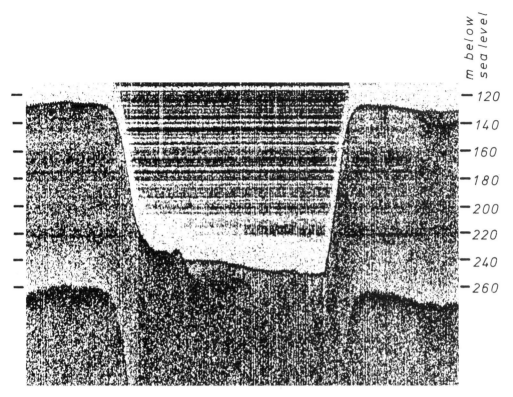

Fig. V-7. Acoustic reflection record of a tunnel valley at 58°58'N 0°12'W, the Fladen Ground Deep (from Jansen 1976).

pavement and gravel lag deposits are found between (and as a continuous layer below) the Hinder Banks, the Flemish Banks, the Gabbard Banks, and around Sandettie (Veenstra 1964, Houbolt 1968, Kirby and Oele 1975). Local sorting of gravel was observed by Veenstra (1969) south of Dogger Bank; the shallower parts contain about twice as much coarse gravel as the deeper parts. Since nearly all gravel is found in water of 30 m or deeper, sorting must have taken place when sea level was lower.

Present current pattern and depositional features

The distribution of both relict and subrecent or recent depositional features reflects the present current pattern in the North Sea. The tidal waves enter the North Sea around Scotland and the Shetlands and through the Channel; a residual current that is related to the windfield over the North Atlantic Ocean and the North Sea, causes the surface water in the North Sea to circulate anticlockwise, while in the Southern Bight a resultant current is directed towards the North (Lee and Ramster 1968). Strong tidal currents with an average maximum velocity of >75 cm/s at the surface are found between Scotland and the Shetlands, along the Scottish-English coast, in the Southern Bight and off the Wadden Sea up to Esbjerg (Sager 1963). In the Oyster Grounds the average maximum tidal-surfacecurrent velocity decreases to less than 40 cm/s and in the

Skagerrak to less than 12.5 cm/s. Tidal currents near the sea-floor are usually different from those át the surface in velocity and direction, but they are generally weaker. In the Norwegian Channel and Skagerrak a subsurface current below a depth of about 50 m moves southward along the western side, turns into the Skagerrak and moves northward along the Norwegian coast. Owing to variations in wind strength and topographic conditions, non-tidal currents in the Norwegian Channel may increase to 45 cm/s and in the Skagerrak to more than 75 cm/s (Stride and Chesterman 1973, Enger 1976). An increasing waterdepth renders the effect of surface waves on the seafloor to a minimum in the northern North Sea, and bottom material is reworked on a limited scale only (Bratteland and Bruun 1974).

Deposition of suspended material supplied from the South occurs mainly in the Skagerrak/Kattegat area and in the Norwegian Channel where current velocities are low. The strongest sediment movements take place in the Southern Bight and adjacent areas, where the waterdepths are generally less than 35 m and the maximum tidal currents are in the order of 75—125 cm/s. In spite of the strong currents and wave action, however, erosional features are scarce in the Southern Bight. Recent erosion is mainly restricted to the coastal areas. Offshore erosional features are developed temporarily during storms when sand waves are eroded, and locally where the seafloor is eroded. In the Straits of Dover-Calais strong tidal currents prevent the sand from being deposited on the rocky seafloor.

The Young Sea-sands forming the present sea-floor nearly everywhere in the southern North Sea have been deposited as nearly planar beds, as megaripples with heights of 0.3—2.0 m, as sand waves of 2—15 m height and as linear sand ridges up to 40 m high and 65 km long. The largest megaripple fields are present on top of the sand waves (Terwindt 1971, Langhorne 1973). There is no consensus of opinion about the relation of the large depositional features in the southern North Sea to the present hydraulic conditions. Nearshore sand transport and sedimentation as a result of tidal currents and wave action are well known, and complicated systems of nearshore shallow banks have been charted along the East Anglia coast, in the Thames estuary, off Ramsgate, off Blaavandshuk, at the tidal inlets of southern Holland, the Wadden Sea and the Wash, and along the sandy beaches, where up to three nearshore sandbanks are present parallel to the coast. Farther offshore, tidal currents and waves are strong enough to transport sand and to form small ripples as well as megaripple systems, but to what extent sand waves and linear sand ridges are relict features is still under discussion.

Sand waves

Two large groups of sand waves are found in the southern North Sea, one covering most of the Southern Bight, the other occupying the area southwest of Dogger Bank. Smaller sand wave groups are known in the Southern Bight outside the main sand wave field, off Norfolk and further north along the Scottish coast, around the Orkney Islands, and off northwest Denmark. Sand waves are apparently formed where a) sufficient sand is available, b) current strength is >60 cm/s at mean spring tide (but this is not the case off northwest Denmark), and c) wave action is weak to moderate. The absence of sand waves down to depths of about 18 m along the Dutch coast is attributed to wave action during storms (Terwindt 1971, McCave 1971): small sand waves present during periods of calm weather disappear during a storm period. McCave (1971) suggested a relation between sand waves and the shape of the tidal ellipse, but Stride (1973) proved this to be incorrect. Sand waves are not found in areas with coarse sands (median diameter >500 μm) containing much gravel or in sands with a large admixture (> 15%) of fine-grained material (< 50 μm).

The sand waves in the Southern Bight (which are the most intensively studied) are up to 15 m high and several hundreds of metres long. The crests are oriented normal to the principal directions of ebb and flood, while the slopes are flat or concave and steepening towards the top. The sand wave height gradually decreases from a central sand wave area off the southern Dutch coast to the North as well as to the South. In the central area the sand waves are symmetrical but outside this area they become asymmetrical: in the South the steep slope faces southwest, in the North north-northeast.

Stride (1963) and McCave (1971) found a rough relation between the sand wave orientation and the direction of the maximum tidal current velocity and, presumably, the net total sediment transport. The shape of the sand waves has therefore been used as an indicator of the direction of sand transport (Stride 1963, 1973, Houbolt 1968, McCave 1971). Terwindt (1971), however, pointed out that the formation of sand waves has not yet been adequately explained, and that the asymmetry of the sand waves has not been proved to be a result of the present hydraulic conditions. Asymmetric sand waves therefore cannot be unconditionally used as indicators for sand transport directions. The same is implied by Lekahena (1966), Winkelmolen (1969), Veenstra (1971) and Nio (1976), as well as by van Weering (1975) for the sand waves off NW-Denmark; they considered the sand waves to be partly or mainly formed under different flow conditions during earlier periods of the Holocene when sea level was lower. Another complication is the presence of megaripple systems on top of the sand waves (Houbolt 1968, McCave 1971, Terwindt 1971, Langhorne 1973). These seem to be restricted to the higher sand waves (height >5 m), and the direction of their crests makes an angle of up to 60° with the sand wave crests; this suggests that during maximum sand transport the flow is usually oblique to the general direction of the sand waves. This would be in support of the assumption that sand waves have at least partly been formed under different conditions of flow.

Most investigators (van Veen 1935, 1938, Stride 1963, 1973, McCave 1971, Dingle 1965) consider the sand waves to move in the direction of the steeper slope. Langeraar (1966) found the movement — if any — to be less than 24 m/y; McCave (1971) estimated a net advance of about 15 m/y. Johnson and Stride (1969) calculated a net northward sand transport of 8 x 10^6 m^3/y over a distance of about 40 miles. If it is assumed that the sand waves maintain their shape, the net advance must be about 20 m/y, considering that some coarse material will remain behind and some fine material will be removed in suspension. Stride (1970), however, found a decrease in height off the Dutch coast from south to north, but not a concurrent decrease in length. Allowing for the concave slopes of the sand waves in the South, the volume of the sand waves is not noticeably smaller: some individual waves in the North even have a larger cross-sectional volume than those in the South. This suggests that the sand waves do not maintain their identity while travelling north but may break up and fuse with parts of nearby waves.

The sand waves off northwestern Denmark, which are up to 10 m high, are abnormal, as they occur in an area where tidal currents are weak (average maximum 12.5 cm/s). Bruun and Vollen (1972) suggested that these sand waves are formed by internal waves or as lee waves behind a barrier on the sea-floor. According to Stride and Chesterman (1973) they have been formed by strong wind-induced coastal currents along the Danish coast. Van Weering (1975) considered them as probable relicts from a period when sea level was lower and bottom current velocities higher.

Linear sand ridges

Groups of linear sand ridges are found off the Norfolk coast and in the Southern Bight: the Flemish Banks, the Hinder, Brown and Gabbard Banks, the Zeeland ridges, the Thames estuary banks, the Falls Banks and Sandettie (Houbolt 1968, Caston 1970) (Fig. V-5 and V-8). The length may be up to 65 km and the width up to 2 km. The maximum height above the surrounding sea-floor is 40 m. In cross-section they are usually asymmetric and they are covered with megaripples and sand wave systems, on both sides directed towards the crest. Most linear sand ridges lie on a flat erosion surface with a thin cover of lag deposits (mostly gravel) and are accumulation forms, but the Brown Bank and the Zeeland ridges, that have a resistant core of older deposits, are partly erosional remnants. The ridges are aligned parallel to the main ebb and flood current directions: on one

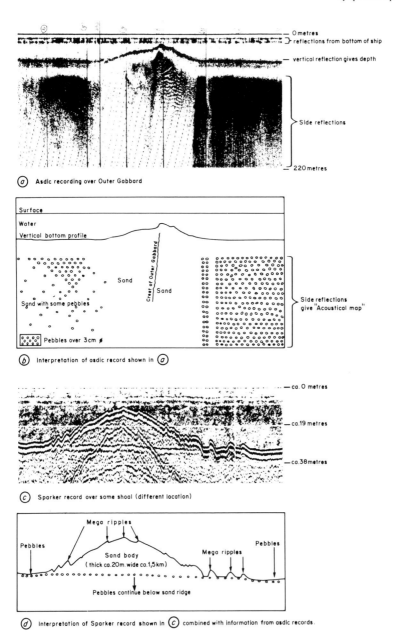

Fig. V-8. Asdic and sparker records over a linear sand ridge (the Outer Gabbard) in the Southern Bight, and their interpretation (from Houbolt 1968).

side the sand moves mainly in the flood direction, on the other mainly in the ebb direction. Most ridges are almost stationary. A comparison of the present situation with old maps shows that the Flemish Banks have only slightly changed in 300 years, and that the Hinder Banks have been stationary for 40 years (Houbolt 1968). The Norfolk Banks are moving very slowly to the NNE, i.e. in the direction of the steep side, while they are also slowly gaining in lengt to the NW, which is the dominant sand transport direction. Caston (1970) proposed a cyclic development of linear ridges which are gradually split up by ebb and flood channels.

The linear sand ridges are clearly related to and in dynamic equilibrium with present flow conditions but it probably took a long time to reach their present shape. Swift (1975) therefore suggested that former elongate estuarine sand banks, such as found today in the Thames estuary, gradually became sand banks on the open shelf during the rise of sea level and developed into linear sand ridges, which are virtually closed systems of sand movement. The flow pattern to maintain the linear ridges, however, is very complicated and not well understood. Rotating currents and lag effects in the entrainment of sand during the tidal cycle have been mentioned in order to explain the accumulation of sand in ridges parallel to the main tidal current axis (Houbolt 1968, Stride 1974, Swift 1975).

Transport and deposition of suspended matter

At present only fine-grained material is being moved in suspension in appreciable quantities and over large distances. There is no sharp upper size limit of the material that can be moved in suspension. McCave (1971) calculated that in the Southern Bight larg quantities of grains of 200—300 μm are moved in this way. It has also been observed that during storms (force 9—10 Beaufort) sand from the seafloor at Dogger Bank was thrown onto the deck of fishing-vessels (Stride 1973). Coulter Counter measurements of particle size in the Southern Bight indicated, however, that only few grains >70 μm were present in suspension throughout the watercolumn down to about 1 m above the sea-floor at windspeeds up to force 7 Beaufort (Eisma and Gieskes 1977, Eisma 1976 and unpubl. data). This indicates that large quantities of sand are moved in suspension only during storms.

The fine-grained material in suspension is supplied by rivers running towards the North Sea (mainly in the South), by erosion of the coasts and of the sea-floor, by inflow from the North Atlantic Ocean, the Channel and the Baltic, by organic production in the North Sea and by inflow of dust from the atmosphere. Allowing for uncertainties concerning erosion of the sea-floor and the amount of deposition in eastuaries and fjords, at least 25 million tons of suspended matter is supplied to the North Sea annually (McCave 1973, Eisma, unpubl. data). About 5 million tons is deposited in the Outer Silver Pit area (chiefly in the Botney Cut and Markham's Hole) and southeast of Helgoland in the German Bight. Probably several million tons are deposited in the Wadden Sea. This leaves at least 17 million tons to be deposited in the Skagerrak/Kattegat and the Norwegian Channel or to be transported into the Atlantic Ocean, since north of Dogger Bank deposition is hardly of importance (Jansen, unpubl. data).

The suspended material in the North Sea is highly concentrated in the coastal waters of the southern North Sea, whereas lower values are found offshore in the Skagerrak and north of Dogger Bank. The concentrations range from more than 100 mg/l in nearshore waters in the Southern Bight (Eisma 1976) to 0.2—0.9 mg/l in the northern North Sea (Hagmeier 1962). The high concentrations can be attributed to river supply and coastal erosion, mainly occurring in the southern North Sea, but also to the water circulation that tends to concentrate suspended matter in the nearshore areas (Dietrich 1955, Ramster 1965). Additional concentration of suspended matter takes place in the Wadden Sea (Van Straaten & Kuenen 1957, Postma 1961) but some suspended matter follows the general anticlockwise circulation reaching the Skagerrak and the Norwegian Channel or flowing out into the Atlantic Ocean.

Pockmarks

Along the western slope of the Norwegian Channel and in the Fladen and Witch Ground area, shallow circular or oval depressions are known op to 10 m deep and 200 m wide (van Weering et al. 1973, Caston 1974, Eden 1975, Holmes et al. 1975, Jansen 1976, Rokoengen and Bugge 1976). Such depressions were first observed on the eastern Canadian shelf by King and McLean (1970) who named them pockmarks (Fig. V-9). They are present in areas with soft clayey sediments and in the Fladen and Witch Grounds there is a close relation between the density of pockmarks (up to $36/km^2$) and the thickness of these soft sediments. Buried pockmarks have been found also in this area.

The formation of pockmarks is generally ascribed to escaping gas or water. The soft sediment is probably brought in suspension and is then carried off by bottom currents, leaving circular or oval depressions, which are found parallel to the main current direction. The gas in unconsolidated sediments may originate from fermentation processes, or from seepage from deeper strata. By fermentation, mainly methane is formed as a result of anaerobic decomposition of organic matter. Gas originating from deep layers normally contains, next to methane, considerable amounts of ethane, propane and butane. In the pockmark area of the Norwegian Channel recent measurements of interstitial gas indicate methane to be the most common component (Enger 1976). However, there is a certain correlation with the presence of hydrocarbon-bearing Tertiary sediments wedging out in this area that are absent farther east (Heybroek et al. 1967, Ronnevik et al. 1975). For the Fladen and Witch Grounds, interstitial gas measurement data are not available, but the pockmarks are found where the underlying Swatchway Beds contain bright spots (Holmes, pers. comm.), which inidicate the presence of gas. Also a slight disturbance of the soft deposits below the pockmarks is in support of a deep origin of the escaping gas. If fermentation in relatively shallow deposits is responsible for the gas, this process may have taken place soon after the deglaciation, when sedimentation rates must have been high and organic material was deposited in large quantities. Comparable cir-

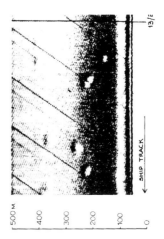

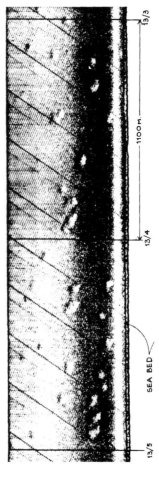

Fig. V-9. Side scan sonar record showing pockmarks at the Witch Ground (from Eden 1975).

cumstances caused methane to be formed in recent sediments on the Labrador shelf (Vilks et al. 1974).

Berkson and Clay (1973) suggested syneresis as an alternative explanation of pockmarks. In that case soil mechanical processes cause water to be expelled, and this ultimately results in shrinkage of the sediments.

Acknowledgements. − Permission given by J. J. H. C. Houbolt, R. A. Eden and R. H. Belderson to reproduce figs. V-8, V-9 and V-6 is gratefully acknowledged.

REFERENCES

Belderson, R. H., Kenyon, N. H. & Wilson, J. B., 1973: Iceberg plough marks in the northeast Atlantic. *Palaeogeogr., Palaeoclim., Palaeoecol.* 13, 3, 215−224.

Belderson, R. H. & Wilson, J. B., 1973: Iceberg plough marks in the vicinity of the Norwegian Trough. *Nor. Geol. Tidsskr.* 53, 3, 323−328.

Berkson, J. N. & Clay, C. S., 1973: Possible syneresis origin of valleys on the floor of Lake Superior. *Nature* 245, 5420, 89−91.

Bratteland, E. & Bruun, P., 1974: Tracer tests in the middle North Sea, *14th Coast. Eng. Conf. Copenhagen,* II, 56, 978−990.

Bruun, P. & Vollen, Ø., 1972: Sand waves on the bottom of the sea, with special reference to conditions in the North Sea. *Schiff und Hafen* 24, 3, 139−142.

Caston, V. N. D., 1970: Linear sandbanks in the southern North Sea. *Sedimentol.* 18, 1/2, 63−78.

Caston, V. N. D., 1974: Bathymetry of the northern North Sea: Knowledge is vital for offshore oil. *Offshore* 34, 2, 76−84.

Destombes, J.-P., Shephard-Thorn, E. R., Redding, J. H. & Morzadec-Kerfourn, M. T. 1975: A buried valley system in the Strait of Dover. *Philos. Trans. R. Soc. London A* 279, 243−256.

Dietrich, G., 1955: Ergebnisse synoptischer ozeanographischer Arbeiten in der Nordsee. *Tagungsber. und wissensch. Abh. Dtsch. Geografentag., Hamburg* 1955, 30, 376−383, Steiner, Wiesbaden.

Dingle, R. V., 1965: Sandwaves in the North Sea mapped by continuous reflection profiling. *Mar. Geol.* 3, 391−400.

Donovan, D. T., 1973: The geology and origin of the Silver Pit and other closed basins in the North Sea. *Proc. Yorks. geol. Soc.* 39, 2, 267−293.

Eden, R. A., 1975: North Sea environmental geology in relation to pipelines and structures. Oceanology International 75, Conf. Pap., 302−309.

Eisma, D., 1976: Deeltjesgrootte van dood gesuspendeerd materiaal in de Zuidelijke Bocht van de Noordzee. *NIOZ Publ. en Versl. 1976−19,* 1−7.

Eisma, D. & Gieskes, W. W. C., 1977: Particle size spectra of non-living suspended matter in the southern North Sea. *NIOZ Publ. en Versl. 1977-7,* 1−7·

Enger, Th., 1976. Design of a deepwater pipeline across the Norwegian Trench. Offshore North Sea Symp., Stavanger 1976, Paper T-1/15, 1−28.

Flinn, D., 1967: Ice front in the North Sea. *Nature* 215, 5106, 1151−1154.

Flinn, D., 1973: The topography of the sea floor around Orkney and Shetland and in the northern North Sea. *J. geol. Soc.* 129, 1, 39−59.

Flodén, T. & Sellevoll, M. A., 1972: Two seismic profiles across the Norwegian Channel west of Bergen. *Stockholm Contr. Geol.* 24, 2, 25−33.

Hagmeiez, E., 1962: Das Seston und seine Komponenten. *Kieler Meeresforsch.* 18, 2, 189−197.

Hansen, K., 1971: Tunnel valleys in Denmark and Northern Germany. *Bull. geol. Soc. Denmark* 20, 3, 295−306.

Harris, I. M. & Jollymore, P. G., 1974: Iceberg furrow marks on the continental shelf northeast of Belle Isle, Newfoundland. *Can. J. Earth Sci.* 11, 1, 43−52.

Heybroek, P., Haanstra, U. & Erdman, D. A., 1967: Observations on the geology of the North Sea area. *Proc. 7th World Pet. Congr. Mexico* 2, 905−916.

Holmes, R., Fannin, N. G. T. & Tully, M. C., 1975: Geological report on the Forties pockmark detailed survey area, M. V. "Sea-lab" drill sites and Forties to Piper, Forties to engineering study area reconnaissance lines. *Inst. Geol. Sci. CSU Rep.* 75/13, 1−12.

Holtedahl, O., 1929: On the geology and physiography of some Antarctic and sub-Antarctic islands. With notes on the character of fjords and strandflats of some northern lands. *Sci. Results Norw. Antarct. Exped.* 3, 1−172.

Holtedahl, O., 1956: Junge Blockverschiebungstektonik in der Randgebieten Norwegens. Geotektonisches Symposium zu Ehren von Hans Stille, 55−63, Stuttgart.

Holtedahl, O., 1964: Echo soundings in the Skagerrak. *Norg. geol. Unders.* 223, 139−160.

Houbolt, J. J. H. C., 1968: Recent sediments in the Southern Bight of the North Sea. *Geol. Mijnbouw* 47, 4, 245−273.

Jansen, J. H. F., 1976: Late Pleistocene and Holocene history of the northern North Sea, based on acoustic refelction records. *Neth. J. Sea Res. 10*, 1, 1–42.

Jansen, J. H. F., Weering, Tj. C. E. van, & Eisma, D., 1977: Late Quaternary sedimentation in the North Sea. this vol.

Johnson, M. A. & Stride, A. G., 1969: Geological significance of North Sea sand transport rates. *Nature 224*, 5223, 1016–1017.

Kent, P. E., 1975: Review of North Sea basin development. *J. geol. Soc. 131*, 435–468.

King, L. H., & Maclean, B., 1970: Pockmarks on the Scotian shelf. *Bull. geol. Soc. Am. 81*, 3141–3148.

Kirby, R. & Oele, E., 1975: The geological history of the Sandettie-Fairy Bank area, southern North Sea. *Philos. Trans. R. Soc. London A. 279*, 257–267.

Kolp, O., 1974: Submarine Uferterrassen in der südlichen Ost- und Nordsee als Marken eines stufenweise erfolgten holozänen Meeresanstiegs. *Baltica 5*, 11–40, Vilnius.

Kolp, O., 1976: Submarine Uferterrassen in der südlichen Ost- und Nordsee als Marken des holozänen Meeresanstiegs und der Überflutungsphasen der Ostsee. *Petermanns Geogr. Mitt. 120*, 1, 1–23, Gotha.

Langeraar, W., 1966: Sand waves in the North Sea. *Hydrogr. Newsletter 1*, 5, 243–246.

Langhorne, D. N., 1973: A sandwave field in the Outer Thames estuary, Great Britain. *Mar. Geol. 14*, 2, 129–143.

Lee, A. & Ramster, J. W., 1968: The hydrography of the North Sea. *Helgoland. Wiss. Meeresunters. 17*, 1/4, 44–63.

Lekahena, E. G., 1966: Megaribbels en hun relatie tot zandtransport in de zuidelijke Nordzee. TNO-nieuws 21, 345–352.

McCave, I. N., 1971: Sandwaves in the North Sea off the coast of Holland. *Mar. Geol. 10*, 3, 199–225.

McCave, I. N., 1973: Mud in the North Sea. In E. D. Goldberg (editor), North Sea Science, 75–100, MIT Press, Cambridge, Mass.

Nio Swie-djin, 1976: Marine transgressions as a factor in the formation of sandwave complexes. *Geol. Mijnbouw. 55*, 1/2, 18–40.

Oele, E., 1969: The Quaternary geology of the Dutch part of the North Sea, north of the Frisian Isles. *Geol. Mijnbouw 48*, 5, 467–480.

Oele, E., 1971: Late Quaternary geology of the North Sea south-east of the Dogger Bank. *Inst. Geol. Sci., Rep. No. 70/50*, 25–34.

Postma, H., 1961: Transport and accumulation of suspended matter in the Dutch Wadden Sea. *Neth. J. Sea Res. 1*, 1/2, 148–190.

Pratje, O., 1951: Die Deutung der Steingründe in der Nordsee als Endmoränen. *Dtsch. Hydrogr. Z. 4*, 3, 106–114.

Ramster, J. W., 1965: Studies with the Woodhead sea-bed drifter in the southern North Sea. *Lab. Leafl. Fish. Lab. Lowestoft (New Ser.) 6*, 1–4.

Reimnitz, E. & Barnes, P. W., 1974: Sea ice as a geologic agent on the Beaufort Sea shelf of Alaska. In J. C. Reed & J. E. Sater (editors), The coast and shelf of the Beaufort Sea, 301–353, Arctic Inst. North Am., Arlington, Va.

Rokoengen, K. & Bugge, T., 1976: Grunnforholdene på den norske kontinentalsokkel sør for 72° N. Continental Shelf Inst. Trondheim, Publ. 81, 1–44.

Ronnevik, H., Bergsager, E. I., Moe, A., Ovrebø, V., Navrestad, T. and Stangenes, J., 1975: The geology of the Norwegian continental shelf. In A. W. Woodland (editor). Petroleum and the continental shelf of North-Western Europe, 1, Geology, 117–129, Applied Sci. Publ. Barking.

Sager, G., 1963: Atlas der Elemente des Tidenhubs und der Gezeitenströme für die Nordsee, den Kanal und die Irische See. Inst. für Meereskunde Warnemünde, 45 pp.

Sellevoll, M. A. & Aalstad, I., 1971: Magetic measurements and seismic profiling in the Skagerrak. *Mar. geophys. Res. 1*, 3, 284–302.

Sellevoll, M. A. & Sundvor, E., 1974: The origin of the Norwegian Channel – a discussion based on seismic measurements. *Can. J. Earth Sci. 11*, 2, 224–231.

Sindowski, K. H., 1970: Das Quartär im Untergrund der Deutschen Bucht (Nordsee). *Eiszeitalter Gegenwart 21*, 33–46.

Straaten, L. M. J. U. van, & Kuenen, Ph. H., 1957: Accumulation of fine grained sediments in the Dutch Wadden Sea. *Geol. Mijnbouw, Nieuwe Ser. 19*, 8, 329–354.

Stride, A. H., 1963: Current-swept sea floors near the southern half of Great Britain. *Q. J. geol. Soc. Lond. 119*, 175–199.

Stride, A. H., 1970: Shape and size trends for sand waves in a depositional zone of the North Sea. *Geol. Mag. 107*, 469–477.

Stride, A. H., 1973: Sediment transport by the North Sea. In E. D. Goldberg (editor), North Sea Science, 101–103, MIT Press, Cambridge, Mass.

Stride, A. H., 1974: Indications of long term, tidal control of net sand loss or gain by European coasts. *Estuarine Coastal Mar. Sci. 2*, 1, 27–36.

Stride, A. H. & Chesterman, W. D., 1973: Sedimentation by non-tidal currents around northern Denmark. *Mar. Geol. 15*, 5, M53–M58.

Swift, D. J. P., 1975: Tidal sand ridges and shoal retreat massifs. *Mar. Geol. 18*, 2, 105–134.

Terwindt, J. H. J., 1971: Sandwaves in the Southern Bight of the North Sea. *Mar. Geol. 10*, 1, 51–67.

Valentin, H., 1957: Die Grenze der letzten Vereisung im Nordseeraum. *Tagungsber. un wissensch. Abh. Dtsch. Geografentag., Hamburg 1955, 30*, 359–366. Steiner, Wiesbaden

Veen, J. van, 1935: Sand waves in the souther North Sea. *Int. Hydrogr. Rev. 12*, 21–29.
Veen, J. van, 1938: Die unterseeische Sandwüste in der Nordsee. *Geol. Meere Binnengewasser 2*, 62–86.
Veenstra, H. J., 1964: Geology of the Hinder Banks, southern Sea.*Hydrogr. Newsl. 1*, 72–80.
Veenstra, H. J., 1969: Gravels of the southern North Sea. *Mar. Geol. 7*, 5, 449–464.
Veenstra, H. J., 1971: Sediments of the southern North Sea. *Inst. Geol. Sci., Rep. No. 70/15*, 9–23.
Vilks, G., Rashid, M. & Linden, W. J. van der, 1974: Methane in recent sediments of the Labrador shelf. *Can. J. Earth Sci. 11*, 10, 1427–1434.
Weering, Tj. C. E. van, 1975: Late Quaternary history of the Skagerrak; an interpretation of acoustical profiles. *Geol. Mijnbouw 54*, 3/4, 130–145.
Weering, Tj. C. E. van, in prep: On the distribution and character of late- and postglacial sediments in the Skagerrak and the Norwegian Channel.
Weering, Tj. C. E. van, Jansen, J. H. F. & Eisma, D., 1973: Acoustic reflection profiles of the Norwegian Channel between Oslo and Bergen. *Neth. J. Sea Res. 6*, 1/2, 241–263.
Winkelmolen, A. M., 1969: Experimental rollability and natural shape sorting of sand. Thesis Groningen Univ., 1–141.
Woodland, A. W., 1970: The buried tunnel-valleys of East Englia. *Proc. Yorks. geol. Soc. 37*, 4, 521–578.
Wright, H. E., 1973: Tunnel valleys, glacial surges, and subglacial hydrology of the Superior Lobe, Minnesota. *Mem. geol. Soc. Am. 136*, 251–276.
Ziegler, P. A., 1975: North Sea basin history in the tectonic framework of North-Western Europe. In A. W. Woodland (editor), Petroleum and the continental shelf of North-Western Europe, 1, Geology, 131–148, Applied Sci. Publ. Barking.
Ziegler, P. A. & Louwerens, C. J., 1977: this volume.

V-c

Sea-level changes in the North Sea basin

SASKIA JELGERSMA

Jelgersma, S., 1979: Sea-level changes in the North Sea basin. In E. Oele, R.T.E. Schüttenhelm & A.J. Wiggers (editors), The Quaternary History of the North Sea, 233–248. *Acta Univ. Ups. Symp. Univ. Ups. Annum Quingentesimum Celebrantis: 2*, Uppsala. ISBN 91-554-0495-2.

In this paper the Late Quaternary changes in sea level will be discussed, and the maximum sea level stands of the Eemian interglacial around the North Sea basin. In the Netherlands, evidence is available of sea-level changes during the Eemian interglacial proper. Relatively little is known of the Weichselian glacial period. One of the salient points is the rapid regression of the sea from the greater part of the Southern Bight during the later part of the Eemian and the beginning of the Weichselian. Changes in sea level during the Holocene ar based on evidence from the coastal plain of the Netherlands and the adjacent North Sea. It must be mentioned that all sea-level data from this area are relative, i.e. they reflect the combined effect of eustatic and tectonic movements.

Dr. S. Jelgersma, Rijks Geologische Dienst, Spaarne 17, Haarlem, the Netherlands.

Introduction

This chapter deals with evidence of sea-level changes since the Saalian in the North Sea area that determine the position of the shorelines. Those studying sea-level changes agree that the variations are the combined effect of tectonic, eustatic and isostatic movements. Due to local conditions the changes in the North Sea basin must be looked upon as relative changes. Fig. V-

10 shows the present uplift and subsidence in NW-Europe. From the North Sea proper no data are available.

Ziegler & Louwerens (this vol.) define the geological position of the present North Sea area as part of the intratectonic NW-European basin, which is surrounded by the Fennoscandian Shield, the Caledonides and the Variscan massifs (Fig. III-1). All over the world such subsiding basins are the sites of significant accumulation of sediments. It is supposed that the tectonic subsidence is maximal in the region of the Tertiary basins with their huge loads of sediments.

It is, however, not always possible to distinguish the tectonic movements in the North Sea basin from the "eustatic" movements of sea level. As a result of the alternating glacial and interglacial periods during the Quaternary, the eustatic movements became a controlling factor for the position of the shorelines. Recent information on eustatic movements prove these changes not to be so simple. Daly (1925) already stated the eustatic changes might cause a deformation of the earth. It is assumed that after a glacial period the redistribution of ice and water results in a redistribution of matter in the interior of the earth. This is thought to cause the distortion of the ocean surface (the geoid), calculated by Walcott (1972).

In the Stockholm 1977 Symposium on Earth Rheology and Late Cenozoic Isostatic Movements, Mörner divided eustatic movements in glacio-eustacy, tectonoeustacy and geoidal eustacy. The glacio-eustacy is controlled by changes in glacial ice volume, the tectono-eustacy by vertical and horizontal tectonic movements caused by the load of water masses. The combined effect of these changes in ocean level is thought to be of equal magnitude all over the globe. Geoidal eustacy, however, should not be the same all over the globe, as it is dependent on gravity.

In the North Sea area another important factor is that during the Quaternary, part of the region was several times covered by expanding land ice. The load of the Scandinavian and British ice caps caused a depression of the crust, followed by a slow updoming after the withdrawal of the ice. This special case of isostasy is proved by the height of numerous raised beaches in the Scandinavian and Scottish areas (see Hafsten, this vol., Eriksson, this vol. and Jardine, this vol.), recognized already by Jamieson (1865). Most geophysicists agree that depressions of the crust by ice loads should cause an uplift in the marginal areas, the so-called peripheral bulge. The isostatic recovery after the retreat of the ice results in a subsidence of adjacent areas (Vening Meinesz 1954, Farrand 1968, Walcott 1972). The above-mentioned ideas have never been confirmed by field studies, and the effects in the marginal areas are still a much disputed subject.

Following the above-mentioned isostatic evidence for the surrounding shields, the question arises how the ice-covered part of the North Sea basin proper reacted to the loading and unloading. There are two important differences between this area and the above-mentioned shields. Firstly, the density of the sedimentary basins in this area is different from that of the shields. Secondly, the area was at least during the late Middle Weichselian glacial maximum situated on the outer fringes of the ice sheet, where the thickness of the ice must have been less than 300 m. As the degree of depression due to ice loading is related to the thickness of the ice cap and to the density of the crust, the present author is of the opinion that in the area of the sedimentary basins of the North Sea region, the amount of isostatic downwarping is questionable. The loading effect of the ice cap on compaction of unconsolidated sediments may be more significant. Especially in areas where the Tertiary is more than 2500 m thick (Ziegler & Louwerens, this vol.), this effect should not be overlooked. In treating the sea level data from the North Sea basin proper the isostatic movements connected to loading and unloading of ice sheets will be neglected, as these movements are highly speculative.

The deposits indicating the stand of sea level must be found in sedimentation zones closely related to sea level, e.g. in tidal flat zones, salt marshes and littoral environments. When using peat as indication of former sea level, it is essential to determine the ground-water table that controlled the growth of peat, and the relation between the ground-water table and sea level The deposits can be dated by relative and ab solute methods. For Holocene and Late Weichselian deposits the radiocarbon method gives excellent results. Older deposits can be

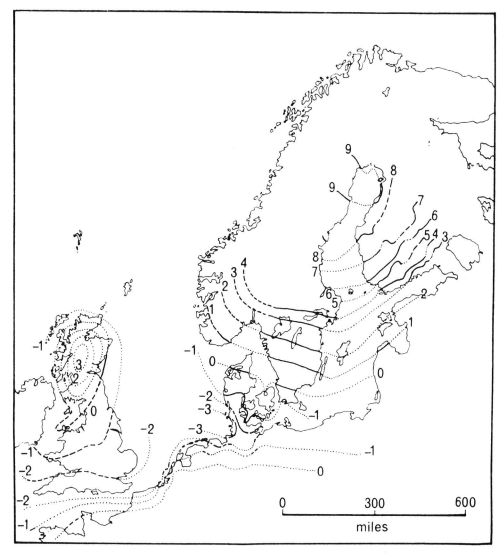

Fig. V-10. Present uplift and downwarping in northwest Europe. The isobases represent the rate of change (+ or −) in mm/year, determined from tide-gauge records. Dashed isobases are less certain, the dotted isobases are based on interpolation (from West 1968).

dated only by palaeontological methods. In the North Sea region investigations by pollen analysis showed a succession of vegetational zones, permitting actual correlations in and around the North Sea basin. For this reason, relative dating by pollen analysis will be preferred.

Sea level during the Eemian in the southern North Sea basin

During the maximum extension of the Saalian land ice, sea level is considered to have been more than 130 m lower than the present level (Emery 1969). Accordingly, the greater part of the present North Sea emerged and became land. After the retreat of the land ice, sea level rose and the North Sea region became sea again. The maximum extension of the Eemian interglacial sea is shown on the colour chart.

The absolute and world-wide position of the maximum Eemian sea level is much disputed. Bonifay and Mars (1959) and Guilcher (1969) advocated levels between 8 and 10 m above the present sea level, others prefer a level of 20 metres above the present one (Woldstedt 1969). Because of our presently much more dynamic view of the earth, a discussion about the absolute maximum Eemian sea level is rather speculative. In view of the evidence obtained from pollen analysis of Eemian deposits and from oxygene-isotopic records in deep sea sediments presumably of Eemian age, the climate must have been slightly warmer than during the Holocene, and the sea level must have been higher than the present level.

The maximum marine Eemian level in the countries surrounding the southern part of the North Sea basin is reported at rather different heights. A few outlines on the maximum height of the marine deposits will be discussed below. See Oele et al. (this vol.) for more details. In Germany Sindowski (1965) indicated in onshore Lower Saxony the height of the maximum Eemian sea level at 9 m below the present one. On the Isle of Juist this level is found at 6.5 m below the present sea level, in Schleswig-Holstein Dittmer (1954) reported 5 m lower than the present sea level. See Behre et al. (this vol.) for more details. In the Netherlands a salt marsh clay in pollen zone E5 (hornbeam zone) represents the maximum elevation of the Eemian sea level, about 8 m below the present sea level (Zagwijn 1961, 1977, Jelgersma 1961). In Belgium, near Gent, De Moor & De Breuck (1973) described tidal flat deposits about 1 m below the present sea level; molluscs and pollen analysis indicate an Eemian age. Paepe et al. (1972) reported marine Eemian deposits near Brugge, the top of which is close to the present sea level.

In England the maximum height of the Eemian sea level was discussed by West (1972) and by Jardine (this vol.). The highest sea level is found in Ipswichian pollen zone III (equivalent to zones E4/E5 of Zagwijn (1961)), as published by West & Sparks (1960) and Sparks & West (1972). In the Lower Thames valley, near Aveley, aggradation amounted to about 15 m above the present sea level. Pollen analysis indicates these sediments to coincide with Ipswichian zone III. In the English Channel, Selsey Sussex, the same age is concluded for a deposit at 8 metres above M.S.L. In Stutton, Suffolk, brackishwater deposits from Ipswichian pollen zone III are situated 1 m above present sea level. The corresponding sea level must be 1 m lower than the present level. The March Gravels in the western part of East Anglia, containing a rich marine fauna, are situated between 10 and 12 m above the present sea level. On the basis of the above-mentioned evidence, West (1972) concluded to a tilting of about 9 metres since the Eemian between the western and the eastern part of East Anglia.

Summarizing the above data the elevation of the maximum Eemian sea level from Belgium to the Danish border varies from around present sea level to −8 m, −6 m and finally to −5 m. Along the North Sea coast of SE England, from south to north, elevations have been reported from 8 to 15 m above to 1 m below the present sea level. The above-mentioned differences in elevation might be caused by differences in tectonic downwarping in and around the North Sea basin.

In the Netherlands and the adjacent North Sea basin, relative sea level changes are known from the Eemian interglacial. The area contains a complete section of Eemian deposits, partly formed in a marine or brackish environment. The relative age of the Eemian deposits is based on pollen zones in this part of the North Sea basin (Zagwijn 1961). In general, the Eemian deposits show a transgressive and a regressive phase. The relative dating of these phases and their relation to sea level provides evidence of relative sea level changes during the Eemian. The results of the investigations have been presented by Zagwijn during the Xth Inqua Congress at Birmingham 1977.

Late Saalian and Early Eemian marine deposits are not yet known. They may be present in the northern part of the North Sea where the depth increases. The oldest Eemian deposits of marine origin in the Netherlands and the adjacent North Sea basin were formed during pollen zone E3, the so-called oak zone. These E3 marine deposits are found in Saalian glacial basins. The base of these deposits is found between 60 and 80 m below the present sea level. These depths do not represent, however, a sea level stage. The glacial basins were depressions in a plateau-like area. This plateau is situated about 40 m below M.S.L.; accordingly, sea level during E3 must have been about 40 m below the present level. Subsequently, sea level rose to about 19 m below the present level, and this coincided with the transition from pollen zone E3b to E4a (hazel zone). The above-mentioned evidence is based on the relative age and the depth of the beginning of the Eemian transgression in the Eemian type locality.

During the transition from pollen zone E4a to E4b (taxus zone), sea level reached a height of 16 m below the present level, as indicated by the height on which the beginning of the Eemian transgression was found in Friesland (ter Wee 1976). Finally, the maximum level of 8 m below the present one was reached during pollen zone E5 (hornbeam zone). This level is represented by a marsh clay overlying older Eemian sandy marine deposits in the Eem valley. After that time, a regression of the Eemian sea set in, as indicated by a peat layer on top of the marsh clay in the Eem valley and in Friesland. On the northwest coast of the Netherlands, however, a series of borings proved that marine sediments were formed during the Late Eemian pollen zone E6. According to the results in these holes, the relative sea level during pollen zone E6a must have been about 16 m lower than the present level and dropped to 20—30 m below the present level in pollen zone E6b. These findings indicate a considerable drop in sea level during the Late Eemain. This is supported by the evidence collected in the sediments of the Southern Bight (see below).

Weichselian sea levels in the North Sea basin

There has been much discussion about the position of sea level during the Early Weichselian, the interstadials of the Middle Weichselian and during the Late Weichselian. In this paper the Middle to Late Weichselian boundary will be placed at 13000 B.P. The reason for the uncertainties is the scarcity of information on Weichselian shorelines, located on the shelf and presently submerged. From the North Sea region scarce data give a rough indication of the position of sea level during the Weichselian (Oele et al., this vol.). According to the bathymetric contour lines the present North Sea area can be divided into two parts: one south of Dogger Bank where the depth is less than 50 metres and one north of Dogger Bank where the bottom slopes to more than 100 m below the present sea level. In the mentioned northern part some data are available on marine deposits that must have been formed during the Early and Middle Weichselian. The exact age, the environmental conditions, and information on the corresponding sea levels, are, however, extremely scarce (Jansen 1976, Holmes 1977). South of the Dogger Bank, material of marine Weichselian age has not been found, and the known Weichselian deposits are of terrestrial origin.

During the Late Eemian and Early Weichselian, the sea retreated from a large area in the Southern Bight of the North Sea. This regression is interpreted as a eustatic lowering of sea level and not as a lowering due to tectonic movement; it has most likely been caused by the growth of the Laurentide ice sheet, as demonstrated by investigations on the North American continent (Dreimanis 1960). It is accepted here that in the North Sea region during the Early Weichselian, sea level must have been more than 40 m lower than the present level, even during the interstadials. This is indicated by the sediments in the Southern Bight of the North Sea. A fresh-water clay, overlying marine Eemian deposits in the central part of the Southern Bight, was formed at the beginning of the Early Weichselian (pollen zone EW Ia) and is located from 37 to 42 m below mean sea level.

For part of the Middle Weichselian, a sea level of about 10 m lower than the present level is generally accepted for the so-called 30000 B.P. interstadial (Curray 1965, Millima & Emery 1968). This is highly unlikely, as the Weichselian vegetation is indicative of a very cold climate: in our latitude a polar desert alternating with a shrub tundra, indicating a major land ice cap related to a low sea level elsewhere. See Oele et al. (this vol.) for more information on this subject.

In general, most authors agree that the lowest position of the Weichselian sea level was reached in the late Middle Weichselian, about

Fig. V-11. Hypothetical Weichselian shorelines and relicts of linear ridges in the North Sea basin.
1. shoreline at 18000 B.P., sea level about 130 m below the present level.
2. Linear ridges formed around 18000 B.P.
3. Shoreline at 12000 B.P., sea level about 90 m below the present level.
4. Linear ridges formed around 12000 B.P.
5. Shoreline at 10300 B.P., sea level about 65 m lower than the present level.
6. Linear ridges from around 10300 B.P.

18000 B.P., which coincides with the maximum extension of the land ice caps. At that time sea level must have reached a level of 120 to 130 m lower than the present level. This is based on data from the United States shelves (Curray 1965 and Emery 1969) and on calculations o the volume of land ice caps (Bloom 1971). Leaving tectonic and glacio-isostatic movements in the North Sea basin out of consideration, a shoreline near the 130 m depth contour line (see Fig. V-11) might be accepted during the maximum extension of the land ice.

Around 18000 B.P., the land ice started to retreat and sea level again began to rise. Calculations based on the extension of the ice sheet and on the ice volume during the late Middle Weichselian and Late Weichselian (after 13000 B.P.) indicate an enormous reduction of the ice sheet; the shoreline at $\pm$ 12000 B.P. must have been located at $\pm$ 90 m lower than present mean sea level, as indicated in Fig. V-11. The calculations, summarized by Bloom (1971), indicate that at the transition from the Late Weichselian to the Holocene 50% of the total ice sheet had disappeared. This means a considerable rise of sea level again. Assuming the maximum lowering of sea level to have been 130 m, and accepting the calculations based on the reduction of the land ice, sea level must have been between 70 and 60 m at the end of the Late Glacial. Accordingly, the shoreline in the North Sea basin at that time must have been situated close to the north side of Dogger Bank; this is indicated in Fig. V-11.

It should be stressed that part of the present North Sea submarine topography is assumed to represent a submerged glacial landscape, and that during the Late Glacial and Holocene transgression only local erosion and sedimentation played a role (Jelgersma 1961). Based on this assumption, the hypothetical coastlines represented in Figs. V-11, 14, 16 have been constructed. Differences by tectonic movements have been omitted.

Summarizing the scarce evidence of the Weichselian sea level changes, the following conclusions can be presented. Already during the Early Weichselian the sea retreated from the Southern Bight of the North Sea. During the maximum extension of the ice caps the shoreline may have been located near the 130 m depth contour line. After the late Middle Weichselian land ice started to melt, the sea transgressed over the western part of the northern North Sea as indicated in the map (Fig. V-11). It should be mentioned that during the Late Weichselian the Norwegian and Swedish west coasts were partly submerged by transgression of the North Sea through the Norwegian Channel.

Sea level changes during the Holocene in the southern North Sea basin, with the emphasis on the Netherlands

Introduction

Much more information is available on sea level changes during the Holocene than during the Pleistocene. In the North Sea area the combined effects of eustatic rise in sea level, isostatic rebound due to uplift after the retreat of the ice sheet, and tectonic subsidence of the North Sea basin are expressed in locally varying relative rises in sea level. In parts of the northern area where the former Weichselian ice sheet had reached a large thickness, Early Holocene shorelines are represented by raised beaches. A regression in these areas in younger Holocene time was most likely caused by an updoming of the land that surpassed the eustatic rise of sea level. Sea level changes in these areas are discussed by Mörner (1969) and Jardine, Eriksson and Hafsten (all in this volume). In the southern area of the North Sea basin, however, a continuous rise of sea level during the Holocene might partly have been caused by tectonic downwarping, but may be largely the result of a eustatic sea level rise (Jelgersma et al., this vol.).

Geological setting

Holocene sea level data from the Netherlands and the Dutch sector of the North Sea can be divided into two categories. Early Holocene evidence, covered by the Holocene transgression of the North Sea, can be traced only by offshore surveys, and is scarce (Jelgersma et al., this vol.). Atlantic and later data can be collected in the subsoil of the coastal plain of the Netherlands and is easier to be obtained (Jelgersma et al., this vol.). These will be treated first, and completed later with the scarce data from the North Sea proper.

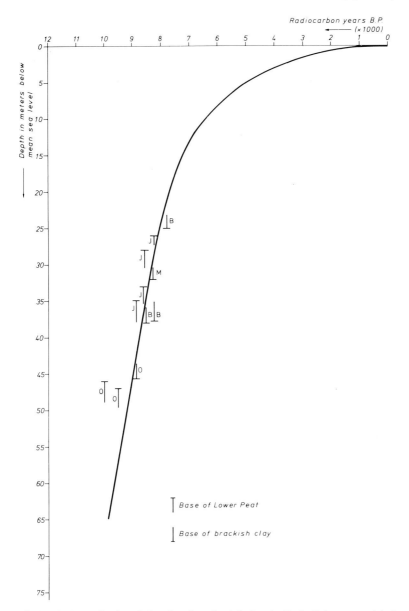

Fig. V-12. Curve for the relative rise of sea level during the Early Holocene, mainly based on data from the Dutch sector. B — Behre et al. (this vol.), J — Jelgersma (1961), M — Morzadec et al. (1972), O — Oele in Jelgersma et al. (this vol.).

The Holocene deposits in the coastal plain of the Netherlands are situated on top of the seaward sloping Pleistocene surface. In an east-west section through the coastal plain three types of sedimentation can be distinguished; in the East peat deposited in a fresh water environ ment, more to the West a clayey lagoonal zone of tidal flats, salt marshes and other brackish-water environments, and towards the sea a zone of beach ridges and dunes. These three zones shifted with time. To a large extent this was caused by differences in the rate of sea-level rise and in the amounts of sediments. A fourth sedimentation zone, called the perimarine area, is present in the valleys of the large rivers Rhine and Meuse. During the Holocene rise of sea level the riverplain was gradually filled up with sand, clay and layers of fen wood peat (Jelgersma et al., this vol.).

Methods of investigation and sources of error

Most sea level data in the Netherlands have been collected in the coastal zone, the peat zone and the perimarine area. In the lagoonal zone, excellent markers of former sea level are available also. Interpretation of the sedimentary structures gives a rather close approximation of the former mean sea level, which can be dated with shells in their natural position or with an overlying peat layer. An important source of error is the compaction of the underlying sediments. In the lagoonal area the Holocene sequence generally consists of alternating sands, clays and peats. In the last two the rate of compaction can be very high. Accordingly, the height of sea levels in the lagoonal area can only be used if the underlying deposits have not been subjected to compaction. Until now, the factor compaction could be satisfactorily eliminated from the data for the lagoonal area only in a few localities.

As mentioned before, nearly all data for constructing the sea level curve have been collected in the area of peat formation (Jelgersma 1961, 1966). The evidence for changes in sea level is based on ^{14}C datings of the peat layer directly overlying the inclined Pleistocene surface at various depths in the coastal region of the Netherlands. The peat layers (the so-called Lower Peat) are considered to have been formed by rising ground water, which in turn was controlled by rising sea level. Firstly, information is required on the ground-water table at the time of peat formation. Secondly, the relation between ground water and sea level has to be ascertained.

To solve the first problem, fen-wood peat were used (Jelgersma 1961), as the plant associations connected with these peats flourish when the ground-water table is close to the surface. Sphagnum peats e.g. have been omitted as they are not dependent on the general ground-water table. The second question, the relation between ground water and sea level, is in many cases debatable, as will be discussed. The dating of the base of peat layers resting directly on Pleistocene sand has both advantages and disadvantages. By using this method the compaction in underlying sediments can be eliminated, because the Pleistocene sands show very little compaction. The disadvantage is the questionable reliability of the datings for sea level changes. The assumption is made that the Lower Peat was formed in close relation to sea level, but in places with upwelling water from the hinterland or stagnant ground water in local depressions this need not be so, which may represent a possible source of error (Jelgersma 1961).

Another source of error in using peat is contamination by older material causing the samples to be too old; penetration by roots, however, causes them to be too young. Peat samples selected for radiocarbon dating require careful inspection. Contamination, however, cannot always be noticed, and the obtained ages had to be checked with other methods of investigation, e.g. pollen analysis. Peat datings obtained from the so-called "Donken" (aeolian dunes on top of levee deposits in the perimarine area) are considered as the most reliable indicators of ground-water movements during the Holocene. See Behre et al. (this vol.) and Jardine (this vol.) for more information on former sea level stands and sources of error.

High tide vs. sea level. — Jelgersma (1961, 1966) discussed the relation between ground water in a river plain and sea level. At that time it was assumed that the highest ground-water level in a river plain approximately coincides with high tide. Accordingly, it was accepted that her curve of the rise of ground water represents the changes in high tide. However, the results of the study of the high tide levels in the coastal barrier complex, dated by ^{14}C with shells, clearly indicate that former ground-water levels as summarized by Jelgersma (1961, 1966) correlate with mean sea level rather than with high tide

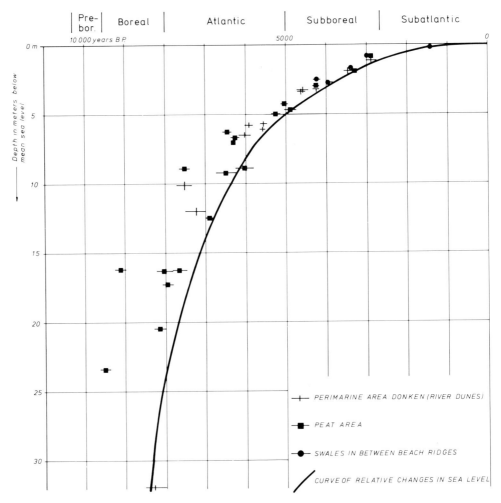

Fig. V-13. Curve for the relative rise of sea level during the Holocene, based on data from the Dutch coastal plain.

(Roep et al. 1975). The results of the study of former sea levels in the coastal barrier complex of the western Netherlands and the ^{14}C datings of shell beds will be published (Roep et al., in prep.). In view of the age of the coastal barriers the paper will cover the fluctuations of high tide during the last 4800 years only.

The dating of moorlogs. — As mentioned before, evidence of sea level changes during the Early Holocene, 10000 to 8000 B.P., can be found only on the bottom of the North Sea. Jelgersma (1961) published pollen-analytical data from "moorlogs", Lower Peat dredged from the bottom of the North Sea. The moorlogs, found in large areas of the southern North Sea, are looked upon as relicts of the peat layer formed by the rising ground water during the submergence of the present North Sea area. As mentioned, the reliability of the Lower Peat in the coastal plain of the Netherlands (see below as an indicator of former sea level is questiona ble. This applies especially to the southern North Sea where the recent topography indicates that closed depressions must have been present. In these depressions, inland lakes could be formed, related to peat layers. In our opinion correct conclusion based on datings of the base of the Lower Peat is that sea level must have been lower than the depth of the sample.

Another problem connected with the Lower Peat is that erosion undoubtedly played a role during the submergence of the North Sea. Especially for dredged samples, the depth is not certain. During the last ten years several new data on the Lower Peat of the southern North Sea have become available, especially from the area SE of Dogger Bank. These data were obtained from cored borings. Oele in Jelgersma et al. (this vol.) mentioned two radiocarbon datings for the base of the Lower Peat: 47 m −M.S.L.: 9445 + 80 B.P. (GRN 5759) and 46 m −M.S.L.: 9935 ± B.P. (GRN 5758). Behre & Menke (1969) made pollen analyses of the Lower Peat on the SE slope of Dogger Bank. The peat layer was located at 45 m −M.S.L., and pollen analysis indicates a Preboreal age. Oele (1969), Oele in Jelgersma et al., (this vol.) and Behre & Menke (1969) also mention the Lower Peat to be overlain by a thin layer of fresh-water clay passing into brackish-water clay. The latter is dated by pollen analysis as Early Boreal. Accordingly, the transgression of the North Sea in this area is dated between 9000 and 8700 B.P. In the central part of the German sector the Lower Peat is found in shallower places, about 38 m −M.S.L. Behre et al (this vol.) estimates the age of the overlying brackish clay as 8190 ± 140 B.P. (HV 7095) and 8485 ± 125 B.P. (HV 7094). This indicates that this area was submerged in the Late Boreal. The depth of the brackish clay deposits only indicates that the contemporary sea level was above the mentioned depth.

The curves of sea level changes

As mentioned above, Jelgersma 1961 published pollen-analytical data from the North Sea. A curve of the relative sea level changes in the southern North Sea basin was based on the relative datings and on the depths. Together with the curve, 4 hypothetical maps show the submergence of the southern North Sea area during the Early Holocene transgression. More data are however available at the present time.

The Dutch sector.—The above-mentioned data on the age of the base of the Lower Peat in the offshore area and the overlying brackish-water deposits have been used to establish a new curve of the relative rise in sea level during the Early Holocene (Fig. V-12). The construction of the curve is based on the assumption that sea level was below the observed depth of the base of the Lower Peat but above the observed depth of the brackish clay deposits. This produces a rather steep curve that shows a relative rise in sea level of 2 m/century during this period. It is clear that fluctuations of the relative sea level are not evident by this method.

The Dutch coastal plain.—A tentative new curve (see Fig. V-13) is presented on the relative sea level movements in the coastal plain of the Netherlands during the Holocene. The method is based on the assumption that in several places the formation of the Lower Peat may have started with a ground-water level that was independent of sea level (i.e. at different elevations above sea level). In that case only the points located in the lowest places for a given age represent a ground-water table that coincided with sea level. In this way a smooth curve is obtained. All aberrations must be considered as errors and not as fluctuations of sea level. The curves show a continuously rising sea level during the last 8000 years, with a gradual levelling-off after 6000 B.P. Next to the curve for the Netherlands, two sea level curves for eastern England can be mentioned, one for the Thames estuary drawn by Greensmith & Tucker (1973), and one for the Humber area by Gaunt & Tooley (1974). These two curves show large fluctuations of sea level, which cannot be accepted since the presented data can be differently interpreted. In our opinion, the fluctuations may be attributed to several sources of error, e.g. compaction in underlying deposits. The datings on shell deposits do not mention the use of bivalves in natural position, which in our opinion is to be preferred. Sedimentological

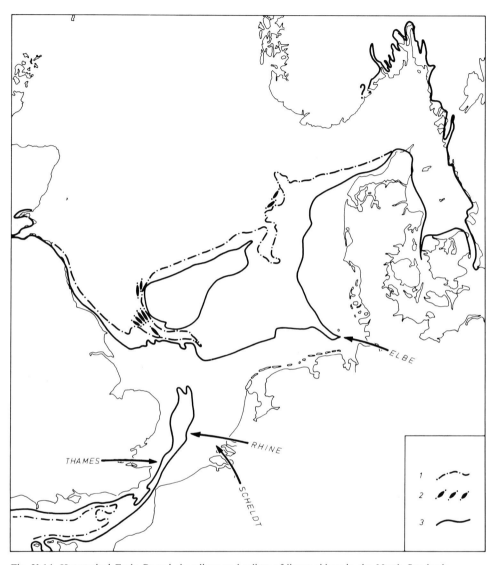

Fig. V-14. Hypotetical Early Boreal shorelines and relicts of linear ridges in the North Sea basin.
1. Shorelines at 9000 B.P., the beginning of the Boreal. Sea level about 50 m below the present level.
2. Linear ridges formed around 9000 B.P.
3. Shoreline at 8700 B.P. Sea level about 36 m lower than the present level.

structures to indicate the exact position in relation to the mean sea level have not been given. The fluctuations presented in the two curves are open to question.

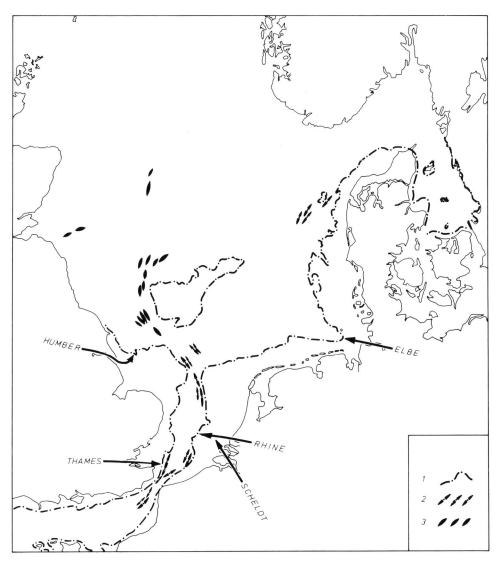

Fig. V-15. Hypotetical Late Boreal shorelines and relicts of linear ridges in the North Sea basin. Sea level around 30 m below the present level.
1. Shorelines at 8300 B.P. d2. Linear ridges formed around 8300 B.P.
3. Older linear ridges.

The Early Holocene shorelines

In addition to the sea level curve, 5 hypothetical shorelines for the Early Holocene period are presented (Figs. V-14, 15, 16). The 65 m contour line is thought to represent the position of

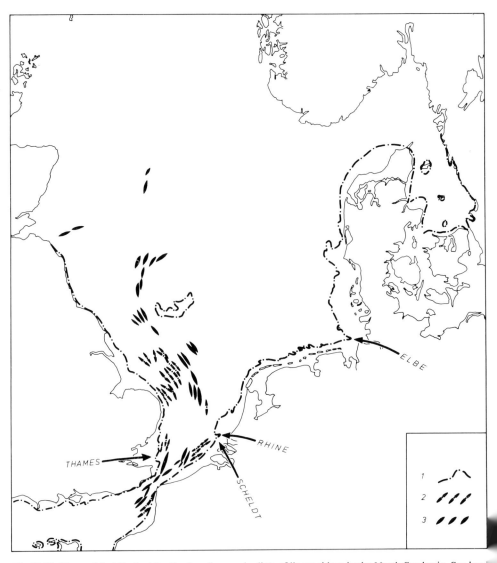

Fig. V-16. Hypotethical Early Atlantic shorelines and relicts of linear ridges in the North Sea basin. Sea level around 20 m below the present level.
1. Shorelines at 7800 B.P.
2. Linear ridges formed around 7800 B.P.
3. Older linear ridges.

the North Sea shoreline during the transition from the Weichselian to the Holocene (Fig. v-11). The 50 m contour line is the shoreline just before the submergence of the area SE of Dogger Bank: about 9000 B.P. just before the deposition of the Elbow clay in this area (Fig. V-14). At 8700 B.P., sea level reached a position of about 36 m below the present level. Th southern part of the North Sea and the northern part were separated only by a narrow land bridge. The 30 m depth contour line shows the situation about 8300 B.P., when the southern and the northern North Sea were recently connected. This period saw two tidal waves, one from the Channel and the other coming south around Scotland (Fig. V-15), and the current movement in this area was rather turbulent. The 20 m depth contour is considered to represent the shoreline position about 7800 B.P., which is somewhat similar to the present position (Fig. V-16).

Erosion and deposition during the Early Holocene transgression

During the Early Holocene transgression, deposition and erosion played a role. In the area southeast of Dogger Bank, the latter probably was of minor importance as the Lower Peat is overlain by a thin layer of brackish clay belonging to the so-called Elbow deposits. In the western part of the North Sea the situation was different, especially in the Southern Bight with more erosion and deposition, as a result of strong tidal currents from the Channel and from Scotland. The linear ridges and the closed depressions in our opinion were formed by tidal scour. Donovan (1973) in his study of the Silver Pit area mentioned that these closed basins could have been formed either by subglacial stream erosion or by tidal scour. See discussion in Eisma et al., (this vol.). Especially for the Southern Bight we are inclined to assume that tidal scour has been responsible for the typical morphology. On the shoreline maps (Figs. V-11, 14, 15, 16) relicts of linear ridges have been indicated.

Acknowledgements. — The critical comments of Dr. W.H. Zagwijn are gratefully acknowledged.

REFERENCES

Behre, K.-E. & Menke, B., 1969: Pollenanalytische Untersuchungen an einem Bohrkern der südlichen Doggerbank. *Beiträge zur Meerskunde 24/25,* 123—129, Deutsche Akademie der Wissenschaften zu Berlin.

Behre, K.-E., Menke, B. & Streif, H., 1979: The Quaternary geological development of the German part of the North Sea, this volume.

Bloom, A.L., 1971: Glacial-eustatic and isostatic controls of sea level since the last glaciation. In K.K. Turekian (editor), Late Cenozoic glacial ages, 355—379, Yale University Press.

Bonifay, E. & Mars, P., 1959: Le Tyrrhénien dans le cadre de la chronologie quaternaire méditerranéenne. *Bull. Soc. Géol. France, Ser. 7, 1,* 62—78.

Curray, J.R., 1965: Late Quaternary history, continental shelves of the United States. In H.E. Wright & D.G. Frey (editors). The Quaternary of the United States, 723—735, Princeton University Press.

Daly, R.A., 1925: Pleistocene changes of level. *Am J. of Sci. 5th ser., 10,* 281—313.

Dittmer, E., 1954: Interstadiale Torfe in würmzeitlichen Schmelzwassersanden Nordfrieslands. *Eiszeitalter und Gegenwart, 4/5,* 172—175.

Donovan, D.T., 1973: The geology and origin of Silver Pit and other closed basins in the North Sea. *Proc. Yorks. geol. Soc. 39,* 267—293.

Dreimanis, A., 1960: Pre-classical Wisconsin in the eastern portion of the Great Lake Region North America. 21st Intern. Geol. Congress Copenhagen, 4, 108.

Eisma, D., Jansen, J.H.F. & van Weering, Tj.C.E., 1979: Sea-floor morphology and recent sediment movement in the North Sea, this volume.

Emery, K.O., 1969: The continental shelves. *Sci. Am. 221,* 107—126.

Eriksson, K. Gösta, 1979: Late Pleistocene and Holocene shorelines on the Swedish West Coast, this volume.

Farrand, W.R., 1968: Postglacial Isostatic Rebount. In R.W. Fairbridge (editor), The Encyclopedia of Geomorphology, Earth Science Ser. 3, 884—888, Reinhold Book Corp., New York.

Gaunt, G.D. & Tooley, M.J., 1974: Evidence for Flandrian sea-level changes in the Humber estuary and adjacent areas. *Bull. Geol. Surv. G.B. 48,* 25—41.

Greensmith, J.T. & Tucker, E.V., 1973: Holocene transgressions and regressions on the Essex Coast outer Thames estuary. *Geol. Mijnbouw 52,* 193—202.

Guilcher, A., 1969: Pleistocene and Holocene Sea Level Changes. *Earth-Sci. Rev. 5,* 69—97.

Hafsten, U., 1979: Late and Post-Weichselian shore level changes in South Norway, this volume.

Holmes, R., 1977: Quaternary deposits of the central North Sea, 5. The Quaternary Geology of the U.K. sector of the North Sea between 56° and 58° N. *Rep. Inst. Geol. Sci.* 77/14m 50 pp.

Jamieson, T.F., 1865: On the history of the last geological changes in Scotland. *Quart. J. Geol. Soc. London 21,* 161—203.

Jansen, J.H.F., 1976: Late Pleistocene and Holocene

history of the northern North Sea, based on acoustic reflection records, *Neth, J. Sea Res. 10,* 1–43.
Jardine, W.G., 1979: The western (United Kingdom) shore of the North Sea in Late Pleistocene and Holocene times, this volume.
Jelgersma, S., 1961: Holocene sea level changes in the Netherlands. *Meded. Geol. Sticht., C, VI,* 7, 100 pp.
Jelgersma, S., 1966: Sea Level Changes During the Last 10,000 Years. In World Climate from 8000 to O B.C. Proceedings of the International Symposium Held at Imperial College, London, 18 and 19 April 1966, 54–71, Royal Meterological Society, London.
Jelgersma, S., Oele, E. & Wiggers, A.J., 1979: Depositional history and coastal development in the Netherlands and the adjacent North Sea since the Eemian, this volume.
Milliman, J.D. & Emery, K.O., 1968: Sea levels during the past 25,000 years. *Science 162,* 1121–1123.
Mörner, N.A., 1969: The Late Quaternary history of the Kattegatt Sea and the Swedish West Coast. Deglaciation, shorelevel displacement, shronology and eustacy. *Sver. Geol. Unders. C. 640,* 487 pp.
De Moor, G. & De Breuck, W., 1973: Sedimentologie en stratigrafie van enkele pleistocene afzettingen in de Belgische Kustvlakte. *Natuurwet. Tijdschr. 55,* 3–96.
Morzadec-Kerfourn, M.T. & Delibrias, G., 1972: Analyses polliniques et dations radiocarbone des sédiments quaternaires prélevés en Manche centrale et orientale. In Colloque sur la Géologie de la Manche. *Mém. Bur. Rech. Géol. Min. 79,* 160–165.
Oele, E., 1969: The Quaternary geology of the Dutch part of the North Sea, north of the Frisian Islands. *Geol. Mijnbouw 48,* 467–480.
Oele, E. & Schüttenhelm, R.T.E., 1979: Development of the North Sea after the Saalian glaciation, this volume.
Paepe, R., Vanhoorne, R. & Deraymaker, D., 1972: Eemian sediments near Bruges (Belgian Coastal Plain). Prof. Paper 1972/9, Belg. Geol. Survey, Brussel, 11 pp.
Roep, Th.B., Beets, D.J. & Ruegg, G.H.J., 1975: Wavebuilt structures in subrecent beach barriers of the Netherlands. Proc. IXth International Congress of Sedimentology, 6, 141–145, Nice.
Sindowski, K.-H., 1965: Das Eem im Ostfriesischen Küstengebiet. *Z. dtsch. geol. Ges. 115, (Jahrg. 1963),* 163–166.
Sparks, B.W. & West, R.G., 1972: The Ice Age in Britain., Methuen, London.
Vening Meinesz, F.A., 1954: Crustal warping in the Netherlands. *Geol. Mijnbouw 16,* 207 ff.
Walcott, R.I., 1972: Past Sea levels, Eustacy and Deformation of the Earth. *Quaternary Res. 2,* 1–14.
Wee, M.W. ter, 1976: Toelichtingen bij de Geologische Kaart van Nederland 1:50.000. Blad Sneek (10 W, 10 0), Rijks Geologische Dienst, Haarlem, 131 pp.
West, R.G., 1968: Pleistocene geology and biology. XIII + 377 pp., Longmans, London.
West, R.G., 1972: Relative land-sea-level changes in South-eastern England during the Pleistocene. *Philos. Trans. R. Soc. London A272,* 87–98.
West, R.G. & Sparks, B.W., 1960: Coastal interglacial deposits of the English Channel. *Philos. Trans. R. Soc. London 243,* 95–133.
Woldstedt, P, 1969: Quartär. In F. Lotze (editor), Handbuch der Stratigraphischen Geologie, II, 263 pp., F. Enke, Stuttgart.
Zagwijn, W.H., 1961: Vegetation, climate and radiocarbon datings in the Late Pleistocene of the Netherlands. Part I: Eemian and Early Weichselian. *Meded. Geol. Sticht., Nieuwe Ser. 14,* 15–45.
Zagwijn, W.H., 1977: Sea level changes during the Eemian in the Netherlands. Xth INQUA Congress, Birmingham 1977, abstracts, 509.
Ziegler, P.A. & Louwerens, C.J., 1979: Tectonics of the North Sea, this volume.